PRACTICAL SCANNING ELECTRON MICROSCOPY

Electron and Ion Microprobe Analysis

AUTHORS

J. I. Goldstein — Metallurgy and Materials Science Department, Lehigh University, Bethlehem, Pennsylvania

H. Yakowitz — National Bureau of Standards, Washington, D.C.

D. E. Newbury — National Bureau of Standards, Washington, D.C.

E. Lifshin — General Electric Corporate Research and Development, Schenectady, New York

J. W. Colby — Bell Telephone Laboratories, Allentown, Pennsylvania

J. R. Coleman — Department of Radiation, Biology, and Biophysics, University of Rochester Medical Center, Rochester, New York

With the assistance of
R. B. Bolon and M. F. Ciccarelli — General Electric Corporate Research and Development, Schenectady, New York

PRACTICAL SCANNING ELECTRON MICROSCOPY

Electron and Ion Microprobe Analysis

Edited by

Joseph I. Goldstein

Metallurgy and Materials Science Department
Lehigh University
Bethlehem, Pennsylvania

and

Harvey Yakowitz

U.S. Department of Commerce
National Bureau of Standards
Washington, D.C.

Foreword by T. E. EVERHART

PLENUM PRESS • NEW YORK AND LONDON

Library of Congress Cataloging in Publication Data

Goldstein, Joseph, 1939-
    Practical scanning electron microscopy.

    Includes bibliographical references and index.
    1. Scanning electron microscope. 2. Microprobe analysis.
I. Yakowitz, H., joint author. II. Title.
QH212.S3G64   502′.8   74-34162
ISBN 0-306-30820-7

© 1975 Plenum Press, New York
A Division of Plenum Publishing Corporation
227 West 17th Street, New York, N.Y. 10011

United Kingdom edition published by Plenum Press, London
A Division of Plenum Publishing Company, Ltd.
4a Lower John Street, London W1R 3PD, England

Printed in the United States of America

FOREWORD

In the spring of 1963, a well-known research institute made a market survey to assess how many scanning electron microscopes might be sold in the United States. They predicted that three to five might be sold in the first year a commercial SEM was available, and that ten instruments would saturate the marketplace. In 1964, the Cambridge Instruments Stereoscan was introduced into the United States and, in the following decade, over 1200 scanning electron microscopes were sold in the U.S. alone, representing an investment conservatively estimated at $50,000–$100,000 each. Why were the market surveyors wrong? Perhaps because they asked the wrong persons, such as electron microscopists who were using the highly developed transmission electron microscopes of the day, with resolutions from 5–10 Å. These scientists could see little application for a microscope that was useful for looking at surfaces with a resolution of only (then) about 200 Å. Since that time, many scientists have learned to appreciate that information content in an image may be of more importance than resolution *per se*. The SEM, with its large depth of field and easily interpreted images of samples that often require little or no sample preparation for viewing, is capable of providing significant information about rough samples at magnifications ranging from 50× to 100,000×. This range overlaps considerably with the light microscope at the low end, and with the electron microscope at the high end. The SEM became available just at the time that man attempted to make integrated circuits with dimensions measured in microns, to tolerances of fractions of microns. Applications in one field led to experiments in other fields, and suddenly the scientific community realized how much time and effort could be saved by "seeing" objects that previously had been out of reach of the light microscope and unsuited for the transmission electron microscope. Although these objects were "seen" in vacuum, they were seen with a depth and clarity unprecedented in the light microscope, and the information provided by the SEM images was great indeed. Requiring no preparation, they were free of artifacts produced by preparation techniques. More recently, biologists have developed a variety of techniques to fix or replicate biological surfaces so that they can be "seen" in the SEM. The new morphological detail "seen" in this way, when correlated with parallel work in the light microscope or transmission electron microscope, provides valuable insights into the biological world.

In a sense, the SEM has come of age. Many scientists are applying it to solve their problems as a well-understood tool, while others are probing the physics, chemistry, and biology on which a sound interpretation of SEM images are based. This need to understand quantitatively how signals are generated is even more important in the case of the specialized SEM, the electron probe microanalyzer, that uses x-rays as the signal. This older relative of the SEM is invaluable to the materials scientist because it determines what elements are present in the first micron or so of a solid surface. As most electron probes for microanalysis now scan and produce an image, and many produce secondary electron images to complement the x-ray image, I consider them to be SEM's that use x-rays as the information signal.

Many scientists and engineers wish to learn how to use the SEM intelligently (which means they need to understand the fundamental operation of the instrument, including how images are produced). Short courses such as the one which led to the development of this book have provided a valuable service to the scientific community. This book places the information available to the attendee of a short course at the disposal of any person who can use a microscope or microprobe as he explores the book, so the experience of using the instrument can reinforce the knowledge gained from the book. I would have been delighted to have had such a book at my disposal when I first became acquainted with the SEM some two decades ago; the book would have saved me many hours of study and puzzlement, just as the SEM itself has saved many hours of puzzlement for countless scientists by providing new "sights" and insights. Not all readers will need to study all topics presented. Almost all readers will learn something from the book because of the breadth of coverage. In order that the foreword not keep the interested reader from the body of knowledge contained in the book any longer, it ends here. Good reading.

T. E. Everhart

Berkeley, California

PREFACE

In our rapidly expanding technology, scientists and engineers are often required to observe and correctly explain phenomena occurring on a micrometer (μm) and submicrometer scale. The scanning electron microscope (SEM) and electron probe microanalyzer (EPMA) are two relatively new and powerful instruments which permit the characterization of heterogeneous materials and surfaces on such a fine scale. Among the major types of information obtained are compositional information of both a qualitative and quantitative nature, and topographic information contained in images obtained at relatively high resolution.

The SEM and EPMA are, in reality, one and the same instrument and can be treated as such. The use of the SEM–EPMA has increased markedly over the last few years and sales of these major instruments (cost $20,000–$150,000 per instrument) have expanded at a rapid rate. The number of scientific papers dealing with the theory and practical application of the two instruments has also increased, perhaps even more rapidly. Excellent review articles about various aspects of the SEM–EPMA have appeared in the literature. Nevertheless, the user of these instruments is often faced with a bewildering list of references on various topics of interest. The novice is at an even greater disadvantage.

Clearly, with the expanding use of the SEM–EPMA, there arises a need for material which is presented in a tutorial style. This book was written in order to provide a straightforward means for the reader to learn the concepts of scanning electron microscopy and electron probe microanalysis. We have attempted to prepare the material in a unified form and to present it in tutorial fashion. This method of presentation was suggested by our cumulative experience in teaching the intensive short courses in SEM and EPMA methods given at Lehigh University in the summers of 1971, 1972, and 1974 as well as various tutorials given at the Electron Probe Analysis Society of America national meetings and the

Scanning Electron Microscopy Symposiums sponsored by IITRI. Prof. Goldstein has also used much of this material in an undergraduate course on electron metallography which he has taught at Lehigh for the past five years. Since no single author is an expert in the details of every phase of SEM–EPMA analysis, a multiauthor approach was adopted. The experience of the Lehigh Summer Course has allowed us to fully integrate the chapters.

The book is set out in the following manner. Chapters I–V and VII give the basic information necessary for understanding the theory and practice of scanning electron microscopy–electron microprobe analysis. No previous experience or specific educational background beyond the first two years of college is assumed. Chapters VI and XI discuss specimen preparation and give practical examples of the applications of these techniques. The applications illustrate the uses of the instrument and are taken from the various fields of interest. Chapters IX and X discuss in detail the various methods for quantitative x-ray analysis and how to apply them successfully. Several specific techniques, such as thin-film analysis, x-ray statistics, x-ray source size, and light element analysis are discussed in Chapters VIII and XII. The application of the SEM and EPMA to biological samples is discussed separately in Chapter XIII since specimen preparation is so critical to analyses of this type. Finally, Chapter XIV discusses the use of the ion microprobe mass analyzer. This book will hopefully provide an important complement to the SEM–EPMA instruments. It is suitable for use as a text in intensive short courses or in regular undergraduate or graduate university courses. Technicians, scientists, and engineers should find this book useful for their introduction to the field, and experts in the fields of SEM and EPMA should find it useful as a reference.

The editors wish to thank M. F. Ciccarelli and R. B. Bolon for their contributions to Chapters VII and VIII. Discussions with students of the Lehigh Short Course on SEM–EPMA and particularly with Lehigh students using our own SEM–EPMA instruments have been very helpful in improving the quality of the book. One of the editors (J.I.G.) wishes to acknowledge the research support from NASA (Grant #s NGR 39-007-056, NGR 39-007-043) and from NSF (Grant #GA 15349), which has encouraged him to develop his own interests in the SEM–EPMA technique. In addition, the encouragement of Dr. G. P. Conard, Chairman of the Metallurgy and Materials Science Department, Lehigh University, to develop a book on this topic is greatly appreciated. Special thanks go to Barbara Hayes, Louise Valkenburg of Lehigh, Barbara Goldstein, Barbara Newbury, and Goldie Yakowitz, without whose efforts the book could not have been completed. The wholehearted support of the National Bureau of Standards is also gratefully acknowledged.

J. I. Goldstein

H. Yakowitz

ACKNOWLEDGMENTS

We would like to express our appreciation to the following: at the *National Bureau of Standards*: K. F. J. Heinrich, C. E. Fiori, R. L. Myklebust, S. D. Rasberry, A. W. Ruff, Jr., D. B. Ballard, P. A. Boyer, D. L. Maske, K. Loraski, R. A. Hormuth, W. J. Keery, K. O. Leedy, B. W. Christ, and R. Bowen; at *Cambridge University*: J. Agrell, W. C. Nixon, L. R. Peters, and G. R. Plows; at *Southwest Research Institute*: D. Davidson; at *ETEC Corporation*: N. C. Yew; at *Oxford University*: D. C. Joy, R. Booker, D. W. Tansley, and A. M. B. Shaw; at *Technion*, Haifa: Professor B. Z. Weiss; at *Bell Laboratories*: J. P. Ballantyne; at the *University of Illinois*: C. A. Evans; at *Lehigh University*: D. Bush and R. Hewins.

Chapter XIII is based in part on work performed under contract with the U.S. Atomic Energy Commission at the University of Rochester Atomic Energy Project and assigned report # UR-3490-561 and in part on work supported by the U.S. Public Health Service research grants 1RO 1AM 14272 and 1RO 1AM 17074.

CONTENTS

Chapter V

Contrast Mechanisms of Special Interest In Materials Science

D. E. Newbury and H. Yakowitz

Chapter VI

Specimen Preparation, Special Techniques, and Applications of the Scanning Electron Microscope

D. E. Newbury and H. Yakowitz

Chapter VII

X-Ray Spectral Measurement and Interpretation

E. Lifshin, M. F. Ciccarelli, and R. B. Bolon

Chapter VIII

Microanalysis of Thin Films and Fine Structure . .   299

R. B. Bolon, E. Lifshin, and M. F. Ciccarelli

Chapter IX

**Methods of Quantitative X-Ray Analysis Used
in Electron Probe Microanalysis
and Scanning Electron Microscopy . .**   327

H. Yakowitz

Chapter X

Computational Schemes for Quantitative X-Ray Analysis: On-Line Analysis with Small Computers

H. Yakowitz

Chapter XI

Practical Aspects of X-Ray Microanalysis

H. Yakowitz and J. I. Goldstein

Chapter XII

Special Techniques in the X-Ray Analysis of Samples 435

J. I. Goldstein and J. W. Colby

Chapter XIII

Biological Applications: Sample Preparation and Quantitation 491

James R. Coleman

INTRODUCTION

J. I. Goldstein, H. Yakowitz, and D. E. Newbury

In our rapidly expanding technology, the scientist today is more frequently required to observe and correctly explain phenomena occurring on a micrometer (μm) and submicrometer scale. The electron microprobe and scanning electron microscope are two powerful instruments which permit the characterization of heterogeneous materials and surfaces on such a local scale. In both instruments, the area to be examined is irradiated with a finely focused electron beam, which may be static, or swept in a raster across the surface of the specimen. The types of signals which are produced when the focused electron beam impinges on a specimen surface include secondary electrons, backscattered electrons, characteristic x-rays, Auger electrons, and photons of various energies. They are obtained from specific emission volumes within the sample and are used to measure many characteristics of the sample (composition, surface topography, crystallography, etc.).

In the electron probe microanalyzer (EPMA), also frequently referred to as the electron microprobe, the primary signal of interest is the characteristic x-rays which are emitted as a result of the electron bombardment. The analysis of the characteristic x-radiation, emitted from the region at which the electron beam impinges, yields compositional information of both a qualitative and quantitative nature. In the scanning electron microscope, the primary signal of interest is the variation in secondary

J. I. GOLDSTEIN—Metallurgy and Materials Science Department, Lehigh University, Bethlehem, Pennsylvania. H. YAKOWITZ and D. E. NEWBURY—Institute for Materials Research, National Bureau of Standards, Washington, D.C.

electron emission that takes place as the electron beam is swept in a raster across the surface of a specimen due to differences in surface topography. The secondary electron yield is confined near the beam impact area, which permits images to be obtained at relatively high resolution. The three-dimensional appearance of the images is due to the large depth of focus of the scanning electron microscope. Other signals are available from these two instruments, which prove quite useful in many cases, and will be the subject of subsequent chapters.

Historically, the scanning electron microscope (SEM) and electron microprobe (EPMA) have evolved as separate instruments. It is obvious on inspection, however, that these two instruments are quite similar but differ from each other primarily in the way in which they are utilized. The development of each of these instruments (SEM and EPMA) and the similarities of modern commercial instruments are discussed in this chapter.

I. EVOLUTION OF THE SCANNING ELECTRON MICROSCOPE

The scanning electron microscope (SEM) is one of the most versatile instruments available for the examination and analysis of the microstructural characteristics of solid objects. The primary reason for the SEM's usefulness is the high resolution which can be obtained when bulk objects are examined; values of the order of 10 nm (100 Å) are usually quoted for commercial instruments. Advanced research instruments have been described which have achieved resolutions of about 2.5 nm (25 Å).[1] A high-resolution micrograph is shown in Figure 1. This micrograph was taken with a commercial SEM under typical operating conditions.

Another important feature of SEM images is the three-dimensional appearance of the specimen, which is a direct result of the large depth of focus. Figure 2a shows the skeleton of a small sea animal (radiolarian) viewed optically, and Figure 2b as viewed with the SEM. The greater depth of focus of the SEM provides much more information about the specimen. In fact, examination of the SEM literature indicates that it is this feature which is of the most value to the SEM user. Most SEM micrographs have been produced with magnifications below 8000 diameters. At these magnifications, the resolution capability of the SEM is not taxed. The SEM is also capable of examining objects at very low magnification. This feature is useful in forensic studies as well as other fields. An example of a low-magnification micrograph of a familiar subject is shown in Figure 3.

The variety of signals, e.g., x-rays, Auger electrons, available in the

FIGURE 1. High-magnification micrograph of National Bureau of Standards Research Material 100 consisting of Al–W dendritic material. Micrograph taken using a commercial SEM under typical operating conditions.

SEM can provide useful information about the composition at the specimen surface. Other phenomena which can be studied through electron–specimen interactions are crystallographic, magnetic, and electrical characteristics of the specimen.

The basic components of the SEM are the lens system, electron gun, electron collector, visual and recording cathode ray tubes (CRT's), and the electronics associated with them (see Figure 8). The first successful commercial packaging of these components was offered in 1965, the Cambridge Scientific Instruments Mark I instrument. Considering the present popularity of the SEM, the fact that 23 years passed between the time Zworykin, Hillier, and Snyder[2] published the basis for a modern SEM and this development seems incredible. The purpose of this brief historical introduction is to point out the pioneers of scanning electron microscopy and in the process trace the evolution of the instrument.

FIGURE 2a. Optical micrograph of radiolarian.

The first published work describing the use of scanning in electron microscopy appears to be that of von Ardenne in 1938.[3,4] In fact, von Ardenne added scan coils to a transmission electron microscope (TEM) and in so doing produced what amounts to the first scanning transmission electron microscope (STEM). Both the theoretical bases and practical aspects of the STEM were discussed in fairly complete detail, even by today's standards. The first STEM micrograph was of a ZnO crystal. The operating voltage was 23 kV. The original magnification was 8000×, and the resolution was between 500 and 1000 Å. The photograph contained 400 × 400 scan lines and took 20 min to record[3]; the film was scanned in synchronism with the beam. The instrument had two electrostatic condenser lenses, with the scan coils placed between the lenses. The instrument possessed a CRT but it was not used to photograph the image.[4]

The first SEM used to examine thick specimens was described by Zworykin *et al.* in 1942.[2] These authors recognized that secondary electron emission would be responsible for topographic contrast and accordingly constructed the design shown in Figure 4. The collector was biased positive to the specimen by 50 V and the secondary electron current collected on it

FIGURE 2b.   SEM micrograph of same radiolarian shown in Figure 2a. The depth of focus and superior resolving capability in this micrograph are apparent.

produced a voltage drop across a resistor. This voltage drop was sent through a regular television set to produce the image; however, resolutions of only 1 μm were attained. This was considered unsatisfactory since better resolution than that obtainable with optical microscopes (2000 Å) was being sought.

Therefore Zworykin *et al.* decided to reduce spot size, as well as improve signal-to-noise ratio, and so produce a better instrument. These authors considered all of the required contributions to the relationship of lens aberrations, gun brightness, and spot size, and they obtained the appropriate relation for minimum spot size as a function of beam current.[2] Next they sought to improve gun brightness by using a field emitter source, but instability in these cold cathode sources forced a return to thermionic emission sources. Nevertheless, as early as 1942 the field emitter tip was used to produce high-magnification, high-resolution images.[2] The next contribution was the use of an electron multiplier tube as a preamplifier for the secondary emission current for the specimen. In the Zworykin *et al.* version, the secondaries struck a fluorescent screen in front of the electron multiplier. The resulting photocurrent was suitably amplified

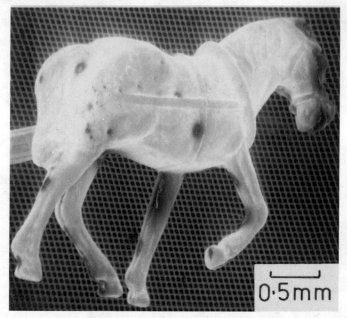

FIGURE 3.   Very low-magnification photograph of a toy horse showing depth of field and surface details on horse.

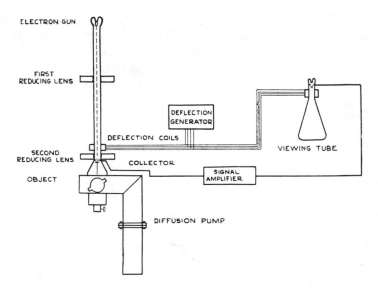

FIGURE 4. Schematic diagram of 1942 SEM (after Ref. 2).

to form the image, which was viewed on the CRT. Care was taken to match the sensitivity peak of the phosphor scintillator with that of the photocathode of the multiplier. Useful but noisy (by today's standards) photomicrographs were obtained. The final instrument consisted of three electrostatic lenses with scan coils placed between the second and third lenses. The electron gun was at the bottom so that the specimen chamber was at a comfortable height for the operator. Resolutions of at least 500 Å were demonstrated with this, the first modern SEM.[2] World War II caused a hiatus in the development of the SEM. Zworykin *et al.* were unable to fully develop their instrument. In fact the team was split up and U.S. advances ceased until the 1960's. In 1948, C. W. Oatley at the University of Cambridge became interested in building electron microscopes. He and McMullan built the first SEM at Cambridge. By 1952, this unit had achieved a resolution of 500 Å.[5] McMullan was followed by Smith, who recognized that signal processing could improve micrographs. Hence Smith introduced nonlinear signal amplification (γ-processing). He also replaced electrostatic lenses with electromagnetic lenses and improved the scanning system by introducing double deflection scanning. Smith was also the first to insert a stigmator into the SEM.[6]

The next step forward was the modification of the detector described by Zworykin *et al.* This was accomplished by Everhart and Thornley, who attached the scintillator to a light pipe directly at the face of the photomultiplier.[7] This improvement increases the amount of signal collected and results in an improvement in signal-to-noise ratio. Hence, weak contrast mechanisms could be better investigated. Oatley and Everhart were able to observe, for the first time, the phenomenon known as voltage contrast.[8]

Pease, under Nixon's guidance, built a system, known as SEM V, with three magnetic lenses, the gun at the bottom, and the Everhart–Thornley detector system. This instrument amounts to the prototype of the Cambridge Scientific Instruments Mark I and, in many ways, is similar to the 1942 instrument. SEM V, of course, embodied all of the improvements outlined above which were introduced after 1952.[9,10]

A. D. G. Stewart and co-workers at the Cambridge Scientific Instrument Company carried out the commercial design and packaging of the instrument.* In the ensuing decade, more than 1000 SEM units have been sold by several manufacturers representing the U. S., U. K., France, Holland, Japan, and Germany who are actively developing new, improved

---

*The best source data on this early instrument is found in the commercial descriptive brochures.

instruments. Yet even today the basic SEM is not far removed from the one described in 1942.

Since 1965, many advances have been made. One of these was the development of the LaB$_6$ electron source by Broers.[1] This source is a high-brightness electron gun; hence more electron current can be concentrated into a smaller beam spot. In this way, an effective improvement in resolution can be obtained (see Chapter II for details). The field emission tip electron source first used in the SEM in 1942 was revived by Crewe and developed to the point where it can be used successfully in obtaining high-resolution images.[11] The reason that the field emission gun is excellent for high resolution is that the brightness is very high and the source is very small. Hence, current densities of thousands of amperes per square centimeter are available even when the beam current is on the order of 10^{-11} A. Field emission sources have two potential drawbacks: (1) if one draws currents exceeding a few nanoamperes from them, resolution deteriorates rapidly, and (2) the source is currently not as stable as desirable; hence rapid scanning is almost always needed to provide an image free from defects.

Other advances involve contrast mechanisms not readily available in other types of instrumentation. Crystallographic contrast, produced by crystal orientation and lattice interactions with the primary beam, was found by Coates[12] and exploited initially by workers at Oxford University.[13] Magnetic contrast in certain noncubic materials was observed, simultaneously but independently, by Banbury[14] (Cambridge) and Joy[15] (Oxford). Magnetic contrast in cubic materials was first observed by Philibert and Tixier[16]; the contrast mechanism was explained later by Fathers *et al.*[17,18]

Often the contrast of the features one is examining is so low as to be invisible to the eye. Therefore contrast enhancement by processing of the signal is needed. Early signal processing included nonlinear amplification, as noted, and differential amplification (black level suppression), both incorporated into SEM's at Cambridge University. Derivative signal processing (differentiation) to enhance small details was introduced later.[19,20] Most commercial SEM units today are provided with these signal processing capabilities.

The image itself can be processed either in analog or digital form. Image storing circuits have been developed so that one can observe the image and/or operate on it off-line.[21,22] These devices are extremely useful, not prohibitively expensive, but not as versatile as full computer image processing. White and co-workers have developed a series of image processing computer programs known as CESEMI (Computer Evaluation, SEM Images) which can provide a great deal of information such as grain size, amount of phase present, etc.[23] Digital scanning in which the coordinates of the picture point and the signal intensity of the point are delivered to the computer is needed to take full advantage of such programs. In fact, computer interaction and control of the SEM has been accomplished.[24]

The large depth of focus available in the SEM makes it possible to observe three-dimensional objects in stereo. Equipment has been developed which allows quantitative evaluation of surface topography making use of this feature.[25] Provisions for direct stereo viewing on the SEM have been described as well.[26]

The addition of an energy-dispersive x-ray detector to an electron probe microanalyzer[27] signaled the eventual coupling of such instrumentation to the SEM. Today, a majority of SEM facilities are equipped with x-ray analytical capabilities. Thus, topographic, crystallographic, and compositional information can often be obtained rapidly and efficiently.

II. EVOLUTION OF THE ELECTRON PROBE MICROANALYZER

The electron probe microanalyzer (EPMA) is one of the most powerful techniques for the microanalysis of materials. The primary reason for the EPMA's usefulness is that compositional information, using characteristic x-ray lines, with a spatial resolution of the order of 1 μm can be obtained from a sample. The sample is analyzed nondestructively, and quantitative analysis can be obtained in many cases with an accuracy of the order of 2% of the amount present for a given element. Figure 5 shows a Ni and Co microprobe analysis of a ferrite (bcc) phase which has nucleated and grown by a diffusion-controlled process in an austenite (fcc) phase matrix during cooling of a lunar metal particle on the moon's surface.[28] The Co and Ni data were taken with a commercial EPMA under typical operating conditions and illustrate the resolution of the quantitative x-ray analysis which can be obtained.

Another important feature of the EPMA is the capability for obtaining x-ray signal scanning pictures. The x-ray pictures show the elemental distribution in the area of interest. Figure 6 shows the distribution of Ag, Hg, and Sn among the various phases in a complex dental alloy.[29] The areas scanned are 275 $\times$ 220 μm. Magnifications up to 2500$\times$ are possible without exceeding the resolution of the instrument. The attractiveness of this form of data gathering is that detailed microcompositional information can be directly correlated with optical metallography. The EPMA literature

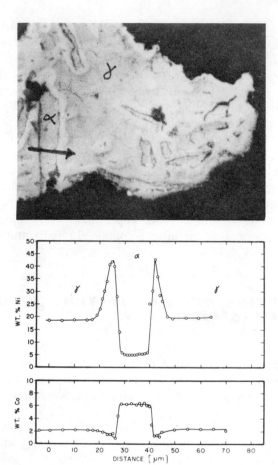

FIGURE 5. Lunar particle F28.8 from the Apollo 14 soil 14003. The photomicrograph shows the microstructure of the particle. The α-ferrite appears as laths or rhombs in the γ-austenite matrix. The field of view is $145 \times 110 \ \mu$m. The EPMA scan shows the Ni and Co across the largest ($\sim$10 μm) kamacite plate shown in the photomicrograph. The trace was taken normal to the α/γ boundaries as shown. Co is enriched in the α-phase and the Ni distribution is similar to that observed across α-ferrite plates in iron meteorites. (Adapted from Goldstein and Axon.[28])

indicates that this feature is of great value to the user. In addition, the variety of signals available in the EPMA (emitted electrons, photons, etc.) can provide useful information about surface topography and composition in small regions of the specimen.

In 1913, Moseley found that the frequency of emitted characteristic x-ray radiation is a function of the atomic number of the emitting element.[30] This discovery led to the technique of x-ray spectrochemical analysis by which the elements present in a specimen could be identified by the examination of the directly or indirectly excited x-ray spectra. The area analyzed, however, was quite broad (>1 mm²). The idea of the electron microanalyzer, in which a focused electron beam was used to excite a small area on a specimen (~1 μm²), and including an optical microscope for locating the area of interest, was first patented in the 1940's.[31] It was not until 1949, however, that Castaing, under the direction

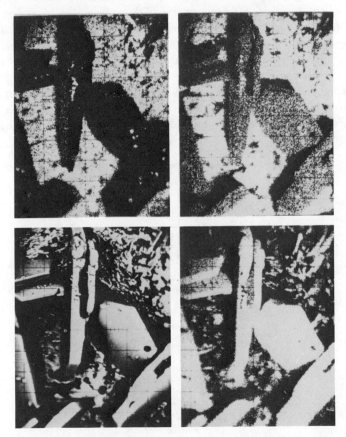

FIGURE 6.   Recrystallized dental alloy, unprepared surface. Upper left: Ag $L\alpha$, x-ray scan. Upper right: Hg $M\alpha$, x-ray scan. Lower left: target current scan. Lower right: Sn $L\alpha$, x-ray scan. Area scanned: 275 μm $\times$ 220 μm. (Adapted from Heinrich.[29])

of Guinier, described and built an instrument called the "microsonde electronique" or electron microprobe.[32-34] In his doctoral thesis Castaing demonstrated that a localized chemical analysis could be made on the surface of a specimen. Concurrently with the work of Castaing in France, Borovskii in the USSR developed an EPMA quite dissimilar in design.[35]

During the early 1950's several EPMA instruments were developed in laboratories both in Europe and the United States.[36-40] The first EPMA commercial instrument (Cameca 1956) was based on the design of an EPMA built by Castaing in the laboratories of the Recherches Aeronautiques.[34] Figure 7 shows diagrammatically the principle of the apparatus. The electron optics system consists of an electron gun followed by reducing lenses which form an electron probe with a diameter of approximately 0.1–1 μm on the specimen. Since electrons produce x-rays from a volume often exceeding 1 μm in diameter and 1 μm in depth, it is usually unnecessary to use probes of very small diameter. An optical microscope for accurately choosing the point to be analyzed and a set of spectrometers for analyzing the x-radiation emitted are also part of the instrument.

Cosslett and Duncumb designed and built the first scanning electron microprobe in 1956 at the Cavendish Laboratory in Cambridge, England.[41]

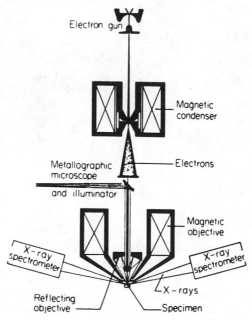

FIGURE 7.   Schematic diagram of the French microanalyzer (adapted from Castaing[34]).

Whereas all previous electron microprobes had operated with a static electron probe, Duncumb and Cosslett swept the beam across the surface of a specimen in a raster, as is done in current SEM's. They used the backscattered electron signal to modulate the brightness of a cathode ray tube sweeping in synchronism with the electron probe. They also used the x-ray signal to modulate the brightness, permitting a scanned image to be obtained showing the lateral distribution of a particular element. Although the concept of a local x-ray analysis is in itself a strong incentive for the utilization of a microprobe, the addition of the scanning concept was an extremely significant contribution, and this probably accounted for the subsequent increased popularity of the electron microprobe.

In his doctoral thesis, Castaing developed the physical theory so that the analyst could convert measured x-ray intensities to chemical composition.[33] Castaing proposed a method of analysis based on comparisons of intensity, i.e., the comparison of the intensity N_1 of a characteristic line of a given element emitted by a specimen, under one set of conditions of electron bombardment, and the intensity N_2 of the same characteristic radiation, when emitted by a standard containing the same element, under the same electron bombardment conditions. The ratio of the two readings is proportional to the mass concentration of a given element in the region analyzed. Recognition of the complexity of converting x-ray intensities to chemical composition has led numerous investigators (for example, Refs. 42–44) to expand the theoretical treatment of quantitative analysis proposed by Castaing.

In the years since the development of the first scanning EPMA instrument many advances have been made. Of particular importance was the development of diffracting crystals having large interplanar spacings.[45,46] These crystals enable long-wavelength x-rays from the light elements to be measured. The ability to detect fluorine, oxygen, nitrogen, carbon, and boron enabled users of the EPMA to investigate many new types of problems with the instrument. Techniques of soft x-ray spectroscopy have now been applied in the EPMA to establish how each element is chemically combined in the sample. The chemical effect may be observed as changes in wavelength, shape, or relative intensity of emission and absorption spectra.

When the application of the EPMA was extended to nonmetallic specimens, it became apparent that other types of excitation phenomena might also be useful. For example, the color of visible light (cathodoluminescence) produced by the electron probe has been associated with the presence of certain impurities in mineral specimens,[47] and photon radiation produced from the recombination of excess hole–electron pairs in a semiconductor can be studied.[48] Measurement of cathodoluminescence in the EPMA has now been developed as another important use of this instrument.

Increased use of the computer in conjunction with the EPMA has greatly improved the quality and quantity of the data obtained. Many computer programs have been developed to convert the x-ray intensity ratios to chemical compositions, primarily because some of the correction parameters are functions of concentration and hence make successive approximations necessary.[49] Some of these programs can be run on a minicomputer[50] and compositions are calculated directly from recorded digital data. The advantage of rapid calculation of chemical compositions is that the operator has greater flexibility in carrying out analyses. In addition, computer automation of the EPMA has been developed to varying degrees of complexity. Dedicated computers and the accompanying software have been developed in several laboratories to control the electron beam, specimen stage, and spectrometers. The advantages of automation are many but in particular it greatly facilitates repetitive-type analysis, increases the efficiency of quantitative analysis performed, and leaves the operator free to concentrate on evaluating the analysis and designing experiments to be performed.

The addition of an energy-resolving x-ray detector to the EPMA[27] has provided a means for obtaining rapid element identification in the instrument. This type of detector has increased the capability of the EPMA and has also provided, for the first time, the possibility of x-ray analysis in the SEM. The next section discusses the similar capabilities of the two instruments and the combined SEM–EPMA instrument.

III. COMBINATION SEM–EPMA

The realization that secondary electron images could be obtained in the electron microprobe has induced several manufacturers to improve their electron beam resolutions to the range 100–1000 Å and to provide both secondary electron detectors and tilting stages. A complete conversion is not simple because many of the EPMA's are not properly shielded from stray magnetic fields, and their specimen stages are susceptible to vibration.

In recent years high-resolution, solid state x-ray detectors have become available which separate the x-ray spectrum by energy rather than by wavelength. The appearance of the energy-dispersive detector has caused a flurry of activity because it can be easily added to an SEM and possesses enough sensitivity to provide x-ray spectral data at the low SEM beam currents. Therefore it is now possible to do x-ray scanning in a high-resolution SEM with good imaging capabilities. However, this does not mean that chemical concentrations can be obtained for specimen constituents of this size (≤ 100 Å), since the electron scattering volume and there-fore the x-ray source is still approximately 1–5 μm in diameter. In addition, the resolution of the energy-dispersive detector is substantially worse than that of the wavelength-dispersive detector, allowing the possibility of over-lapping x-ray peaks and decreased sensitivity. The energy-dispersive detector is also not usually capable of measuring the presence of the light elements below about Na, although recently a description of a system capable of observing elements down to carbon was given.[51] If desired, wavelength dispersive spectrometer (WDS) facilities can be added to an SEM.

The SEM and EPMA are in reality one and the same instrument. Because of the obvious similarity of the two instruments, several manu-facturers have constructed instruments capable of being operated as an electron microprobe or as a high-resolution scanning electron microscope. Figure 8 shows a schematic of the electron and x-ray optics of such a com-bination instrument. Both a secondary electron detector and x-ray de-tector(s) are placed below the final lens. For quantitative x-ray analysis and for the measurement of x-rays from the light elements, at least two wavelength-dispersive spectrometers are desirable. In addition, provision for a broad useful current range, 10^{-12}–10^{-6} A, must be made. The scanning unit allows both electron and x-ray signals to be measured and displayed on the CRT. Such combination instruments are still in the developmental stage. Several books and monographs in the general area of SEM and EPMA have been prepared; a list of such texts is given as a bibliography at the end of this chapter.

IV. OUTLINE AND PURPOSE OF THIS BOOK

Because the EPMA and SEM can be considered as one instrument, it is useful to discuss these two instruments and their capabilities together. The electron optics system and the signals produced during electron bombard-ment are very similar for the EPMA–SEM and are discussed in Chapters II and III. The remainder of the book is devoted to the details of measuring the available signals and the techniques of using these signals to determine particular types of information about a given sample. The emphasis is on the selection and utilization of techniques appropriate to the solution of problems often presented to the SEM and EPMA analysis staff by their clients.

Chapters IV, V, VII, and VIII consider the detection and processing of secondary electron, backscattered electron, cathodoluminescence, and

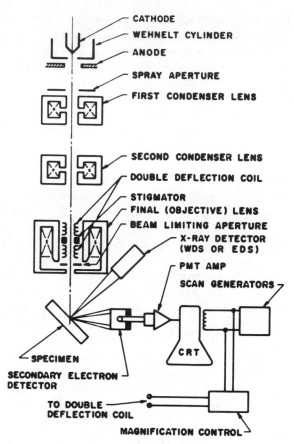

FIGURE 8.   Schematic drawing of the electron and x-ray optics of a combined
                  SEM–EPMA.

x-ray signals as obtained from the SEM–EPMA. Chapters IX and X discuss various methods for quantitative x-ray analysis. Applications of these instruments, as well as specimen preparation, are considered in Chapters VI and XI. Several specific x-ray techniques, such as light element analysis, soft x-ray analysis, analysis at interfaces, etc., are discussed in detail in Chapter XII. Thin-film analysis with the film either free standing or on a substrate constitutes an important class of investigations. Techniques for carrying out such analyses are discussed in detail in Chapters VIII and XII; applications to biological specimens are outlined in Chapter XIII. The application of the SEM and EPMA to biological samples is discussed separately in Chapter XIII since specimen preparation is so critical to analyses of this type. Finally, Chapter XIV discusses the use of the ion microprobe mass analyzer. This tool provides an important complement to the SEM–EPMA instruments but is still in the developmental stage.

REFERENCES

1. A. N. Broers, in *Scanning Electron Microscopy/1974*, IITRI, Chicago, Illinois (1974) p. 9.
2. V. K. Zworykin, J. Hillier, and R. L. Snyder, *ASTM Bull.* **117**, 15 (1942).
3. M. von Ardenne, *Z. Phys.*, **109**, 553 (1938).
4. M. von Ardenne, *Z. Techn. Phys.*, **19**, 407 (1938).
5. D. McMullan, Ph.D. Dissertation, Cambridge University (1952).
6. K. C. A. Smith, Ph.D. Dissertation, Cambridge University (1956).
7. T. E. Everhart and R. F. M. Thornley, *J. Sci. Instr.*, **37**, 246 (1960).
8. C. W. Oatley and T. E. Everhart, *J. Electron.*, **2**, 568 (1957).
9. R. F. W. Pease, Ph.D. Dissertation, Cambridge University (1963).
10. R. F. W. Pease and W. C. Nixon, *J. Sci. Instr.*, **42**, 81 (1965).
11. A. V. Crewe, in *Scanning Electron Microscopy/1974*, IITRI, Chicago, Illinois (1969), p. 11.
12. D. G. Coates, *Phil. Mag.*, **16**, 1179 (1967).
13. G. R. Booker, in *Modern Diffraction Techniques in Materials Science* (S. Amelincx, ed.), North Holland, Amsterdam (1970), p. 647.
14. J. R. Banbury, Ph.D. Dissertation, Cambridge University (1970).
15. D. C. Joy, Ph.D. Dissertation, Oxford University (1969).
16. J. Philibert and R. Tixier, *Micron*, **1**, 174 (1969).
17. D. J. Fathers, J. P. Jakubovics, D. C. Joy, D. E. Newbury, and H. Yakowitz, *phys. stat. sol. a*, **20**, 535 (1973).
18. D. J. Fathers, J. P. Jakubovics, D. C. Joy, D. E. Newbury, and H. Yakowitz, *phys. stat. sol. a* **22**, 609 (1974).
19. K. F. J. Heinrich, C. E. Fiori, and H. Yakowitz, *Science*, **167**, 1129 (1970).
20. C. E. Fiori, H. Yakowitz, and D. E. Newbury, in *Scanning Electron Microscopy/1974*, IITRI, Chicago, Illinois (1974), p. 167.
21. C. J. D. Catto, Ph.D. Dissertation, Cambridge University (1972).
22. N. C. Yew and D. E. Pease, in *Scanning Electron Microscopy/1974*, IITRI, Chicago, Illinois (1974), p. 191.
23. J. Lebiedzik, K. G. Burke, S. Troutman, G. G. Johnson, Jr., and E. W. White, in *Scanning Electron Microscopy/1974*, IITRI, Chicago, Illinois (1974), p. 121.
24. R. F. Herzog, B. L. Lewis, and T. E. Everhart, in *Scanning Electron Microscopy/1974*, IITRI, Chicago, Illinois (1974), p. 175.
25. A. Boyde, in *Scanning Electron Microscopy*/1974, IITRI, Chicago, Illinois (1974), p. 91.
26. A. R. Dinnes, in *Scanning Electron Microscopy: Systems and Applications 1973*, Conference Series 18, Institute of Physics, London (1973), p. 76.
27. R. Fitzgerald, K. Keil, and K. F. J. Heinrich, *Science*, **159**, 528 (1968).
28. J. I. Goldstein and H. J. Axon, *Naturwiss.*, **60**, 313 (1973).
29. K. F. J. Heinrich, in *Fifty Years of Progress in Metallographic Techniques*, ASTM Special Tech. Publ. No. 430 (1968), p. 315.
30. H. Moseley, *Phil. Mag.*, **26**, 1024 (1913).

31. J. Hillier, U.S. Pat. 2,418,029 (1947); L. L. Marton, U.S. Pat. 2,233,286 (1941).
32. R. Castaing and A. Guinier, in *Proceedings of the 1st International Conference on Electron Microscopy, Delft, 1949*, (1950), p. 60.
33. R. Castaing, Thesis, University of Paris (1951), ONERA publ. No. 55.
34. R. Castaing, in *Advances in Electronics and Electron Physics* Vol. 13 (1960), p. 317.
35. I. B. Borovskii and N. P. Ilin, *Dokl. Akad. Nauk SSSR*, **106**, 655 (1953).
36. M. E. Haine and T. Mulvey, *J. Sci. Instr.*, **26**, 350 (1959).
37. L. S. Birks and E. J. Brooks, *Rev. Sci. Instr.*, **28**, 709 (1957).
38. R. M. Fisher and J. C. Schwarts, *J. Appl. Phys.*, **28**, 1377 (1957).
39. D. B. Wittry, Thesis, California Institute of Technology (1957).
40. J. R. Cuthill, L. L. Wyman, and H. Yakowitz, *J. Metals*, **15**, 763 (1963).
41. V. E. Cosslett and P. Duncumb, *Nature*, **177**, 1172 (1956).
42. D. B. Wittry, in *ASTM Spec. Techn. Publ. 349*, (1963), p. 128.
43. J. Philibert, in *X-Ray Optics and X-Ray Microanalysis* (H. H. Pattee, V. E. Cosslett, and A. Engström, eds.), Academic Press, New York (1963), p. 379.
44. P. Duncumb and P. K. Shields, in *The Electron Microprobe* (T. D. McKinley, K. F. J. Heinrich, and D. B. Wittry, eds.), Wiley, New York (1966), p. 284.
45. B. L. Henke, in *Advances in X-Ray Analysis*, Vol. 7, Plenum Press, New York (1964), p. 460.
46. B. L. Henke, in *Advances in X-Ray Analysis*, Vol. 8, Plenum Press, New York (1965), p. 269.
47. J. V. P. Long and S. O. Agrell, *Min. Mag.*, **34**, 318 (1965).
48. D. F. Kyser and D. B. Wittry, in *The Electron Microprobe* (T. D. McKinley, K. F. J. Heinrich, and D. B. Wittry, eds.), Wiley, New York, (1966), p. 691.
49. D. R. Beaman and J. A. Isasi, *Anal. Chem.*, **42**, 1540 (1970).
50. H. Yakowitz, R. L. Myklebust, and K. F. J. Heinrich, Nat. Bureau of Standards Tech. Note 796 (1973).
51. N. C. Barbi, A. D. Sandborg, J. C. Russ, and C. E. Soderquist, in *Scanning Electron Microscopy/1974*, IITRI, Chicago, Illinois (1974), p. 151.

BIBLIOGRAPHY OF TEXTS AND MONOGRAPHS IN SEM AND EPMA

1951   R. Castaing, "Application of Electron Probes to Local Chemical and Crystallographic Analysis," Ph. D. Thesis, University of Paris.
1966   H. A. Elion, *Instrument and Chemical Analysis Aspects of Electron Microanalysis and Macroanalysis*, Pergamon, New York.
1968   K. F. J. Heinrich (ed.), *Quantitative Electron Probe Microanalysis*, NBS Spec. Publ. 298.
1968   P. R. Thornton, *Scanning Electron Microscopy*, Chapman and Hall, London.
1969   A. J. Tousimis and L. L. Marton (eds.), *Electron Probe Microanalysis*, Advances in Electronics and Electron Physics (Suppl. 6), Academic Press, New York.
1971   L. S. Birks, *Electron Probe Microanalysis*, 2nd ed., Wiley–Interscience, New York.
1971   V. H. Heywood (ed.), *Scanning Electron Microscopy—Systematic and Evolutionary Applications*, Academic Press, London.
1972   D. R. Beaman and J. A. Isasi, *Electron Beam Microanalysis*, ASTM Spec. Tech. Publ. 506, 80 pp.

1972   J. W. S. Hearle, J. T. Sparrow, and P. M. Cross, *The Use of the Scanning Electron Microscope*, Pergamon Press, Oxford.

1972   C. W. Oatley, *The Scanning Electron Microscope*, University Press, Cambridge.

1973   C. A. Andersen (ed.), *Microprobe Analysis*, Wiley, New York.

1973   R. Woldseth, *X-Ray Energy Spectrometry*, Kevex, Burlingame, California.

1974   D. B. Holt, M. D. Muir, P. R. Grant, and I. M. Boswarva (eds.), *Quantitative Scanning Electron Microscopy*, Academic Press, London.

1974   O. C. Wells, *Scanning Electron Microscopy*, McGraw-Hill, New York.

1975   S. J. B. Reed, *Electron Microprobe Analysis*, University Press, Cambridge.

The proceedings of the annual SEM conference sponsored by the Illinois Institute for Technology Research Institute (IITRI) each year since 1968 constitute the best source for individual research papers on SEM. Compiled and edited each year by Dr. O. Johari, this set of proceedings is necessary to any SEM laboratory. Current and past proceedings can be obtained from Dr. Johari at IITRI.

ELECTRON OPTICS

J. I. Goldstein

The amount of current in a finely focused electron beam impinging on a specimen determines the magnitude of the signals (x-ray, secondary electrons, etc.) emitted, other things being equal. In addition, the size of the final probe spot determines the resolution of the scanning electron microscope (SEM) and electron microprobe (EPMA) for many mechanisms of contrast formation. Therefore, the electron optical system in these instruments is designed so that the maximum possible current is obtained in the smallest possible electron probe. In order to use the instruments intelligently, it is important to understand how the optical column is designed, how the various components of the optical system function, and which components are most important in determining the final current and spot size. In this chapter, we will discuss the various components of the electron optical system, develop the relationships between electron probe current and spot size, and discuss the factors which influence this relationship.

I. ELECTRON GUNS

A. Tungsten Filament Cathode

As shown in Chapter I, electron optical columns for the EPMA and SEM consist of the electron gun and two or more electron lenses. The elec-

J. I. GOLDSTEIN—Metallurgy and Materials Science Department, Lehigh University, Bethlehem, Pennsylvania.

tron gun provides a stable source of electrons which is used to form the electron beam. These electrons are usually obtained from a source by a process called thermionic emission. In this process, at sufficiently high temperatures, a certain percentage of the electrons become sufficiently energetic to overcome the work function E_w of the cathode material and escape the source. The emission current density J_c coming from the cathode is expressed by the Richardson law,

$$J_c = AT^2 \exp(-E_w/kT) \quad A/cm^2 \tag{1}$$

where A is a constant which is a function of the material, T is the emission temperature, and E_w is the work function.

The cathode is a wire filament, usually of tungsten, approximately 100 μm in diameter, bent in the shape of a hairpin with a V-shaped tip which is about 100 μm in radius. The filament is heated directly as current i_f from the filament supply is passed through it and is maintained at a high negative voltage (1–50 kV) during operation. For tungsten, at a typical operating temperature of 2700°K, J_c is equal to 1.75 A/cm² as calculated from the Richardson equation, where $A = 60$ A/cm² °K² and $E_w = 4.5$ eV. Filament life is a function of the tungsten evaporation rate and decreases with increasing temperature. At an emission current of 1.75 A/cm², filament life should average around 40–80 hr in a vacuum of 10^{-5} Torr. Although raising the operating temperature would have the advantage of increasing J_c, this would be accomplished only at the loss of filament life. At normal filament temperatures (2650–2900°K), the emitted electrons leave the V-shaped tip from an emission area of about 100 $\times$ 150 μm and have a Maxwellian distribution of velocities. At 2900°K, for example, the electrons have a maximum in the energy distribution at about 0.25 eV and a range of energies from 0 to about 2 eV.[1] The electrons obtained from the filament (cathode) are then accelerated to ground (anode) by a 1000–50,000 V potential between the cathode and anode. The configuration for a typical electron gun[2] is shown in Figure 1.

Surrounding the filament is a grid cap or Wehnelt cylinder with a circular aperture centered at the filament apex. The grid cap is biased negatively between 0 and 500 V with respect to the cathode. The effect of the electric field formed in such a gun configuration, the filament, Wehnelt cylinder, and anode causes the emitted electrons from the filament to converge to a crossover of dimension d_0. This configuration is equivalent to having an electrostatic immersion lens in the system. Figure 1 also shows the equipotential field lines which are produced between the filament grid cap and the anode. The constant field lines are plotted with respect to the filament (cathode) voltage and vary between zero at the filament

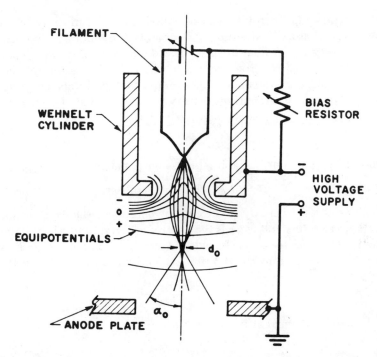

FIGURE 1. Configuration of self-biased electron gun (adapted from Hall[2]).

to a negative potential (up to 500 V) at the grid cap to the large positive potential (1–50,000 V) at the anode.

The emitted electrons are accelerated through this field and attempt to follow the maximum field gradient which is perpendicular to the field lines. When the constant field lines are negative, however, the electrons are repelled. Figure 1 shows the paths of electrons through the field lines. Note the focusing action as the electrons approach the negatively biased grid cap. By the use of the grid cap, the electrons are focused to a crossover of dimension d_0 and divergence angle α_0 below the Wehnelt cylinder. The intensity distribution of the electrons at crossover is usually taken to be Gaussian. The condenser and probe forming lenses demagnify this electron image, at crossover, to obtain the final electron probe.

The current density in the electron beam at crossover represents the current that could be concentrated into a focused spot on the specimen if no aberrations were present in the electron lenses. This current density, J_b (A/cm²), is the maximum intensity of electrons in the electron beam at crossover, and can be defined as

$$J_b = 4i_b/\pi d_0^2 \tag{2}$$

where i_b represents the total beam or emission current measured from the filament. It is usually desirable, in practice, to obtain the maximum current density in the final image. Since the maximum usable divergence angle of the focused electron beam is fixed by the aberrations of the final lens in the imaging system, the most important performance parameter of the electron gun is the current density per unit solid angle. This is called the electron beam brightness β and is given as

$$\beta = 4i_b/(\pi d_0\alpha_0)^2 \tag{3}$$

As shown by Langmuir,[3] the brightness has a maximum value given by

$$\beta = J_c eE_0/\pi kT \quad \text{A/cm}^2\text{ster} \tag{4}$$

for high voltages, where J_c is the current density at the cathode surface, E_0 is the accelerating voltage, e is the electronic charge, and k is Boltzmann's constant. The brightness can be calculated as $\beta = 11{,}600 J_c E_0/\pi T$, where the units are A/cm^2 for J_c and V for E_0. The current density can then be rewritten as

$$J_b = \pi\beta\alpha_0^2 \tag{5}$$

and the maximum current density is

$$J_b = J_c eE_0\alpha_0^2/kT \tag{6}$$

The theoretical current density per unit solid angle (brightness) for a given gun configuration can be approached in practice provided an optimum bias voltage is applied between cathode and grid cap. In the electron gun the bias is produced by placing a variable "bias resistor" in series with the negative side of the high-voltage power supply and the filament (Figure 1). In this configuration, when the current is supplied to the filament a negative voltage will be applied across the grid cap. As the resistance of the "bias resistor" changes, the negative bias voltage changes in direct relation. The major effect of varying the bias voltage is to change the electrostatic potentials around the cathode.

At low bias, the negative field gradient becomes weak (Figure 1) and the focusing action of the negative equipotentials is relatively ineffective. The electrons see only a positive field gradient toward the anode; and therefore the emission current i_b is high. Since little or no focusing takes place, the crossover d_0 is large and the brightness β obtained is not optimum. If the bias is increased too much, however, the negative equipotentials around the filament will be so strong that the electrons which are emitted from the filament will observe only a negative field gradient and will return to the filament. In this case the emission current as well as the brightness decrease to zero. Figure 2 summarizes the relationship between emission current, brightness, and bias voltage. As shown in Figure 2, there is an optimum bias setting for maximum brightness. This setting can be obtained for each electron gun. If the filament-to-Wehnelt cylinder distance can be changed, then the shape of the electrostatic field lines can also be altered. Depending upon the particular instrument, one may have the freedom to change either the filament-to-grid spacing and/or the bias resistance. These two factors must be properly adjusted to obtain the maximum brightness from the gun.

Typical values of d_0 and α_0 for electron guns used in the EPMA and SEM are $d_0 \sim 25\text{--}100$ μm and $\alpha_0 \sim 3 \times 10^{-3}$ to 8×10^{-3} rad. For a tungsten filament operated at 2700°K at a cathode current density J_c of 1.75 A/cm², the brightness at 25 kV, as calculated from equation (4), is about 60,000 A/cm² ster. The beam current i_b usually varies between 100 and 200 μA. For comparison purposes, in a conventional sealed-off x-ray tube, a considerably higher beam current i_b, typically of the order of 15–25 mA, is produced.

It is also important to obtain a stable, well-regulated beam current. As the current through the tungsten filament i_f is increased from zero, the temperature of the filament is increased and electron emission occurs. Figure 3 shows the emission characteristics of a self-biased gun in which the beam current i_b is plotted vs. the filament current i_f. If the bias resistor is correctly set for maximum brightness, the beam current i_b does not vary as the current through the filament is increased above a certain minimum value. This saturation condition occurs because as the current in the filament i_f is increased above that necessary for emission, the bias voltage also increases, causing the negative field gradient around the fila-

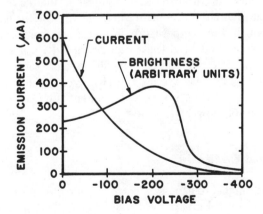

FIGURE 2.   Relationship of emission current and brightness to bias voltage; schematic only.

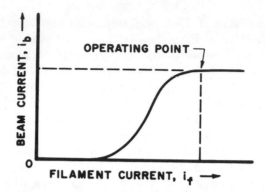

FIGURE 3.   Emission characteristic of the self-biased electron gun. Schematic
                      drawing of beam current i_b versus filament current i_f (adapted from
                      Hall[2]).

ment to increase and therefore limiting the rise in i_b. This balancing condition is called saturation and produces a self-regulating gun and a stable
beam current.

The current density at crossover J_b can be increased if the brightness
of the gun can be increased. Improved brightness would then provide either
increased current for the same beam size, or a reduced beam diameter for
the same current. Therefore, by increasing the gun brightness, either
sensitivity or resolution can be improved. The brightness of the filament
can be increased according to Langmuir's formula [equation (4)] by either
increasing the high voltage E_0 or increasing the current density J_c at the
cathode. Since voltages on the gun are held within relatively narrow limits,
major improvement in brightness for tungsten filaments will be made
by increasing the filament temperature in order to raise the emission current
J_c. For example, if the temperature is increased from 2700 to 3000°K,
the emission current and brightness can be raised by a factor of five or
more from $J_c = 1.75$ A/cm² and $\beta = 60{,}000$ A/cm² ster at 25 kV to $J_c =
14.2$ A/cm² and $\beta = 440{,}000$ A/cm² ster. However, although the brightness
increases by over a factor of seven, the filament life decreases from 40–80
hr to an unacceptably low 1 hr.

B.   LaB₆ Rod Cathode

Recently a high-brightness electron gun has been developed by
Broers[4,5] which uses a LaB_6 rod cathode. The cathode has a much higher
ratio of electron emission density to evaporation rate than the conventional
0.1-mm tungsten hairpin filament. Difficulties that have hindered the

development of the LaB$_6$ gun are: (1) the need for a better vacuum condition (10^{-6} torr) in the vicinity of the gun than is normally available in the EPMA or SEM and (2) the fact that the LaB$_6$ material cannot be directly heated as in the case of the tungsten filament. Typically the cathode is a small rod of solid LaB$_6$ approximately 1 mm square in cross section and 1.6 cm long. The end of the cathode rod from which the emission is drawn is milled to a fine point with a radius of approximately 10 μm. The other end is braced into a heat sink which is maintained at a much lower temperature during operation. Figure 4 shows the gun configuration. The cathode is heated by a coil of tungsten wire surrounding the filament to high temperatures by passing a current through it. Heat is applied to the cathode by radiation or by a combination of radiation and electron bombardment from the heater coil. Heat and evaporation shields are also placed around the heater. The temperature at the tip of the LaB$_6$ filament typically ranges from 1700 to 2100°K. The bias voltage should be 1000 V or greater; 2500 V is typical. [6]

Measurements of lifetimes and emission current densities have been made by Ahmed and Broers. [7] They found that LaB$_6$ produced a very high current density, 65 A/cm^2 at 1600°C and approximately 100 A/cm^2 at 1680°C. This current density is about an order of magnitude higher than that for the conventional W filament. Plotting their data using the Richardson law, equation (1), the constant A is found to be 40 A/cm^2 °K^2 and the work function E_w is 2.4 eV. The voltage field at the point of the cathode can be quite high and is sufficient to reduce the work function E_w by about

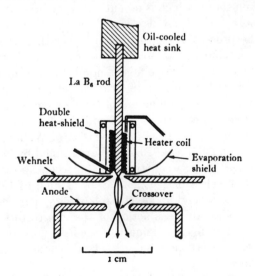

FIGURE 4. Gun configuration for LaB$_6$ cathode (from Broers[5]).

0.1 eV by the Schottky effect.[6] The work function for LaB_6 is much lower than for W, resulting in much higher emission. At less than 10 A/cm^2 the life of the LaB_6 filament exceeds 10,000 hr, two orders of magnitude higher than that of the standard W filament. Even at 50 A/cm^2, the filament lifetime is 200 hr. An emission density of 65 A/cm^2 and a temperature of 1600°C would be sufficient, according to Langmuir's equation, to produce a brightness of about 10^7 A/cm^2 ster at 75 kV or about 3×10^6 A/cm^2 ster at 25 kV. This brightness is at least five times greater than that of a tungsten filament operating at its maximum temperature at 25 kV. Satisfactory LaB_6 cathodes have been shown experimentally to produce a maximum brightness of 10^7 A/cm^2 ster at 75 kV.[8] Maximum brightness often drops off, however, after about 50 hr but satisfactory operation can still continue for several hundred hours at levels greater than 5×10^5 A/cm^2 ster at 75 kV.

A LaB_6 gun has been used in a high-resolution SEM built at IBM[9] and a minimum probe diameter estimated to be between 30 and 50 Å was measured with a beam current of 10^{-11} A at 24 kV. Several manufacturers have recently made available the LaB_6 gun for use on their SEM's. The necessity for vacuums of 10^{-6} torr or better requires improved gun pumping and limits the incorporation of LaB_6 guns in many present instruments.

C. Field Emission Gun

An electron source which is potentially very bright is the field emission gun. This source has been developed by Crewe and his associates.[10] It employs the process of field emission rather than thermionic emission, as discussed previously, to obtain a source of electrons. For the field emission process, a negative voltage is applied to a very sharp metal point and this high, negative field gradient drives electrons out and away from the point. Figure 5 shows a diagram of the field emission gun.[11] A voltage V_1 of around 3000 V between the field emission tip and first anode controls the emission current from the pointed whisker of W. Tips of these cathodes have radii between 200 and 2000 Å and are formed by electrolytic etching. A second voltage V_0 between tip and the second anode determines the final energy of the electrons, which can be as high as 100 keV (Figure 5). An aperture in the second anode controls the angular spread of the beam. Between the anodes, the lens action of the electrostatic field focuses the electron beam and forms a real image of the field emission tip some distance (3–5 cm) beyond the second anode.[11] The two anodes are specially shaped to minimize aberrations.

Current densities as high as 10^6 A/cm^2 can be obtained from the field

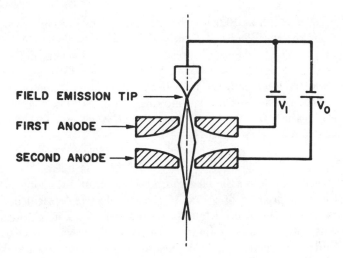

FIELD EMISSION TIP

FIRST ANODE

SECOND ANODE

V_1 V_0

FIGURE 5. Gun configuration for field emission (adapted from Crewe *et al.*[11]).

emitting tip and the apparent source size is even smaller than the tip size. The properties of this electron gun are such that, when used without any auxiliary lens, it is capable of producing a focused spot of electrons smaller than 100 Å and a high beam current of 10^{-10} A. As estimated by Broers,[12] the brightness of this gun at 23 kV is about 2×10^9 A/cm² ster, about two orders of magnitude better than the LaB_6 gun. The field emission source has been used without any magnetic lenses in an SEM to produce a resolution of 100–200 Å at beam currents of 10^{-11}–10^{-10} A.[11]

The major problems with field emission guns are the stringent vacuum required, about 10^{-10} torr, and the relative instability of the cathode tips. As pointed out by Broers,[13] because of the very small source size and the limited source current, the performance of the field emission gun falls below that of thermal emission cathodes for beam sizes above about 300 Å. Above 300 Å beam sizes, more beam current is available at the specimen using thermal emission cathodes.

II. ELECTRON LENSES

A. General Properties of Magnetic Lenses

The condenser and objective lens systems are used to demagnify the electron image formed at crossover ($d_0 \sim 25$–100 μm) in the electron gun to the final probe size on the sample (50 Å–1 μm). This represents a demagnification of as much as 10,000. The condenser lens system, which is composed of one or more lenses, determines the beam current which impinges on the sample. The final lens, often called the objective lens, determines the final spot size of the electron beam. Conventional electromagnetic lenses are used and the electron beam is focused by the interaction of the electromagnetic field of the lens on the moving electrons. The vector equation which relates the force on the electron **F** to the velocity of the electron **v** in a magnetic field of strength **H** is

$$\mathbf{F} = -e(\mathbf{v} \times \mathbf{H}) \tag{7}$$

where e is the charge on the electron, and the multiplication operation is the vector cross-product of **v** and **H**. In the electron optical instruments under discussion, the electron moves with a velocity **v** in a magnetic field **H** which is rotationally symmetric. Figure 6 shows a schematic section of the cylindrical electromagnetic lens commonly used as condenser lenses in these instruments. The lens is drawn so that the windings which are used to induce the magnetic field in the iron core can also be seen. The bore of the electromagnetic lens has a diameter D. The gap located in the center of the iron core is the distance S between the north and south pole pieces of the lens and is parallel to the direction in which the electrons are traveling.

The strength of the magnetic lens, that is, the intensity of the magnetic field in the gap, is proportional to NI, the number of turns in the solenoid winding times the current flowing through the lens. Since the speed v of the electrons is controlled by the high voltage E_0, the lens effect on the electrons is inversely proportional to the accelerating potential of the electron gun. The focal length of a lens f is that point on the z axis where a ray initially parallel to the z axis crosses the axis after passing through the lens. This focal length is inversely proportional to the strength of the lens. The greater the strength of the magnetic lens, the shorter will be its focal length. A beam of electrons which enters such a rotationally symmetric field will be forced to converge and may be brought to a focus. The shape and strength of the magnetic field determine the focal length and therefore the amount of demagnification. Figure 6 shows also that an electron beam which is diverging from an image of diameter d_0 above the lens is brought to a focus and demagnified to a diameter d_i outside the magnetic field of the lens.

A more detailed picture of the focusing action of the electromagnetic lens can also be gained from Figure 6. In Figure 6(a) the magnetic field lines **H** formed within the gap of the lens are depicted for a lens with axial symmetry. The magnetic field lines flow between two parallel iron pole pieces of axial symmetry. **H** is the vector which is parallel to the

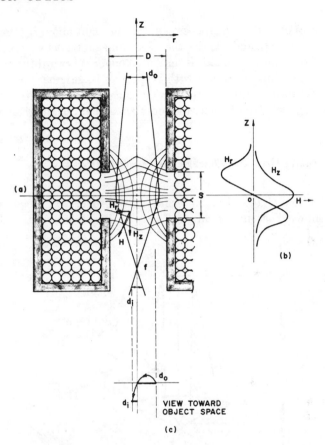

FIGURE 6. Schematic of an axially symmetric electromagnetic lens (adapted from Hall[2]). The magnetic field lines are plotted along with the components of the magnetic field.

magnetic field, H_r is the vector which is perpendicular to the direction of the electron optical axis, and H_z is the magnetic field which is parallel to the electron optical axis. Since the magnetic field H varies as a function of position through the electromagnetic lens, the values of H_r and H_z also vary. In Figure 6(b), a plot of the magnetic field intensity H_z parallel to the electron optical axis is given. The maximum magnetic field is obtained in the center of the gap. H_r, which is the radial magnetic field, has a zero component when the magnetic field is parallel to the z axis, that is, in the center of the gap of the electromagnetic lens, and has a maximum on either side of the center of the gap. The lines of constant magnetic field strength, the equipotential lines, which are normal to the magnetic field lines, stretch across the gap (Figure 6a) and show the converging field produced. Figures

6(a) and 6(c) also show the focusing action brought about by the combined magnetic fields H_r and H_z in the electromagnetic lens. A ray initially diverging from the optical axis is made to converge toward the center of the lens (Figure 6a) and to rotate with respect to its original position (Figure 6c). The detailed motions of the electrons can be determined by applying equation (7), using the two components H_r and H_z of the magnetic field.

B. Production of Minimum Spot Size

Reduction of the electron beam at crossover to a focused electron probe is shown diagrammatically in Figure 7. Ray traces of the electrons as

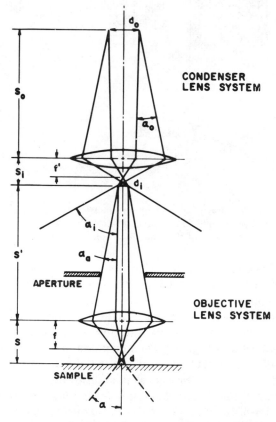

FIGURE 7.   Schematic of ray traces in a typical scanning electron microscope or electron microprobe optical column.

they pass from the crossover point in the electron gun through the condenser lens system and the probe forming lens are given for a typical scanning electron microscope or electron microprobe optical column. The electron image at crossover, of diameter d_0 and divergence angle α_0, passes through the condenser lens and is focused to a diameter d_i with a divergence angle α_i. In most instruments, the entrance aperture of the condenser lens system is greater than the divergence of the beam, enabling almost all of the beam to enter the lens system. The distance from the crossover point to the condenser lens gap is fixed at S_0 while the distance that the electrons are focused on the other side of the condenser lens is S_i and can be varied by changing the strength of the condenser lens system. The demagnification M of such a lens is given by the equation $M = S_0/S_i$, and is greater than one in the SEM–EPMA. The diameter d_i of the electron beam after passing through the condenser lens is equal to the crossover spot size d_0 divided by the demagnification M, and the divergence angle α_i of the electrons from the focused image below the condenser lens at S_i is equal to α_0 times the demagnification M.

If, as often the case, the lens thickness is negligible in comparison to S_0 and S_i, the Gaussian form of the thin lens equation from geometrical optics can be applied to the lens systems in the scanning electron microscope–electron microprobe. This equation has the form[2]

$$(1/S_0) + (1/S_i) = 1/f' \tag{8}$$

where f' is the focal length of the lens. We can apply this equation to the condenser lens system as shown in Figure 7. As discussed previously, as the current I in the condenser lens system increases, the strength of the lens increases and the focal length of the lens decreases. Therefore, according to equation (8), S_i will decrease. Also, as the strength of the lens increases, the demagnification M will increase. However, α_i, the divergence of the electrons from the focused spot, will increase. For the two-lens system as shown in Figure 7, equation (8) can be applied for each lens separately. The total demagnification is the product of the demagnifications of the lenses taken separately.

The distance from the intermediate image to the objective lens gap is S' and the final focused electron beam which impinges on the specimen is a distance S from the objective lens (Figure 7). The distance from the bottom pole piece of the objective lens to the sample surface is called the working distance and is around 5–25 mm in most instruments. Such a long working distance is necessary so that low-energy secondary electrons and magnetic specimens are outside the magnetic field of the lens. Also, in some EPMA's, the longer working distance prevents the x-rays emitted from the sample from being absorbed by the lens pole pieces. An objective aperture (100–300 μm in diameter) is placed in the electron optical column below the intermediate image (see Figure 7). This aperture decreases the divergence angle of the electron beam α_i from the condenser lens system and a smaller divergence angle, α_a is obtained for the electrons entering the objective or final lens. The electrons are then focused by the objective lens to a resultant spot size d on the sample with a corresponding divergence angle α. The demagnification of the objective lens is given by $M = S'/S$ and the final spot size on the sample is equal to d_0 divided by the product of the demagnifications M obtained from the multilens system.

The condenser lens in combination with the objective aperture determines the current in the final probe. As the demagnification of the electron image from the condenser lens increases, α_i increases. However, the amount of current which passes through the objective aperture is given by the ratio $(\alpha_a/\alpha_i)^2$ times the current which is available in the intermediate image d_i. Therefore, as the strength of the condenser lens increases, α_i increases, and the amount of current available in the final focused spot decreases. Figure 8 illustrates how the focusing of the condenser lens controls the electron beam intensity in the electron probe.[14] In one case (Figure 8a), the first lens is focused to allow most of the beam to pass through the aperture of the objective lens, while in the second case (Figure 8b), the first lens is set for a shorter focal length and only a small portion of the beam passes through the second lens aperture. The intermediate image d_i is larger in the first case. Therefore, minimizing probe size d by minimizing d_i can only be done by losing current in the final electron probe.

If no aberrations are inherent in the electron lens system, then the minimum spot size d at the specimen can be calculated. The successive images of the crossover produced by the various reducing lenses are formed at constant beam voltage E_0 so that the brightness of the final spot is again equal to the brightness β. Therefore the current density which is available in the final spot J_A is given by

$$J_A = \pi\beta\alpha^2 = J_c e E_0 \alpha^2/kT \tag{9}$$

and the current i in the spot is given by

$$i = J_A k_0 \frac{\pi}{4} d^2 = \left(\frac{\pi k_0 d^2}{4} \frac{eE_0}{kT}\right) J_c \alpha^2 \tag{10}$$

where $k_0 = 0.62$ if the distribution of the electrons in the focused beam is considered out to a point where the current density has fallen to $\sim J_A/5$. Using these relations, we obtain

$$d = (i/B\alpha^2)^{1/2} = (4i/\beta\pi^2\alpha^2)^{1/2} \tag{11}$$

where

$$B = (0.62\pi/4)(eE_0/kT)J_c$$

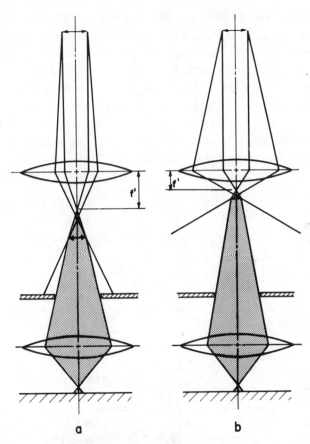

FIGURE 8.   Two condenser lens focusing conditions which control the current in
the final electron probe at the specimen (adapted from Birks[14]).

If there were no aberrations in the system, it would only be necessary to
increase α in order to increase the current at a constant probe diameter.
However, because of the several aberrations present in the electron optical
system, α must be kept small and the current available for a given probe
diameter is limited.

C. *Aberrations in the Electron Optical Column*

1. SPHERICAL ABERRATION

Spherical aberration arises because electrons moving in trajectories
which are further away from the optical axis are focused more strongly than

those near the axis. In other words, the strength of the lens is greater for rays passing through the lens the larger the distance from the optical axis. This aberration is illustrated in Figure 9. Electrons that diverge from point P and follow a path, for example PA, close to the optical axis will be focused to a point Q. Electrons that follow the path PB, which is the maximum divergence by the aperture of the lens, will be focused more strongly. These rays are focused on the axis closer to the lens than point Q. As shown in Figure 9, these rays (path PB) are focused to a point Q' at the image plane rather than to point Q. This process causes a disk of

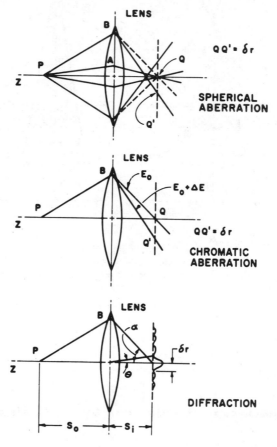

FIGURE 9.   Schematic drawings showing spherical and chromatic aberration, as well as diffraction at a lens aperture (adapted from Hall[2] and Oatley[15]).

confusion in the image plane, $\delta r = QQ'$, of point P. The diameter of the disk of least confusion can be written as

$$d_s = \tfrac{1}{2}C_s\alpha^3 \tag{12}$$

where α is the divergence angle at the image plane formed between BQ and the optical axis, and C_s, the spherical aberration coefficient, is related to the beam energy E_0 and the focal length f of the lens. The contribution of d_s to the final electron probe diameter can be made small by decreasing α. However, to accomplish this, the objective aperture size must be decreased; and, therefore, the current in the final spot will also be decreased. Alternatively, one can decrease the working distance by placing the specimen near the final lens pole piece. This requires an increase in the final lens excitation, thus effectively decreasing f and also C_s.

2. Chromatic Aberration

A variation in the energy E_0 and the corresponding velocity v of the electrons passing through the lens or a variation in the magnetic field H of the lens will change the point at which electrons emanating from a point P are focused. Figure 9 illustrates what may happen if the voltage of the electrons in the diverging ray PB are different, that is, rays of energies E_0 and $E_0 + \Delta E$. Some electrons will be focused to point Q and some to point Q' at the image plane. This process causes a disk of confusion in the image plane, $\delta r = QQ'$, of point P. The diameter d_c of the disk of least confusion is usually written as

$$d_c = (\Delta E/E_0)C_c\alpha \tag{13}$$

where α is the divergence angle at the image plane between BQ and the electron optical axis, $\Delta E/E_0$ is the fractional variation in the electron beam energy, and C_c is the chromatic aberration coefficient. The chromatic aberration coefficient is directly related to the focal length of the lens.

Variations in both E_0 and the magnetic field H (not illustrated in Figure 9) may occur from imperfect stabilization of the various power supplies. If the lens current or high voltage is stabilized to one part in 10^6 per minute, the effect due to the variation in H and E_0 will be unimportant.[15] Nevertheless, we still have a variation ΔE in the energy of the electrons due to the Maxwellian distribution of initial velocities, that is, the spread of the initial velocities (energy) leaving the cathode. Typical values of this velocity spread as expressed in electron energy are 2–3 eV for the tungsten hairpin filament,[13] 2–3 eV for LaB$_6$ depending on the bluntness of the tip of the cathode,[6] and 0.2–0.5 eV for emitted electrons from field emission cathodes.[13] The value of d_c can be minimized by decreasing the divergence angle α at the specimen as is the case for spherical aberration.

3. DIFFRACTION

Even if the two aberrations previously discussed were insignificant, the image of a point P would still be of finite dimensions, owing to the wave nature of electrons and the aperture size of the final lens. This effect is illustrated in Figure 9. The intensity distribution of the point source at the image plane caused by diffraction is shown in this figure. The radius of the first minimum δr subtends an angle θ at the lens. As described by Hall,[2] the effect of diffraction also yields a disk of confusion of diameter d_d which is given by

$$d_d = 1.22\lambda/\alpha \qquad (14)$$

where λ is the wavelength of the electrons, given by $\lambda = 12.26/E_0^{1/2}$ with λ in angstroms, E_0 in electron volts, and α, the angle between the converging ray and the electron optical axis, in radians (Figure 9). For this aberration, the larger the value of α, the smaller will be the contribution of d_d.

4. ASTIGMATISM

Although it has been tacitly assumed that magnetic lenses have perfect symmetry, this is not necessarily so. Machining errors and possible inhomogeneous magnetic fields within the iron, asymmetry in the windings, and uneven buildup of any contamination lead to loss of symmetry. If a lens system has elliptical rather than circular symmetry, for example, electrons diverging from a line focus will converge to two separate line foci at right angles to each other rather than to a line focus. This astigmatism effect then will enlarge the effective size of the final electron probe diameter. Fortunately, we can use a stigmator in the final lens to supply a weak correcting field in order to produce the desired symmetric field in the lens. The stigmator usually has two major controls, one to correct for the magnitude of the asymmetry and one to correct for the direction of the asymmetry of the main field. One can usually correct for the astigmatism in the system by adjusting the stigmator magnitude and orientation alternately and by refocusing on the image at medium magnifications (5000–10,000×).

D. Design of the Final Lens

It is assumed that all the significant aberrations are caused by the final lens. This is justified since the images of the crossover produced by the earlier lenses have much larger diameters than the final spot size, and the effects of aberrations in these lens are relatively small when compared to the size of the intermediate images. Therefore the design of the final lens is very important to producing small electron beam spot sizes in the SEM–EPMA. The significant lens aberrations that have to be considered in the final lens design are spherical aberration, chromatic aberration, and astigmatism.

Other performance characteristics must be also considered for the final lens. For the SEM, secondary electrons emitted over a wide solid angle have energies of only a few electron volts and must reach the detector to produce the necessary signal. Therefore the magnetic field at the specimen must be low enough not to hinder the efficient collection of secondary electrons and not to interfere with the development of voltage contrast, magnetic contrast, beam-induced conductivity contrast, etc. (see Chapter VI) at the surface of the sample. The lens design must also allow a clear path for x-rays produced in the specimen to be collected without hitting parts of the final lens. Also, the bore of the lens should allow enough room to place the scanning coils, stigmator, and beam limiting aperture and will have to avoid obstructing those electrons that are deflected a considerable distance from the electron optical axis during scanning. Since aberrations of the lens increase rapidly with focal length, the focal length should also be as short as possible.

All these performance characteristics suggest that for the best resolution the specimen should be mounted immediately outside the bore of the lens. In order to keep the magnetic field at the specimen low enough to allow efficient collection of secondary electrons and minimize the focal length, the lens is usually of the pinhole variety as shown in Figure 10. In this design, the diameter of the outer pole piece is much smaller than that of the inner pole piece. The lens is highly asymmetric, as opposed

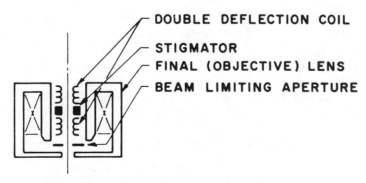

DOUBLE DEFLECTION COIL

STIGMATOR

FINAL (OBJECTIVE) LENS

BEAM LIMITING APERTURE

FIGURE 10. Final lens of the pinhole variety (adapted from Broers[13]).

to the more conventional condenser lenses (Figure 6) and H_z reaches its maximum value quite close to the inner face of the pole piece containing the smaller bore.[15] The limiting aperture is usually placed at this position, where H_z is a maximum. For the design of the final lens of the high-resolution instrument due to Pease and Nixon,[16] the spherical aberration coefficient C_s was 20 ± 5 mm and the chromatic aberration constant C_c was 8 mm at a working distance of 5 mm. For the high-resolution instrument designed by Broers[9] using a LaB$_6$ gun, the final lens had a design spherical aberration coefficient C_s of 18 mm and a chromatic aberration coefficient C_c of 11 mm at a working distance of 7 mm.

III. ELECTRON PROBE DIAMETER d_p VS. ELECTRON PROBE CURRENT i

A. Calculation of d_{min} and i_{max}

Following Smith,[17] it is possible to determine the diameter d_p of an electron probe carrying a given current i. It is assumed that all the significant aberrations are caused by the final lens. The aberrations considered are chromatic and spherical aberration as well as the effect of diffraction. The procedure that is followed is to regard the individual estimates of probe diameters d, d_c, d_s, and d_d as error functions and regard the effective spot size d_p equal to the square root of the sum of the squares of the separate diameters (quadrature), that is,

$$d_p = (d^2 + d_c{}^2 + d_s{}^2 + d_d{}^2)^{1/2} \tag{15}$$

From equations (11)–(14), we obtain

$$d_p{}^2 = \left[\frac{i}{B} + (1.22\lambda)^2\right]\frac{1}{\alpha^2} + \left(\frac{1}{2}C_s\right)^2 \alpha^6 + \left(\frac{\Delta E}{E}C_c{}^2\right)\alpha^2 \tag{16}$$

Pease and Nixon,[16] following Smith,[17] obtained the theoretical limits to probe current and probe diameter by considering only the first two terms of equation (16), spherical aberration and diffraction. They differentiated d_p in equation (16) with respect to the aperture angle α in order to obtain an optimum α. For the optimum α, the current in the beam will be a maximum and the beam size will be a minimum. Using this procedure, values of α_{opt}, d_{min}, and i_{max} were obtained from equation (16). These relations are

$$d_{min} = 1.29 C_s{}^{1/4}\lambda^{3/4}[7.92(iT/J_c) \times 10^9 + 1]^{3/8} \tag{17}$$

$$i_{max} = 1.26(J_c/T)[(0.51 d^{8/3}/C_s{}^{2/3}\lambda^2) - 1] \times 10^{-10} \tag{18}$$

$$\alpha_{opt} = (d/C_s)^{1/3} \tag{19}$$

It can be seen in equation (18) that the incident beam current will vary with the 8/3 power of the probe diameter. Since secondary electron and x-ray emission vary directly with probe current, they fall off very rapidly as the probe diameter is reduced.

There are, however, a few ways in which i_{max} can be increased. As the voltage of the electron beam increases, λ decreases, the value of i_{max} increases, and d_{min} will decrease. However, to keep the x-ray emission volume small as discussed in Chapter III, the maximum voltage that can successfully be applied when x-ray analysis is desired is about 30 kV. The value of i_{max} can also be increased if the spherical aberration coefficient C_s can be reduced by decreasing the focal length of the objective lens. However, because of the need for an adequate working distance beneath the final lens, the focal length cannot be greatly reduced. In addition, significant improvements in C_s, which would require radical changes in lens design, provide only small increases in current since i_{max} is proportional to $C_s^{-2/3}$. Mulvey,[18] in a recent paper, discusses the design of minilenses which are just being developed for SEM and EPMA applications. These lenses have improved coil windings which allow a major reduction in the volume of iron that has to be magnetized in order to produce a given focal property. Indications are that the axial field distribution H_z produced by a coil of small inner diameter will have an appreciably lower spherical aberration coefficient (1–2 mm) than the best iron pole piece lens (10–20 mm). A reduction in C_s by a factor of ten will decrease d_{min} by about a factor of two or increase i_{max} by about a factor of five. These new lenses are still at an early stage but appear to provide interesting possibilities for the future. At the present time, however, any large increases in i_{max} or corresponding decreases in d_{min} will come primarily from improvements in gun current density J_c as discussed previously.

Figure 11 illustrates the relationships between probe current and the size of the electron beam as given by equations (17) and (18). Values of C_s (20 mm) and J_c (4.1 A/cm² for W at 2820°K and 25 A/cm² for LaB$_6$ at 1900°K) that were chosen for the calculation are typical of operational instruments. The relationships between probe current and electron beam size are given for two operating voltages, 15 and 30 kV. The corresponding values of brightness β according to Langmuir's equation [equation (4)] are 8×10^4 and 1.6×10^5 A/cm² ster for W at 15 and 30 kV and 6.9×10^5 and 1.37×10^6 A/cm² ster for LaB$_6$ at 15 and 30 kV.

It can be observed from Figure 11(a) for the EPMA microanalysis range that for the particular C_s and J_c used, the maximum current available in a 1-μm electron beam using a conventional W filament is about 10^{-6} A at 15 kV and 2×10^{-6} A at 30 kV. This amount of current is well above the minimum current (1–5 $\times 10^{-8}$ A) usually needed to perform satis-

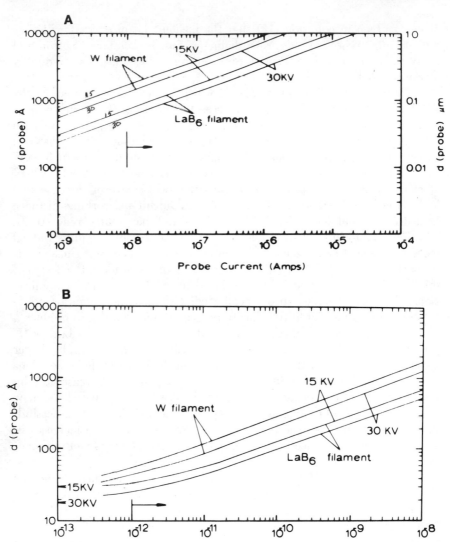

FIGURE 11.   Relationship between probe current i_{max} and the size of the electron beam d_{min}. Calculations consider both the W hairpin filament and the LaB$_6$ gun operating at 15 and 30 kV. (a) EPMA microanalysis range, (b) SEM range.

factory quantitative x-ray analyses with wavelength-dispersive spectrometers (WDS). X-ray analysis can, according to Figure 11(a), be obtained using a W filament with minimum electron beam sizes of the order of 0.2 μm (2000 Å). This spot size is well below the diameter of the region of x-ray emission from the sample ($\sim$1 μm, see Chapter III). A small beam size of this order allows the operator the freedom to take electron scanning images of the analyzed areas as well without changing operating conditions. The LaB$_6$ gun would provide additional advantages in the microanalysis range since it allows the analyst to carry out successful x-ray analysis with an electron beam below 0.1 μm in size. It should be pointed out that at normal SEM beam sizes of approximately 100 Å (Figure 11b), the beam current for W or LaB$_6$ filaments is below 10^{-19} A and is much too low for wavelength-dispersive x-ray analysis. This is, however, just the current range at which energy-dispersive x-ray analysis can be accomplished (see Chapter VII).

Broers[13] has calculated the relationships between probe current and the size of the electron beam for a simple field emission gun and a field emission gun plus magnetic lens as well as for thermal cathodes. The lens added to the field emission gun allows optimum demagnification of the gun crossover to be maintained over a range of gun operating conditions. Figure 12 illustrates these relationships. The curve for $\beta = 10^5$ A/cm^2 ster approximates the behavior of the W filament plotted in Figure 11 and the curve for $\beta = 10^6$ A/cm^2 ster approximates the behavior of the LaB$_6$ filament plotted in Figure 11. As discussed by Broers,[13] because of the very small source size and the limited source current, the performance of

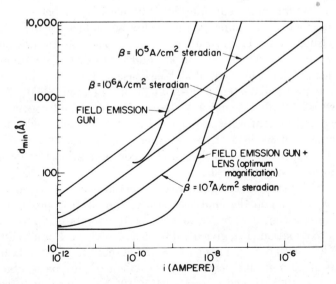

FIGURE 12.   Beam diameter versus current for a simple field emission gun and field emission gun plus magnetic lens. Curves are superimposed on performance for thermal cathodes. (From Figure 6, Broers[13]).

field emission guns falls below that of thermal cathodes (W, LaB_6) for beam sizes above about 300 Å. This fact may not be important in the SEM mode but field emission cathodes do not currently offer a performance advantage when larger beam sizes are required, as in the EPMA.

B. High-Resolution Scanning Electron Microscopy

In the limit when $0.51d^{8/3}/C^{2/3}\lambda^2$ [equation (18)] equals one, $i_{max} = 0$ and there is no current in the electron probe. Therefore one can calculate the minimum probe size d_{min} [equation (17)] as

$$d_{min} = 1.29C_s^{1/4}\lambda^{3/4} \tag{20}$$

This equation is similar to the formula for the limit of the resolving power of a conventional transmission electron microscope. At 15 and 30 kV the ultimate resolution is 30 and 18 Å, respectively, if $C_s = 2$ cm (Figure 11b). If the specimen is placed within the objective lens, as in the scanning transmission microscope (STEM), C_s is reduced because of the shorter focal length and the value of d_{min} will approach that of the standard transmission instrument. Since it is necessary to have the probe current i appreciably greater than zero to obtain useful secondary electron or x-ray signals, the value of d_{min} is greater than the theoretical resolution.

The minimum probe current generally accepted as being sufficient to form a satisfactory scanning picture using secondary electrons is about 10^{-12} A. Using Figure 11(b), it appears that a 50-Å electron beam can be attained using a W filament at 30 kV and a 30-Å electron beam can be attained using a LaB_6 filament at 30 kV. Pease and Nixon[16] have reported a minimum probe diameter of 50-Å at 30 kV in their SEM. The probe current was 8×10^{-13} A and a W hairpin filament and conventional oil diffusion pump system were employed. Broers[6] has reported a minimum probe diameter of 30–50 Å at 24 kV using a LaB_6 electron gun. The beam current was approximately 10^{-11} A and a contamination-free ion-pumped vacuum system was employed.

It is of interest to calculate the importance of the various aberrations d_c, d_s, and d_d on the final spot size of the electron probe. As an example, one can consider the various probe diameters d, d_s, and d_d that, according to equation (15), give a value of d_{min} of 50 Å. For a SEM operating at 30 kV, with a W filament ($J_c = 4.1$ A/cm²), a spherical aberration coefficient of 20 mm, and neglecting the effect of chromatic aberrations, $\alpha_{opt} = 0.63 \times 10^{-2}$ rad and $i_{max} = 1.64 \times 10^{-12}$ A according to equations (18) and (19). From equations (11), (12), and (14), the various contributions to the final 50-Å diameter are $d = 42$ Å, $d_s = 25$ Å, and $d_d = 14$ Å.

These calculations assume that chromatic aberration does not affect the final beam size $(d_c \simeq 0)$. For high-resolution microscopy at low voltages and when tungsten hairpin filaments are used, the effect of chromatic aberration is not trivial. The effect of chromatic aberration can be calculated using equation (13), $d_c = (\Delta E/E_0)C_c\alpha$, with a typical value of C_c, 0.8 cm.[16] For a thermionic cathode, the value of ΔE is 2–3 eV, as discussed previously. Using these values for the 50-Å beam at 30 keV and for $\alpha = 0.63 \times 10^{-2}$ rad previously discussed, the value of d_c is found to be about 40 Å. This is a significant contribution and according to equation (15), will have the effect of increasing d_p from 50 to about 65 Å.

The effect of chromatic aberration for a W hairpin filament at lower voltages is more important. For example, for a 50-Å beam, the value of d_c is 80 Å at 15 kV, using the same calculation scheme as discussed previously. The effect of chromatic aberration leads to an enlarged probe of about 95 Å. Since the energy spread ΔE of the emitted electrons in the LaB$_6$ gun is about the same as for the W filament,[6] significant reductions in the effect of chromatic aberration cannot be expected. The brightness of the LaB$_6$ gun is, however, significantly higher and smaller values of d_{min} are expected. Nevertheless, the effect of chromatic aberrations is quite important in calculating the ultimate electron beam resolution using this electron gun. It is interesting to note that the energy spread in the field emission gun is 0.2–0.5 eV,[13] much lower than the thermionic guns previously discussed.

In any attempt to achieve small probe sizes, ≤ 100 Å, in electron optical instruments, not only must the electron optics be designed to minimize C_s and C_c and maximize J_c, but the instrument must also be correctly aligned and attempts must be made to eliminate the deleterious effects of vibration, ac stray magnetic field interference, and specimen contamination. It is most important to keep the instrument well aligned. Filaments will warp with time and move from the alignment position. Therefore the filament must be recentered during operation. The final aperture which defines the final value of α and the current also requires constant care. It collects much of the beam current and can become easily contaminated. The apertures must be cleaned frequently and carefully positioned when replaced in the instrument. Stray ac magnetic fields from nearby apparatus and power supplies are troublesome to the operation of the instrument at high magnification ranges. These fields have frequenices of between 50 and 200 Hz and must be reduced in magnitude to a value on the order of 5–10 mG in the vicinity of the electron column. Attempts to minimize these effects have been described by Broers.[9] To reduce the effects of contamination as much as possible, an ion pump is used for high-vacuum pumping and the exposure of the system to oil is kept to a minimum. The specimen stage which is used at very high resolution is made in such a way that there is no mechanical contact between the specimen and the base of the chamber during analysis, which reduces vibration below a detectable level. Low-frequency mechanical vibrations (2–10 Hz) can cause the whole instrument to vibrate. The instrument must be isolated from the effect of these vibrations or high resolution will not be achieved. All of these effects can be eliminated by thorough engineering and therefore are not of great importance when considering ultimate resolution. However, they are of importance in the practical operation of the instrument.

IV. DEPTH OF FIELD

One of the most important advantages of the SEM–EPMA instrument is the large depth of field which is available. The depth of field can be defined with the help of Figure 13. A perfect lens is assumed and the specimen is placed a distance Q from the center of the lens. The depth of field defines the distance that the specimen surface can be displaced from point Q without affecting the performance of the instrument, that is, without the specimen going out of electron optical focus. If Q' and Q'' represent the limits of this displacement, then the distance $Q'Q''$ defines the field of view or depth of field. Stated another way, a rough surface can be examined by the SEM and remain in focus as long as the surface features lie between Q' and Q''.

One can consider that at Q' or Q'' a disk of confusion is obtained, rather than a point, whose radius r (Figure 13) is equal to $QQ''\alpha$. If δ is

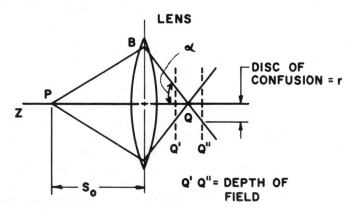

FIGURE 13. Schematic drawing showing depth of field (adapted from Oatley[15]).

the smallest distance that the eye can resolve on the record CRT or on the SEM photograph and M is the total magnification of the image then

$$Q'Q\alpha = QQ''\alpha = \delta/2M \tag{21}$$

and the total depth of field $Q'Q''$ is $\delta/M\alpha$. In the limit as δ/M approaches the electron optical resolution of the instrument d_p, we have

$$Q'Q'' = [(\delta/M) - d_p]/\alpha \tag{22}$$

since any point at position Q must be at least d_p wide. If the resolution of the eye δ is 0.1 mm, the magnification on the photograph is 10,000×, α of the final lens is 0.5×10^{-2} rad, and $d_p = 50$ Å, the depth of field $Q'Q''$ is 1 μm. This is quite a large value at this magnification. The depth of field for a low-magnification SEM photograph is much larger, for example, 2 mm at 10×. At comparable magnifications, the depth of field of the SEM is over 100 times greater than that of the light microscope since the values of α from the objective lens of light microscopes are much larger. This increase in depth of field over the light microscope is extremely important for observing specimens with rough surfaces.

One must often compromise between depth of field and high resolution when viewing a sample with the SEM. The depth of field can be increased by decreasing α. If the final aperture is a fixed size, α can be decreased by increasing the distance of the specimen from the bottom of the final lens, the working distance. However, this change will decrease the resolution of the instrument and increase d_p since the amount of demagnification in the final lens is decreased (see Figure 7). If the working distance is held constant, the depth of field can be increased by decreasing the size of the final aperture used. The smaller α obtained may no longer be α_{opt} and may decrease the resolution of the instrument and increase d_p. In many instruments one can vary both working distance and aperture size. The operator of the SEM must therefore decide what type of information is desired in order to obtain the best compromise.

REFERENCES

1. V. K. Zworykin, G. A. Morton, E. G. Ramberg, J. Hillier, and A. W. Vance, *Electron Optics and the Electron Microscope*, Wiley, New York (1945).
2. C. E. Hall, *Introduction to Electron Microscopy*, McGraw-Hill, New York (1953).
3. D. B. Langmuir, *Proc. IRE*, **25**, 977 (1937).
4. A. N. Broers, *J. Appl. Phys.* **38**, 1991 (1967).
5. A. N. Broers, *J. Physics E* **2**, 273 (1969).
6. A. N. Broers, in *Scanning Electron Microscopy/1974*, IITRI, Chicago, Illinois (1974), p. 10.

7. H. Ahmed and A. N. Broers, *J. Appl. Phys.*, **43**, 2185 (1972).

8. A. N. Broers, *J. Vac. Sci. Technol.*, **10**, 979 (1973).

9. A. N. Broers, *Rev. Sci. Instr.*, **40**, 1040 (1969).

10. A. V. Crewe, D. N. Eggenberger, J. Wall, and L. M. Welter, *Rev. Sci. Instr.*, **39**, 576 (1968).

11. A. V. Crewe, M. Isaacson, and D. Johnson, *Rev. Sci. Instr.*, **40**, 241 (1969).

12. A. N. Broers, in *Scanning Electron Microscopy*/1970, IITRI, Chicago, Illinois (1970), p. 3.

13. A. N. Broers, in *Microprobe Analysis* (C. A. Andersen, ed.), Wiley, New York (1973), p. 83.

14. L. S. Birks, *Electron Probe Microanalysis*, 2nd ed., Wiley–Interscience, New York (1971).

15. C. W. Oatley, *The Scanning Electron Microscope*, Part 1, "The Instrument," Cambridge University Press, Cambridge (1972).

16. R. F. W. Pease and W. C. Nixon, *J. Sci. Instr.*, **42**, 81 (1965).

17. K. C. A. Smith, Ph.D. Dissertation, Univ. of Cambridge (1956).

18. T. Mulvey, in *Scanning Electron Microscopy*/1974, IITRI, Chicago, Illinois (1974), p. 44.

ELECTRON BEAM–SPECIMEN INTERACTION

J. I. Goldstein

A large number of interactions occur when a focused electron beam impinges on a specimen surface. Among the signals produced are secondary electrons, backscattered electrons, characteristic and continuum x-rays, Auger electrons, and photons of various energies. These signals are obtained from specific emission volumes within the sample, and these emission volumes are strong functions of the electron beam energy E_0 and the atomic number of the specimen Z. In fact the resolution for a particular signal in the electron microprobe or scanning electron microscope is primarily determined by its excitation volume and not by the electron probe size. This chapter will discuss scattering and electron penetration in solids in order to provide a basis for understanding how the electron beam interacts with the sample. In addition we will discuss, in turn, each of the signals as well as the spatial resolution that can be obtained in these instruments.

I. ELECTRON SCATTERING IN SOLIDS

Electrons having energies in the range 1–50 keV impinging on the surface of a solid sample exhibit very complex behavior. The primary effects

J. I. GOLDSTEIN—Metallurgy and Materials Science Department, Lehigh University, Bethlehem, Pennsylvania.

on the electrons of the impinging high-voltage electron beam in the solid target are elastic scattering (change of direction with negligible energy loss) and inelastic scattering (energy loss with negligible change in direction). Elastic scattering is caused mainly by interactions with the nucleus and significant deviations from the incident direction occur. Inelastic scattering is caused by two mechanisms, inelastic interaction with the atomic nucleus and inelastic interaction with the bound electrons.

Let us consider inelastic scattering first. Inelastic scattering is primarily responsible for producing signals other than backscattered electrons. If inelastic scattering occurs through interaction with the nuclei of the atoms, the moving electrons lose energy in the Coulomb field of the nucleus and emit white or continuum x-ray radiation. If inelastic collisions occur between the loosely bound outer electrons and the incoming beam, energy is lost from the beam electrons and the loosely bound electrons are ejected. The ejected electrons have an energy typically less than or equal to 50 eV and are called secondary electrons. If these secondary electrons are produced close to the surface and the energy of the secondary electrons is greater than the surface barrier energy (2–6 eV), the secondary electrons have a high probability of escaping from the surface. These electrons are strongly absorbed, however, and if they are produced much below $\geq$ 100 Å of the surface of the sample the probability of escape is extremely small. If the secondary electrons recombine with the holes formed during the scattering process in some materials, a photon of energy is produced which has a wavelength in the visible or near-infrared range. This visible luminescence can frequently be seen optically when examining insulators. In all of these inelastic collisions, the process is combined with the loss of some or all of the incoming energy of the electrons. Inelastic collisions can result in a variety of ionization processes. One of the results of inelastic collisions is the production of characteristic x-ray lines. The primary electron beam loses energy equivalent to the binding energies of the K, L, or M shells, E_K, E_L, or E_M, and the electrons are ejected during the production of the characteristic x-radiation. Occasionally, following ejection of an electron, the deexcitation process may cause another electron called an Auger electron to be ejected without x-ray photon emission.

A thorough treatment of the dissipation of the energy of an electron beam during inelastic scattering in a solid target is necessarily quite complex. From quantum theory, Bethe[1] obtained an expression for the rate of loss of energy of an electron beam in matter in terms of the mean energy E_m and the path length traversed X:

$$- \frac{dE_m}{dX} = 2\pi e^4 N_0 \frac{Z}{A} \frac{\rho}{E_m} \ln \frac{1.166 E_m}{J} \tag{1}$$

where e is the electronic charge, ρ is the physical density, Z is the atomic number, A is the atomic weight of the element concerned, and N_0 is Avogadro's number. The parameter J is called the mean ionization potential of the scattering element, and is the average energy loss per interaction of all the individual energy loss processes (inelastic interactions). The Bethe equation thus allows a discrete energy loss process to be described by a continuous, mean loss. A more convenient form of equation (1) is the stopping power S defined by

$$S = -\frac{1}{\rho}\frac{dE_m}{dX} \tag{2}$$

which indicates the energy loss per unit mass thickness. Wilson[2] obtained a value $J = 11.5Z$ (eV) for aluminum which has been widely used for the entire range of atomic numbers. Empirical data[3] also show that the mean ionization potential J increases with increasing atomic number above $Z = 10$. The density dependence of equation (1) is eliminated by dividing by ρ in equation (2). Therefore the stopping power S increases with decreasing atomic number and is about 50% greater in Al than in Au at 20 keV.[4] It is important to note that the Bethe expression relates the energy loss along the path length X. Therefore equation (1) does not give the rate of energy dissipation with a depth in a target z and will give too rapid a variation with Z and E. One can only obtain energy dissipation with depth if the energy loss is combined with some method which takes into account lateral spread of the beam by elastic scattering. A discussion of elastic scattering follows.

Elastic scattering by the nucleus is by far the most probable large-angle scattering mechanism. The cross section (probability) for inelastic scattering into angles greater than 10^{-2} rad is much smaller than for elastic scattering for all elements except those of low atomic number. It is convenient to think of the elastic scattering process as occurring in two parts: (a) Rutherford scattering, which occurs in the Coulomb field of the nucleus whereby a single scattering act may result in a large change of direction (occasionally greater than 90°), and (b) multiple scattering, which is composed of many small-angle scattering events. In each of these small-angle events, the electron passes through the electron cloud of the atom, which acts as a screening field for the nucleus. Multiple scattering may also result in a large change of direction of the impinging electron beam. Beam electrons may change directions in a series of events, travel back to the surface and escape. This is the process of backscattering. The backscattered electrons leave with reduced energy due to the inelastic processes. At some depth within the target, the original direction of the electron beam is lost and the electrons diffuse through the material at random. The position at which this occurs can be thought of as the depth of complete diffusion x_d. A rather complete discussion of scattering theory and experiments is given in a series of papers by Cosslett and Thomas.[5-7]

The scattering cross section at constant energy varies with Z^2 and the probability of scattering through a given angle varies as Z^2/E^2. As calculated by Murata et al.,[8] the mean free path between scattering events at 30 keV decreases with increasing atomic number from 528 Å for Al to 131 Å for Cu to 50 Å for Au. Since the probability of scattering is low and the mean free path is large for low atomic number samples, there is not much scattering near the surface of the sample as the electrons enter. Only a few electrons are scattered through large angles and leave the sample as backscattered electrons. For low atomic number samples, then, most electrons penetrate deeply into the target before changing direction by more than 90° and hence are absorbed by the sample. In a high atomic number sample, however, there is considerable (singular and multiple) scattering close to the surface and a large fraction of the incoming electrons are backscattered. As the energy of the electrons decreases during passage through the sample, the amount of scattering will increase and a state of complete diffusion will occur. In the case of a heavy element, such as gold, diffusion sets in much nearer the surface than for a light element, and most of the electron backscatter is caused by multiple rather than single scattering. The shape of the electron distribution within the target as a function of voltage and atomic number can therefore be determined qualitatively as discussed by Duncumb and Shields[9]. Figure 1 shows a section through the sample of the electron distribution as a function of depth z. The outside limits of the electron distribution represent zero energy in the electron beam. From this figure it can be observed that, at the same energy, the electrons appear to penetrate more deeply into the low atomic number element and the electron distribution appears to be more pear-shaped. If the energy of the incident electrons is increased, the path length of each electron is lengthened, and the envelope is expanded but retains the same shape. Little change occurs in the fraction backscattered, although the electrons that are backscattered obviously have greater energy.

II. ELECTRON RANGE AND SPATIAL DISTRIBUTION OF THE PRIMARY ELECTRON BEAM

The electron range R is defined as the average total distance (measured from the surface of the sample) that an electron travels in the sample along

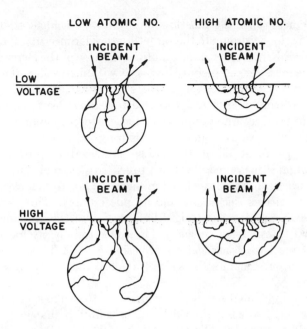

FIGURE 1.   Section through specimen surface illustrating the variation of electron scattering with voltage and atomic number (from Duncumb and Shields[9]).

a trajectory. The spatial distribution is the spread of the electron beam laterally from the center of impact. For the incident electrons, the electron range and the spatial distribution are similar in value since they are both defined by the amount of elastic scattering. There are several ways to define electron range R. One of these is as a function of the incident electron energy E_0 and is given by the following integration:

$$R = \int_{E=E_0}^{E=0} \frac{1}{dE/dX} \, dE \qquad (3)$$

By substituting the Bethe energy loss expression [equation (1)] in equation (3), one can define the Bethe range R_B. This range is inversely proportional to the density of the target and it is convenient to use the Bethe "mass range" ρR_B, the product of R_B multiplied by the density, usually expressed in mg/cm^2:

$$\rho R_B = \int_{E_0}^{E=0} \frac{1}{(dE/dX)(1/\rho)} \, dE = \int_{E_0}^{E=0} \frac{dE}{S} \qquad (4)$$

The Bethe mass range is a function mainly of the incident electron energy but it also varies according to the target material. Since the stopping power S decreases with increasing atomic number because of the increasing density ρ, the Bethe mass range generally increases with increasing atomic number. The value of the Bethe mass range ρR_B is determined along the path length X of the electrons and is not the electron range as measured from the surface. However, as a first approximation, the depth and radial spread of the electron beam can be equated to this range.

From experiments on the transmission of electrons through thin films, one can obtain a measured electron range. Cosslett and Thomas[4] measured the "practical" or extrapolated mass range ρR_x from measurements on thin films of various thicknesses and atomic number. This range is determined by extrapolating the straight line portion of the experimental curve of fractional transmission versus incident electron energy for various film thicknesses until the fractional transmission is zero. These authors found that ρR_x is approximately the same for all elements at a given incident electron energy.

Cosslett and Thomas[4] also measured another range, called the maximum range R_{max}, which is the thickness of thin film that just reduces the transmitted beam to zero. The value of ρR_{max} is approximately the same for all elements measured. The experimental values of range are proportional to the nth power of the incident acceleration voltage, where n varies from 1.2 to 1.7 according to the definition of "range." For the extrapolated mass range ρR_x, n was 1.5 for all elements.

Kanaya and Okayama[10] have derived an expression for the maximum range R from the energy loss equation (3), using a total scattering cross section which takes into account both elastic and inelastic collisions. The mass range ρR derived is expressed as

$$\rho R = 0.0276 E_0^{1.67} A / Z^{8/9} \tag{5}$$

where E_0 is given in keV, ρ in g/cm³, and R in μm. In this equation the mass range increases with atomic number by about 50% from Al to Au. Table I affords a comparison of the experimental range values R_x,[6] R_{max},[6] and the calculated Bethe range R_B for various electron beam energies.

It appears that the ratio R_{max}/R_x is about 1.5 for all the elements. In addition, the two calculated ranges R_B and R are close in value; however, the values of R are less than R_B for the heavier elements. The Bethe range R_B, as discussed previously, cannot be directly compared with values of the three other ranges since it gives the distance along the path length X. As the atomic number of the specimen increases, the amount of high-angle scattering increases. Therefore a significant percentage of the electron path length

TABLE I. Extrapolated Range $R_x^{(6)}$ and Maximum Range $R_{max}^{(6)}$ Expressed in μm Compared with Bethe Range R_B and Maximum Range $R^{(10)}$ for Incident Energies E_0 from 2.5 to 15 keV

	R_x	R_{max}	R_B	R		R_x	R_{max}	R_B	R
		2.5 keV					5 keV		
Al	0.13	0.21	0.1	0.12		0.33	0.48	0.32	0.4
Cu	0.049	0.073	0.043	0.046		0.11	0.18	0.13	0.15
Au	0.021	0.032	0.037	0.027		0.05	0.08	0.10	0.088
		10 keV					15 keV		
Al	0.85	1.1	1.1	1.25		—	—	2.2	2.4
Cu	0.34	0.47	0.42	0.47		0.6	—	0.85	0.9
Au	0.15	0.22	0.31	0.28		0.26	—	0.6	0.54

is no longer perpendicular to the surface. In effect this reduces the maximum depth or range to which the electrons travel; and therefore R_B overestimates the true range for high atomic number elements. On the other hand, one sees good agreement of R_{max}, R_B, and R for the light and medium elements above 5 keV. The values of the extrapolated range R_x and R_{max} are higher than the Bethe range for Al and Cu for low-energy electrons. The calculated maximum range of Kanaya and Okayama[10] appears to agree quite closely with the measured values of R_{max} and can be used in a practical sense to determine the electron range in situations of interest. For a given incident electron energy the electron range varies with atomic number. These range values (Table I) bear out the qualitative impressions of the effects of elastic scattering that were discussed in Section I and illustrated in Figure 1. This table indicates that the spatial resolution of the electron interaction volume can be improved by selecting low voltages for analysis.

Information concerning detailed trajectories of electrons in the target can be obtained in principle by using a Monte Carlo calculation. In the Monte Carlo method[11] an electron of energy E_0 is considered to strike a sample surface at some point P_0 (Figure 2). The electron can undergo elastic and inelastic collisions and can be backscattered out of the sample. The calculation procedure assumes that each electron travels a small distance ΔS_i in a straight line within the sample between random scattering events with an energy E_i after having been scattered at a point P_i with scattering angles ω_i, ϕ_i. The calculation determines the value of ω_i, ϕ_i, $\omega_i E_i$ at the next point P_i, ΔS_i along the path length. Random numbers are used to determine the new direction and energy of an electron after each scattering event by sampling from the appropriate scattering distribution. A calculation of the trajectory and the energy loss for an electron along its path is

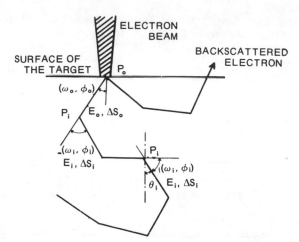

FIGURE 2. Simplified model of electron trajectories (Shinoda *et al.*[11]).

made until the electron energy is such that the electron can no longer cause ionizations, say 500 eV. Trajectories for a large number of electrons are calculated until a statistically valid description of the electron scattering process is available depending on the electron scattering and deceleration model selected.

Such a description can be made as long as the angular distribution of scattered electrons, the energy loss, and the step length can be obtained in a physically valid manner. For these calculations Murata *et al.*[8] used a screened Rutherford-type expression for the single scattering cross section, the Bethe continuous energy loss law [equation (1)], and a step length proportional to the mean free path of the electrons. In calculations for various elements the number of individual tracjectories considered was a high as 2000 and the number of steps per trajectory in some cases was more than 1000. Thus, the number of calculations is very costly in terms of computer time. As will be discussed in a later chapter, the accuracy of the calculations for quantitative x-ray analysis depends directly on the physical constants available for the various physical laws.

Figure 3 shows the electron trajectories calculated for aluminum, copper, and gold with the electron beam incident with 30 keV energy at both 90° and 45°. The number of trajectories drawn for each plot is 100 and the scale is in mass thickness ρx (mg/cm^2). These curves agree with the qualitative ones discussed earlier and shown in Figure 1 in which the electron spread becomes compressed as the atomic number increases. The pear-shaped distribution at low atomic numbers grades into a hemispherical shape as atomic number increases. The electron mass range is similar for all atomic numbers and the depth of complete diffusion decreases with increasing atomic number. If one considers the trajectories in terms of depth z or

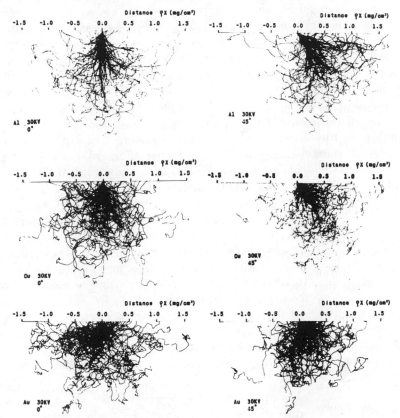

FIGURE 3. Electron trajectories calculated for aluminum, copper, and gold, $E_0 = 30$ keV at (a) normal and (b) 45° incidence (Murata *et al.*[8]).

range R as given in μm (note Figure 3), then the electron range R is much greater for low atomic number elements than for high atomic number elements. For example, $\rho x = 1$ mg/cm² corresponds to $x = 4.3$ μm for Al, $x = 1.2$ μm for Cu, and $x = 0.52$ μm for Au. In addition, the depth of complete diffusion x_d is less for higher atomic number elements.

III. EMITTED ELECTRONS—BACKSCATTERED ELECTRONS

Backscattered electrons are produced by large-angle single elastic scattering events and by small-angle multiple elastic scattering events.

In low atomic number elements, slightly more than 50% of the total back-scattering occurs from multiple scattering events[7] when the incident electrons have energies in the range 10–20 keV. Since the mean free path between events is large, multiple scattering occurs rather deep in the target. In high atomic number elements, slightly less than 50% of the backscattering is caused by multiple rather than single scattering for electrons in the 10–20 keV energy range.[7] The mean free path between scattering events is small, and multiple scattering occurs close to the surface. Because the amount of elastic scattering (single plus multiple) increases with atomic number Z, the fraction η of electrons which are backscattered also increases. This functionality is shown in Figure 4, according to Heinrich,[12] where the electron backscatter coefficients for 30-keV incident electrons are plotted as a function of the atomic number of the target. The curve shown here agrees with those of other authors.[13,14] In addition, little change of η with acceleration voltage has been observed. This result is consistent with the concept that the envelope of electrons scattered within the specimen expands without change in shape as the voltage is raised. However, Bishop[3] found a small decrease in η with increasing voltage and that the backscatter coefficients tend to decrease slightly with decreasing acceleration voltage,[12,13] at least for the heavier elements.[15]

The energy distribution of the backscattered electrons will now be considered. If all the electrons were backscattered before any inelastic collisions occurred, then the backscattered energy would be equal to the beam energy E_0. However, a considerable number of backscattered elec-

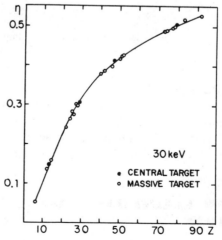

FIGURE 4.   Electron backscatter coefficient η, $E_0 = 30$ keV as a function of the atomic number Z of the target (Heinrich[12]).

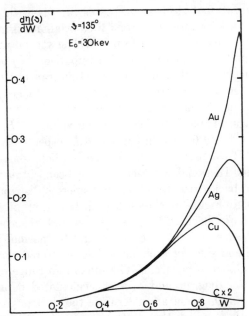

FIGURE 5. Energy distribution of backscattered electrons at a scattering angle of 135° for various elements (Bishop[13]). The normalized energy is equal to E/E_0. Note that the values for carbon have been increased by a factor of two.

trons is produced after some loss of energy occurs. Figure 5 shows the distribution of energy of backscattered electrons at a scattering angle of 135° for various elements. The normalized energy W is equal to E/E_0.[13] The term $d\eta/dW$, which is the ordinate in Figure 5, represents the number of electrons backscattered per incident electron per unit energy interval. The measurement shown in Figure 5 was made for electrons backscattered 45° with respect to the sample surface. For measurements made at smaller angles of incidence, the peak and mean energies of the backscattered electrons approach E_0 more closely. The area beneath the curves of $d\eta/dW$ vs. W for gold, silver, copper, and carbon gives the total number of electrons backscattered. As the atomic number of the sample increases, not only are more electrons backscattered, but the energy of the backscattered electrons is nearer the beam energy E_0. This is consistent with the fact that scattering occurs closer to the surface where the electron has lost little energy. The number of electrons backscattered from carbon is small and peaks at approximately $\frac{1}{2}W$. These energy spectra are almost identical regardless of the beam energy.[16]

As a beam electron penetrates into a solid, the probability that it will escape as a backscattered electron decreases rapidly with distance below the surface. The backscattered electrons thus originate primarily within a small fraction of the electron beam range R. The spatial resolution of the backscattered signal is thus improved. A comparison of the energy distribution in backscattering with that for the fraction transmitted through thin films shows that, in a solid target irradiated with 25 keV electrons, the backscattered fraction comes from a mean depth of about $0.3R$ in Cu to $0.2R$ in Au.[7] From Table I, which contains values of R, these "mean" depths would be 0.4 and 0.1 μm in Cu and Au, respectively.

The Monte Carlo method can also be applied to study the backscattered electrons. Reasonable agreement has been obtained with experimental values of backscatter fraction and energy distribution of backscattered electrons. Figure 6 shows the distribution of the maximum penetration depth for backscattered electrons and absorbed electrons[11] for Cu and Al at several incident energies and for normal incidence electrons. The right-hand side corresponds to backscattered electrons and the left-hand side to absorbed electrons. The maximum range of the electrons R is labeled x_r in the diagram and x_d is the calculated depth of complete diffusion. The maximum penetration depth of the electron as defined in this figure is that depth at which the electron takes a path perpendicular, in the mean, to the direction of incidence. For absorbed electrons the maximum frequency occurs at x_d. These calculations for absorbed electrons further emphasize the relationships previously discussed for the effects of atomic number and incident voltage on values of x_d and the range R.

Figure 6 shows that the backscattered electrons are mainly scattered near the surface; very few penetrate beyond the depth of complete diffusion. For copper irradiated with 25-keV electrons, the mean depth for the backscattered electrons is about 0.35 μm, in close agreement with the data of Cosslett and Thomas.[7] Shimizu and Murata[17] have also obtained the resolving power of the backscattering image by calculating the lateral distribution of the backscattered electrons at normal incidence for Cu and for Al irradiated with 30-keV and 19.76-keV electrons, respectively. The backscattered electrons emerge from a circle on the target surface with a diameter nearly equal to the electron range R. However, the vast majority of electrons are backscattered laterally with a smaller diameter, perhaps about $R/2$; for example, 1 μm for Cu and 2.5 μm for Al. The backscatter fraction η is also a function of the angle of tilt of the specimen surface with respect to the impinging electron beam. As shown in Figure 7a, the measured and calculated values of η for iron and an iron–3.22 wt % Si alloy increase with increasing tilt angle. In this example, 0° tilt is the position where the electron beam is normal to the sample surface. Increasing the

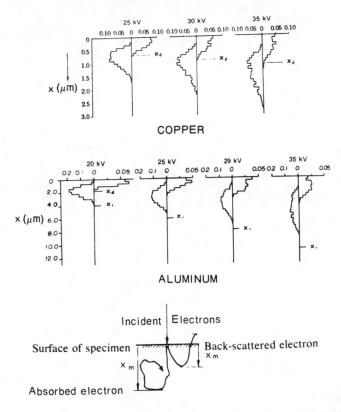

FIGURE 6. Distribution of maximum penetration depth for backscattered electrons (right-hand side) and absorbed electrons (left-hand side) for copper and aluminum (Shinoda et al.[11]).

tilt angle allows the electron beam to strike the sample at a more shallow angle and causes the electrons to penetrate less into the solid (Figure 3).

The angular distribution of backscattered electrons varies markedly with tilt. At normal incidence, the backscattered electrons follow approximately a cosine angular distribution (Figure 7b). The origin of this cosine distribution can be understood from the following argument. If we assume the backscattered electrons originate from a point source located at a depth R below the surface, then the path length to the surface increases as $R/\cos \theta$, where θ is the angle between the normal and a direction of interest. Assuming that the absorption of the electrons is proportional to the path length, the fraction of electrons that escape decreases as $\cos \theta$. At high tilt angles, the angular distribution becomes greatly elongated in

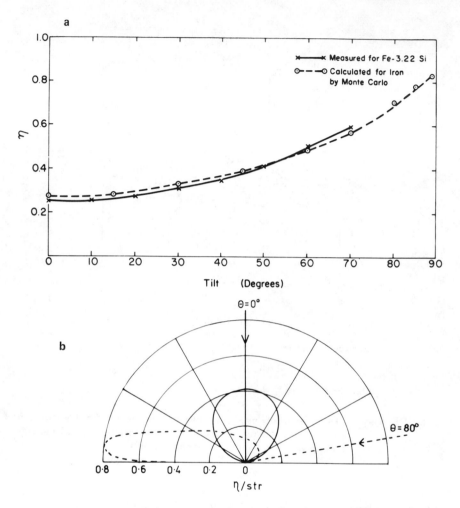

FIGURE 7.   (a) Variation of backscatter fraction η with specimen tilt angle. A tilt of 0° indicates that the beam is normal to the sample surface. (b) Schematic drawing of the angular distribution of backscattered electrons for different beam incidence conditions. Solid line: normal incidence (θ = 0° tilt), ideal cosine distribution. Broken line: high tilt (θ = 80°). (Plot: r vs θ, with values of the backscattering coefficient per unit solid angle along the radius.)

the forward scattering direction. It has been shown experimentally that the most probable scattering angle at high tilt (near grazing incidence) lies close to the surface (Figure 7b). The most probable scattering direction lies in the plane containing the surface normal and the incident beam direction. The backscatter fraction emitted along a line which is not contained within this plane decreases rapidly as the angle between the line and plane increases. Thus, the backscattered electrons emitted from a surface which is tilted with respect to the incident beam are highly directional.

Wells[18] found that the resolution in depth of the backscattered electron image could be improved by as much as an order of magnitude if the electron beam is at a 60° incidence to the sample and if the electrons are collected over an angle that is close to the plane of the specimen surface. This increase in resolution is due to the fact that a significant number of backscattered electrons leave the specimen after very shallow penetration into the sample ($\sim$500Å for 15-keV electrons in Al). It has been proposed that these electrons emerge as a consequence of a single scattering event or at most a few ($\sim$10) events, (plural scattering) and leave the specimen at a shallow angle with only a small loss in energy. These electrons therefore interact with only a small volume of the specimen, in contrast to the electrons that reach 0.3 of the range. The multiply scattered electrons leave the specimen in random directions following penetration to a greater depth. This explanation is consistent with the experimental results on energy distribution curves discussed previously for backscattering angles less than 45° to the specimen surface.[16] In addition, this indicates that there is an increase in single scattering events in which scattering occurs close to the surface in a small volume and that the electrons leave the sample with little loss in energy. Figure 3 shows the electron trajectories for Al, Cu, and Au at 45° incidence as calculated by the Monte Carlo technique. Many of the electrons which are singularly or plurally scattered come from a relatively small volume of the sample.

The backscattering of electrons from a flat, thick target depends uniquely on the atomic number of the target. Direct confirmation of the linearity between sample composition and backscatter yield was obtained by Bishop[13] for a series of Cu–Au samples. Therefore for binary targets, the results can be represented by

$$\eta_{\text{alloy}} = C_A \eta_A + (1 - C_A)\eta_B \qquad (6)$$

where η is the backscatter fraction and C_A is the weight fraction of element A. In a binary system, then, it is possible to perform a quantitative analysis based on backscattered electrons. If the elements are close in atomic number, then η_A and η_B will be quite similar, and the analysis will be difficult to perform. For complex oxides or alloys containing three or more elements quantitative analysis is not possible. However, by measuring electron backscattering or the specimen current, the average atomic number of a complex material can be obtained easily. On a comparison basis, this information can provide a method of identification.

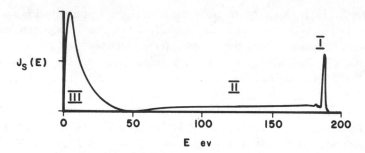

FIGURE 8.   Energy distribution of secondary electrons. $J_s(E)$ is the intensity
of secondary electrons. Note that the beam energy E_0 is 180 eV.
(From Hachenberg and Brauer.[19])

The number of backscattered electrons which are emitted from a
point on a sample in the direction of the backscattered electron (BSE)
detector is also dependent on the angle between the primary beam, the
specimen surface, and the direction of emergence. If the sample surface is
rough, the BSE yield can be selectively absorbed or increased, depending on
the beam position. Therefore, the intensity of the backscattered electrons
is also a function of the topography of the sample. However, if the sample
is polished flat, as in the preparation of samples for quantitative x-ray
analysis, one can obtain the average atomic number of the analyzed area
of the sample. If the sample surface is not polished, a combination of topographical and chemical information will be obtained.

IV. EMITTED ELECTRONS—LOW-ENERGY ELECTRONS

The total energy distribution of emitted electrons, as given by Hachenberg and Brauer,[19] is shown in Figure 8. In this figure, the number of
electrons emitted by a target at a given energy level $J_s(E)$ is plotted against
the energy E of the electrons. Note that the impinging electron beam in
their experiment has a beam energy E_0 of 180 eV, much less than the
beam energies used in scanning electron microscopes. Three groups of
electrons can be distinguished in this distribution curve. Group I represents
elastically scattered electrons of nearly the same energy as the primary
beam. Group II contains, for the most part, multiply scattered electrons
which are reflected to the surface after having passed through a more or less
thick layer of target material. At high beam energies, those used in scan-

ning electron microscopy, the energy of group I and II backscattered electrons is much more widely distributed, as shown in Figure 5.

The third group (III) has an energy which falls below 50 eV and shows a maximum of the distribution curve at a few electron volts. The group III electrons are the low-energy secondary electrons which are formed as a result of excitation by the high-energy primary beam of loosely bound atomic electrons. These are the low-energy secondary electrons which are often used in the scanning electron microscope to image surface topography. In principle, there may be rediffused primaries or inelastically scattered electrons having energies less than 50 eV. However, it has become common practice to arbitrarily separate the two groups at 50 eV. The position of the maximum and the slope of the distribution are strongly influenced by the surface barrier of the material. If there is a low potential barrier, the maximum is expected to occur at low energies. Secondary electron emission is also related to the primary energy of the incident electrons, to the number of outer shell electrons, and to the atomic radius. The angular distribution of the emitted secondary electrons for normal beam incidence follows a cosine law. The origin of this distribution can be understood in terms of the earlier argument for backscattered electron angular distribution (Section III).

The total emitted electron yield contains both the backscattered electron yield η and the low-energy secondary electron yield δ. Values of δ and η for 30-keV electrons are shown in Figure 9 as determined by Wittry [20] for metallic specimens of various atomic numbers. Low-energy secondary

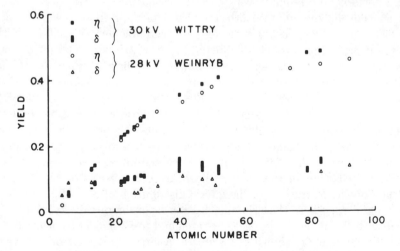

FIGURE 9. Electron backscatter coefficient η and low-energy secondary electron yield δ for $E_0 = 30$ keV; two data sets (from Wittry[20]).

electron yields from pure elements do not vary systematically with atomic number, although the average yield increases somewhat with increasing atomic number. The difference between secondary yields of C and Au is from 6.7 to 13.6% for 30-keV incident electrons. As the energy of the primary electrons increases above 2 keV, for most materials, the number of secondary electrons created increases since secondary electrons are generated throughout the entire volume of primary electron distribution. However, the average depth at which they are formed also increases and the absorption of these true secondary electrons as they travel to the surface becomes quite strong. Therefore, as beam energy increases, the yield of true secondary electrons released from the sample is decreased.[15] Typical secondary yields of 10% for 30-keV electrons and 30–50% for 5-keV electrons have been measured.[20] According to recent measurements by Reimer,[21] typical secondary yields are 5%, 10%, and 40% for aluminum and 10%, 20%, and 70% for gold as determined with 50-keV, 20-keV, and 5-keV electrons, respectively.

The probability that low-energy secondary electrons will escape from the sample decreases exponentially as their point of generation moves away from the surface. This escape probability is equal to $\exp(-z/\lambda)$, where z is the depth and λ is the mean free path of the secondary electrons. According to the work of Seiler,[22] the maximum depth or range of secondary emission is equal to 5λ. For metals $\lambda \approx 10$ Å and for insulators $\lambda \approx 100$ Å. Thus, although primary electrons may penetrate some micrometers into the specimen, nearly all the secondaries which escape through the surface will have originated very close to the surface.

Within this maximum depth, the spread of the primary beam is small. In fact, the primary beam loses only a negligible fraction of its initial energy in traveling through this distance. For a primary electron beam of zero cross-sectional area, as pointed out by Everhart *et al.*,[23] more than half of the secondary electrons are emitted within a distance $\lambda/2$ of the point of entry of the primary beam. For metals this distance is about 5 Å. In practice, the incident beam will have a finite cross section, and the electron distribution over this area will probably be approximately Gaussian. However, the same general principles will apply and the majority of the secondary electrons will be emitted within a circular area whose radius exceeds that of the incident beam by $\lambda/2$. Results of Monte Carlo calculations by Murata[24] of the lateral distribution of secondary electrons for Au and Al, where the primary electron beam having 20 keV energy is assumed to be of zero cross-sectional area and to be incident normal to the specimen surface, are shown in Figure 10. A large value for λ of 100 Å was used which overestimates the actual broadening of the beam. Actually since the mean free path λ of electrons for Au is less than that of Al, the

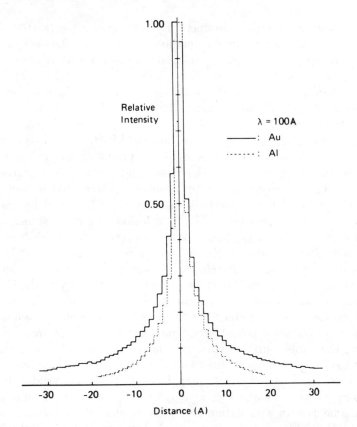

FIGURE 10.   Lateral distribution of secondary electrons from gold and aluminum targets, $E_0 = 20$ keV. The primary electron beam is incident normal to the specimen surface ($\phi = 0°$). (From Murata.[24])

spatial resolution for Au should be somewhat better. Unless the radius of the incident beam is equal to or less than $\lambda/2$, the effect of the spreading of secondary electrons is unlikely to be very important.

Another factor which may affect the spatial resolution of the low-energy secondary image is the production of low-energy secondary electrons by high-energy backscattered electrons as they leave the sample. The backscattered electrons will generate secondary electrons in the same volume in which they are produced.[17] The fraction of low-energy secondary electrons generated by backscattered electrons is significant and increases with the atomic number of the target.[22] Typically, the number of secondaries produced by backscattering electrons is three or four times that produced by incident electrons.[22] Monte Carlo calculations have also been made by Shimizu and Murata[17] to predict the lateral or spatial distribution of secondary electrons produced by high-energy backscattered electrons. The information in these secondary electrons is different from those produced directly at the impact point and normally will contribute to background noise in an SEM image (See Chapter IV). This background will be generated from an area about the size of the lateral distribution of high-energy backscattered electrons (>1 μm). The secondary electron current density at the specimen surface, however, is much greater in the area of impact of the primary beam than outside it. As will be explained in detail in Chapter IV, the secondaries produced by backscattering electrons contribute to noise in the image and limit high-resolution performance. Only in special cases is the resolving power, that is, the spatial resolution of the scanning electron microscope, determined by the spot size of the primary electrons. This conclusion has been shown experimentally to be valid for electron beams with resolutions of 50 Å or below in which surface definition for different samples can be observed.[18,25]

Contrast in the secondary electron image is obtained as a consequence of the variation of the secondary yield and the trajectory of the electrons from one position to another on the sample surface. The major influences on secondary yield and trajectory are the topography of the sample surface and the local electrical potentials or magnetic fields which are present on or above the sample surface. Variations in mean atomic number across the sample have only a minor influence on the secondary yield, although some compounds, such as insulators, may have a very high yield. Therefore secondary electron images can give information on surface topography and local fields with a resolution very close to that of the focused electron beam d_p. How these differences manifest themselves on the final scanning micrographs is discussed in Chapter IV.

The secondary yield is very dependent on the angle of incidence. As the angle α between the incident beam and a line perpendicular to the surface of the sample at the point of incidence increases, the secondary yield increases. This effect is shown in Figure 11. As the angle of incidence α is increased, the envelope of the primary electron beam in which secondary electrons are produced is on the average shifted closer to the surface.

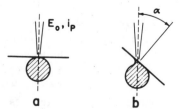

a   b

FIGURE 11.   Topographical configurations to obtain contrast in the scanning electron microscope: α is the angle of incidence.

Therefore, since more secondary electrons are produced close to the surface for the same primary energy E_o and beam current i, the yield of secondary electrons which are emitted from the sample increases. Variations in surface inclination of only a few degrees will generally be sufficient to cause an observable change in the brightness of the final image. However, this is only true if the scale of the variation is larger than the beam diameter.

V. X-RAYS

A. X-Ray Production

There are two electron beam–solid interactions which lead to the production of x-rays: core scattering, which results in the emission of the continuous spectrum, and inner shell ionization, which yields the characteristic spectrum. The continuum x-ray spectrum will be discussed first. The energy of the x-ray photons produced when the impinging electrons are inelastically scattered by the nucleus can take on all values up to the energy of the incident electron E_0. Most electrons give up their energy not in a single step, but rather in numerous unequal increments E. The relation between λ, the wavelength, and the energy of the x-ray photons is

$$\lambda = hc/eE = 12.398/E \qquad (7)$$

where h is Planck's constant, c is the velocity of light, e is the charge on the electron, E is the energy of the x-rays (keV), and λ is the x-ray wavelength given in angstroms. The maximum energy loss occurs when all the energy is absorbed in one collision. Since x-ray wavelength is inversely proportional to energy, the most energetic x-rays will have a minimum wavelength λ_{min}, also called the short-wavelength limit λ_{SWL}. All of the energy of the incoming electron is absorbed and the corresponding energy of the x-rays is E_o, the energy of the incident electron beam. The intensity of the continuum radiation from a molybdenum target as a function of wavelength is shown in Figure 12.[26] In this figure the continuum x-ray intensity is plotted versus the wavelength of the radiation for various electron beam voltages (5, 10, 15, 20, and 25 kV). The more complicated 25-kV spectrum will be discussed later. The minimum wavelength of the continuum, also called the short-wavelength limit, varies as a function of beam voltage, decreasing to shorter and shorter wavelengths as the voltage increases [note equation (7)]. In addition, the intensity of the x-ray continuum reaches a maximum at approximately $1.5\lambda_{min}$ and gradually falls off at longer wavelengths.[27]

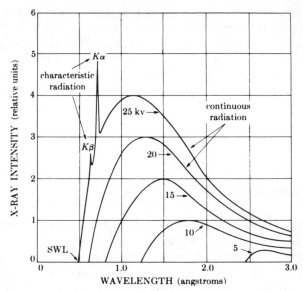

FIGURE 12.   X-ray spectrum of molybdenum as a function of applied voltage (schematic). Linewidths not to scale. (From Cullity,[26] Figure 1-4.)

The intensity of the continuum is a function of both atomic number and accelerating voltage. As the voltage increases, the continuum spectrum moves to shorter wavelengths and increases in intensity. The increase in intensity is due to the fact that, statistically, the beam electrons can undergo more decelerations. In addition, the amount of continuum radiation also increases directly with increasing atomic number since heavy elements have more nuclear scattering and less energy loss by electron–electron interactions. The intensity of the continuum at any one wavelength I_λ has been expressed by Kramers[28] as

$$I_\lambda \sim i\bar{Z}(\lambda/\lambda_{\min} - 1) \sim i\bar{Z}(E_0 - E)/E \qquad (8)$$

where i is the beam current, $\bar{Z}$ is the average atomic number of the sample, and E is the energy corresponding to a particular wavelength of the continuum λ [equation (7)]. For a specific wavelength, I_λ varies directly with electron current i and atomic number. I_λ also varies directly with beam energy E_0 if λ is much greater than $\lambda_{\min}$ or $E_0 - E$ is much greater than one. The continuum radiation forms the background x-ray radiation in the electron microprobe–scanning electron microscope. The amount of continuum background plays an important role in determining the minimum detectability limit for the particular element that is being measured. It is of interest in almost all instances to attempt to keep this continuum or background radiation to a minimum.

Characteristic x-radiation is produced by the interaction of incident electrons with the inner shell electrons of the atoms in the sample. Occasionally, if the incident electron has enough energy, it may dislodge a K, L, or M inner shell electron and leave the atom in an excited or ionized state. The atom returns to its ground state by the transition of an outer electron into the vacancy in the inner shell. When the relaxation of the atom back to its original state occurs, the atom loses energy in the process by the emission of a photon of x-ray radiation. The electrons of the atom are in discrete energy levels described by the quantum numbers of the atom. The restrictions on these numbers allow one energy level for $n = 1$ (K shell), three energy levels for $n = 2$ (L shell), five energy levels for $n = 3$ (M shell), etc., where n is the principal quantum number. Since the electrons are in discrete energy levels, the emitted x-ray photon will also have a discrete energy equal to the energy difference between the initial and final states of the atom. Therefore the wavelengths of the characteristic radiation are specific for atoms of a given atomic number. Detection of the presence of a characteristic x-ray line indicates that the element is present in the sample. These characteristic lines can also be used to obtain the composition of a sample of interest.

Figures 13a and 13b show schematically the process which occurs when the inner electrons of an atom are bombarded by the incoming high-energy electrons. In Figure 13(a) the process of electron excitation is shown,[29] while in Figure 13(b) the electronic transitions for producing various characteristic lines are drawn. Each electron shell K, L, M, etc. is depicted for purposes of simplification as only having one energy level. As an electron impinges upon the inner shells of the atom, a photoelectron is ejected, and the atom is raised to a higher energy level. As the energy of the atom is reduced to lower energy levels, an electron transition occurs. For example, an electron may jump from the L shell to fill the unoccupied site in the K shell. It is also possible for an electron to jump from the M shell to fill the unoccupied site in the K shell. The L- to K-shell transitions produce $K\alpha$ x-ray radiation. The M- to K-shell transitions produce $K\beta$ radiation. Alternatively, if an incoming electron ejects an inner electron from the L shell, then we can produce L radiation. In this process an M-shell electron may jump from the M shell into the unoccupied site in the L shell.

The process for obtaining characteristic radiation can be more easily seen by using an energy level diagram[30] (Figure 14). In Figure 14 the energy of the atom is plotted assuming that the atom is normally at zero energy as a neutral atom. For an incoming electron to ionize the K shell, that is, to eject an electron from the K shell, it requires a certain amount

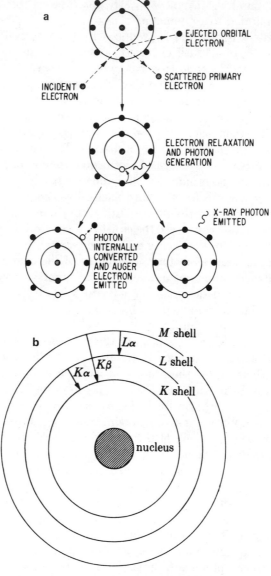

FIGURE 13.   (a) Process of electron excitation producing characteristic x-rays or Auger electrons. (From Lifshin[29].) (b) Electronic transitions in an atom (schematic). Emission processes indicated by arrows. (From Cullity,[26] Figure 1-7.)

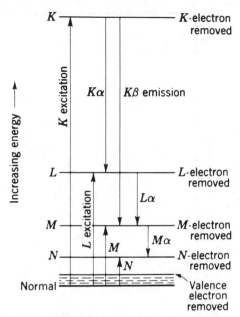

FIGURE 14.   Energy level diagram for an atom (schematic). Excitation and emission processes indicated by arrows. (From Barrett and Massalski,[30] Figure 3-3.)

of energy. This energy value is called the critical excitation energy E_K. This process leaves a hole in the K shell and raises the energy of the atom to E_K. To reduce the energy of the atom, an electron from the L or M shell drops into the vacancy in the K shell. This reduction in energy, which can be seen in Figure 14, from E_K to E_L or E_M produces a photon of characteristic $K\alpha$ or $K\beta$ radiation. The energy of the characteristic lines is given by $E_{K\alpha} = E_K - E_L$ and $E_{K\beta} = E_K - E_M$. The wavelength of the characteristic lines can be calculated from equation (7). Expressed in another way, the critical excitation potentials necessary to eject an electron from the L shell and M shell are E_L and E_M, respectively. The $L\alpha$ and $M\alpha$ radiations are produced in an analogous manner to the $K\alpha$ or $K\beta$ x-rays. It is important to note that if an electron has enough energy to expel a K electron, it can also expel any L- or M-shell electrons. Hence all spectral lines appear simultaneously that result from electron transitions to the innermost electron shell excited and to all shells farther out. In general, all lines of a given type, K, L, or M, appear simultaneously. Even though the L, M, and N shells contain more than one energy level, selection rules limit the number of possible electronic transitions,[31] resulting in rather simple spectra with only a few important lines for each element.

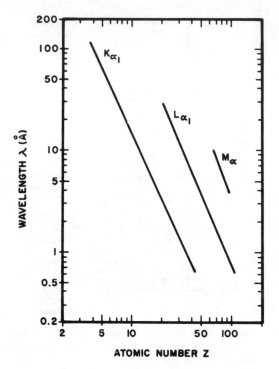

FIGURE 15.   Moseley's relation between λ and Z for the $K\alpha_1$, $L\alpha_1$, and $M\alpha$ characteristic lines.

As shown in Figure 12, no $K\alpha$ characteristic lines are produced for Mo at voltages of 20 kV or less. The critical excitation voltage for Mo_K is 20.01 kV; and therefore, in the 25-kV x-ray spectrum both the $K\alpha$ and $K\beta$ are produced as well as continuum x-rays. Electron transitions originating from sharply defined levels emit spectral lines having a width at half-maximum of ∼0.001 Å. However, the outermost levels in atoms in solids may be broadened under the influence of neighboring atoms. Electron transitions originating from such levels give rise to broadband spectra and examples of this type are the K spectra of the lightest elements.

The wavelength of the x-ray radiation produced decreases with increasing atomic number. This fact was expressed by Moseley as

$$\lambda = K/(Z - \sigma)^2 \tag{9}$$

where K and σ are constants which differ for each series and λ is the characteristic x-ray wavelength. Figure 15 illustrates Moseley's law and the direct relationship between atomic number and wavelength for $K\alpha$, $L\alpha$, and $M\alpha$ radiation. Tables IIA–IIC, which are adapted from a tabulation by Bearden,[32] lists these characteristic wavelengths and the corresponding energies (keV) for $K\alpha_1$, $K\beta$, $L\alpha$, and $M\alpha$ radiation. Tables IIA–IIC also list, for each of the elements, the critical excitation energies E_K, E_{L3}, and E_{M5} and the corresponding wavelengths for K and L x-ray production. The $K\alpha$ radiation lines are usually chosen to analyze elements from beryllium ($Z = 4$) to germanium ($Z = 32$) and the $L\alpha$ radiation lines are usually employed to analyze elements of higher atomic number. Note that the L radiations in Table IIB and Figure 15 overlap some of the same elements that can also be measured with K radiation. A choice is then made as to which characteristic radiation to measure based on the fact that

TABLE IIA. K Series X-Ray Wavelengths and Energies [a]

Element	$K\alpha_1$		$K\beta_1$		K edge	
	λ, Å	E, keV	λ, Å	E, keV	λ, Å	E, keV
4 Be	114.00	0.169	—	—	110.0	0.111
5 B	67.6	0.183	—	—	—	—
6 C	44.7	0.277	——	—	43.68	0.284
7 N	31.6	0.392	—	—	30.99	0.400
8 O	23.62	0.525	—	—	23.32	0.532
9 F	18.32	0.677	—	—	—	—
10 Ne	14.61	0.849	14.45	0.858	14.30	0.867
11 Na	11.91	1.041	11.58	1.071	11.57	1.072
12 Mg	9.89	1.254	9.52	1.302	9.512	1.303
13 Al	8.339	1.487	7.96	1.557	7.948	1.560
14 Si	7.125	1.740	6.75	1.836	6.738	1.84
15 P	6.157	2.014	5.796	2.139	5.784	2.144
16 S	5.372	2.308	5.032	2.464	5.019	2.470
17 Cl	4.728	2.622	4.403	2.816	4.397	2.820
18 A	4.192	2.958	3.886	3.191	3.871	3.203
19 K	3.741	3.314	3.454	3.590	3.437	3.608
20 Ca	3.358	3.692	3.090	4.103	3.070	4.038
21 Sc	3.031	4.091	2.780	4.461	2.762	4.489
22 Ti	2.749	4.511	2.514	4.932	2.497	4.965
23 V	2.504	4.952	2.284	5.427	2.269	5.464
24 Cr	2.290	5.415	2.085	5.947	2.070	5.989
25 Mn	2.102	5.899	1.910	6.490	1.896	6.538
26 Fe	1.936	6.404	1.757	7.058	1.743	7.111
27 Co	1.789	6.930	1.621	7.649	1.608	7.710
28 Ni	1.658	7.478	1.500	8.265	1.483	8.332
29 Cu	1.541	8.048	1.392	8.905	1.381	8.980
30 Zn	1.435	8.639	1.295	9.572	1.283	9.661
31 Ga	1.340	9.252	1.208	10.26	1.196	10.37
32 Ge	1.254	9.886	1.129	10.98	1.17	11.10

[a] Adapted from Bearden.[32]

TABLE IIB. L Series X-Ray Wavelengths and Energies [a]

Element	$L\alpha_1$		L_3 edge	
	λ, Å	E, keV	λ, Å	E, keV
22 Ti	27.42	0.452	—	—
23 V	24.25	0.511	—	—
24 Cr	21.64	0.573	20.7	0.598
25 Mn	19.45	0.637	—	—
26 Fe	17.59	0.705	17.53	0.707
27 Co	15.97	0.776	15.92	0.779
23 Ni	14.56	0.852	14.52	0.854
29 Cu	13.34	0.930	13.29	0.933
30 Zn	12.25	1.012	12.31	1.022
31 Ga	11.29	1.098	11.10	1.117
32 Ge	10.44	1.188	10.19	1.217
33 As	9.671	1.282	9.37	1.324
34 Se	8.99	1.379	8.65	1.434
35 Br	8.375	1.480	7.984	1.553
36 Kr	7.817	1.586	7.392	1.677
37 Rb	7.318	1.694	6.862	1.807
38 Sr	6.863	1.807	6.387	1.941
39 Y	6.449	1.923	5.962	2.079
40 Zr	6.071	2.042	5.579	2.223
41 Nb	5.724	2.166	5.230	2.371
42 Mo	5.407	2.293	4.913	2.523
43 Tc	5.115	2.424	4.630	2.678
44 Ru	4.846	2.559	4.369	2.838
45 Rh	4.597	2.697	4.130	3.002
46 Pd	4.368	2.839	3.907	3.173
47 Ag	4.154	2.984	3.699	3.351
48 Cd	3.956	3.134	3.505	3.538
49 In	3.772	3.287	3.324	3.730
50 Sn	3.600	3.444	3.156	3.929
51 Sb	3.439	3.605	3.000	4.132
52 Te	3.289	3.769	2.856	4.342
53 I	3.149	3.938	2.720	4.559
54 Xe	3.017	4.110	2.593	4.782
55 Cs	2.892	4.287	2.474	5.011
56 Ba	2.776	4.466	2.363	5.247
57 La	2.666	4.651	2.261	5.484
58 Ce	2.562	4.840	2.166	5.723
59 Pr	2.463	5.034	2.079	5.963
60 Nd	2.370	5.230	1.997	6.209
61 Pm	2.282	5.433	1.919	6.461
62 Sm	2.200	5.636	1.846	6.717
63 Eu	2.121	5.846	1.776	6.981
64 Gd	2.047	6.057	1.712	7.243
65 Tb	1.977	6.273	1.650	7.515

TABLE IIB. (contd.)

Element	$L\alpha_1$		L_3 edge	
	λ, Å	E, keV	λ, Å	E, keV
66 Dy	1.909	6.495	1.592	7.790
67 Ho	1.845	6.720	1.537	8.068
68 Er	1.784	6.949	1.484	8.358
69 Tm	1.727	7.180	1.433	8.650
70 Yb	1.672	7.416	1.386	8.944
71 Lu	1.620	7.656	1.341	9.249
72 Hf	1.57	7.899	1.297	9.558
73 Ta	1.522	8.146	1.255	9.877
74 W	1.476	8.398	1.216	10.20
75 Re	1.433	8.653	1.177	10.53
76 Os	1.391	8.912	1.141	10.87
77 Ir	1.351	9.175	1.106	11.21
78 Pt	1.313	9.442	1.072	11.56
79 Au	1.276	9.713	1.040	11.92
80 Hg	1.241	9.989	1.009	12.29
81 Tl	1.207	10.27	0.979	12.66
82 Pb	1.175	10.55	0.951	13.04
83 Bi	1.144	10.84	0.923	13.43
84 Po	1.114	11.13	—	—
85 At	1.085	11.43	—	—
86 Rn	1.057	11.73	—	—
87 Fr	1.030	12.03	—	—
88 Ra	1.005	12.34	0.803	15.44
89 Ac	0.9799	12.65	—	—
90 Th	0.956	12.97	0.761	16.30
91 Pa	0.933	13.29	—	—
92 U	0.911	13.61	0.722	17.17

[a] Adapted from Bearden.[32]

it is desirable to use a line of both high intensity and low continuum background.

The peak intensity I_p obtained from characteristic radiation is a function of both the operating and excitation voltage and the beam current i, where

$$I_p \sim i(E_0 - E_c)^n \tag{10}$$

where E_0 is the operating voltage and E_c is the critical excitation voltage necessary to eject an electron from the inner K, L, or M shell. The exponent n is approximately 1.7.[31] This equation is of importance because it shows that the peak intensity obtained from characteristic radiation is directly proportional to the current in the electron beam and that no radiation is

TABLE IIC. M Series X-Ray Wavelengths and Energies [a]

Element	$M\alpha_1$		M_5 edge	
	λ, Å	E, keV	λ, Å	E, keV
72 Hf	7.539	1.645	—	—
73 Ta	7.252	1.710	7.11	1.743
74 W	6.983	1.775	6.83	1.814
75 Re	6.729	1.843	6.56	1.89
76 Os	6.490	1.910	6.30	1.967
77 Ir	6.262	1.980	6.05	2.048
78 Pt	6.047	2.051	5.81	2.133
79 Au	5.840	2.123	5.58	2.220
80 Hg	5.645	2.196	5.36	2.313
81 Tl	5.460	2.271	5.153	2.406
82 Pb	5.286	2.346	4.955	2.502
83 Bi	5.118	2.423	4.764	2.603
90 Th	4.138	2.996	3.729	3.325
91 Pa	4.022	3.082	—	—
92 Ur	3.910	3.171	3.497	3.545

[a] Adapted from Bearden.[32]

produced if the incoming electron energy is less than E_c. Furthermore, the x-ray intensity is dependent on the difference between E_0 and E_c. In most cases, in order to ensure adequate intensity, the overvoltage U, defined as the ratio E_0/E_c, is made two or greater. The peak to background ratio, i.e., the ratio of the intensity of a characteristic line to the continuum background at the same wavelength, is often greater than 100:1. This peak to background ratio can be written using equations (8) and (10) as

$$P/B = I_p/I_\lambda \sim (E_0 - E_c)^{1.7}/Z[(E_0 - E)/E] \qquad (11)$$

and is independent of beam current.

Since the energy of a characteristic line E is approximately the same as the critical excitation energy for the line E_c, equation (11) can be simplified to

$$P/B \sim (E_0 - E_c)^{0.7}E_c/Z \qquad (12)$$

In summary, then, as the voltage of the electron beam E_0 is increased, the P/B ratio and the sensitivity of the analysis will increase. On the other hand, as E_0 increases, the electron range and the size of the x-ray source increase. As will be discussed later, an optimum operating voltage is obtained when the P/B ratio is maximized and x-ray source size minimized.

B. X-Ray Absorption

The x-rays which are produced at various depths within the target are partially absorbed as they travel from the point of emission to the surface of the sample itself. The absorption of x-rays within the specimen decreases the intensity that is actually measured with respect to the intensity produced within the sample. The fraction of x-rays of a given wavelength λ (characteristic or continuum) transmitted through a sample is given by the equation

$$(I/I_0)_{\text{sample}} = \exp[-(\mu/\rho)_{\text{sample}}(\rho x)] \tag{13}$$

where I_0 is the initial intensity of the x-ray of wavelength λ in the sample, I is the measured intensity of the x-ray of wavelength λ leaving the sample, I/I_0 is the fraction of x-rays transmitted, $(\mu/\rho)_{\text{sample}}$ is the mass absorption coefficient for x-rays of wavelength λ within the sample, ρ is the sample density, and x is the x-ray path length. For a composite material, μ/ρ is equal to the summation of the weight fractions of the mass absorption coefficients of each of the elements present in the sample for the x-ray wavelength of interest. The mass absorption coefficient for the multicomponent material is given by the following equation:

$$(\mu/\rho)_{\text{sample}} = \sum_i (\mu/\rho)_i C_i \tag{14}$$

where $(\mu/\rho)_i$ is the mass absorption coefficient of the x-ray line of wavelength λ measured in element i and C_i is the weight fraction of element i in the sample.

Several mechanisms operate in the absorption process, the most important of which is caused by electronic transitions within the atom. Just as an electron of sufficient energy can knock a K electron out of an atom and thus cause the emission of K radiation, so can an incident x-ray photon. This process can occur just as long as the incident x-rays have an energy greater than the minimum excitation voltage E_c of an element in the sample. In the process, the ejected electron is called a photoelectron and the emitted characteristic radiation is called fluorescent or secondary radiation.

The x-ray intensity of the absorbed radiation decreases while the x-ray intensity of the element which absorbs the x-rays increases. An example of the fluorescence effect is given by examining the effect of the Ni $K\alpha$ radiation produced in a sample containing nickel as well as manganese, iron, and cobalt. Table III shows the atomic numbers, the $K\alpha$ wavelengths, the wavelengths equivalent to the minimum excitation voltage of the various elements in the multicomponent alloy, and the mass absorption coefficients for Ni $K\alpha$ radiation. The Ni $K\alpha$ wavelength is less than the wavelength necessary to produce iron and manganese radiation. Accordingly, the energy of the Ni $K\alpha$ photon is greater than E_K, the energy necessary to form a K-shell vacancy in iron and manganese. Therefore the higher energy of the Ni $K\alpha$ photon enables it to eject a K electron from iron and manganese and produce additional iron and manganese K radiation. On the other hand, Ni $K\alpha$ radiation will be absorbed by the production of iron and manganese fluorescence radiation. Fluorescence radiation is produced by x-ray excitation, rather than electron excitation. The absorption coefficient of x-rays is, however, much less than that for electrons. Therefore this radiation can be produced much deeper in the sample than for electron production since the absorption of x-rays is much less than that for electrons. Mass absorption coefficients for $K\alpha$ lines from Na $(Z = 11)$ to U $(Z = 92)$ and $L\alpha$ lines from Ga $(Z = 31)$ to U $(Z = 92)$ are given by Heinrich,[33] where the absorbers range in order from Li $(Z = 3)$ to U $(Z = 92)$. The mass absorption coefficients for the long-wavelength, low-energy $K\alpha$ lines from Be, B, C, N, O, F, Na, Mg, and Al and $L\alpha$ lines from Ti, Cr, Mn, Fe, Co, Ni, Cu, and Zn have been recently listed by Henke and Ebisu[34] and the absorbers range in order from Li $(Z = 3)$ to U $(Z = 92)$. As discussed in a later chapter on quantitative analysis, these mass absorption coefficients are critical parameters in any calculation scheme meant to obtain accurate chemical compositions.

Figure 16 shows the mass absorption coefficient of a pure Ni absorber plotted as a function of incident x-ray wavelength. In general, the x-ray absorption coefficient, that is, the x-ray stopping power, increases with increasing x-ray wavelength or decreasing x-ray energy. This is to be expected because the shorter the wavelength, the greater the energy and penetrating power. The discontinuities in these curves are called absorption edges; there is one for the K edge (Figure 16). There are three discontinuities for the L, five for the M, etc., at longer wavelengths which are not shown in this figure. The equivalent energies of these edges are equal to the critical excitation potentials for K, L, and M radiation.

If we consider the K edge as an example (Figure 16 at $\lambda > \lambda_K$ edge),

TABLE III. Wavelength and Absorption Coefficient Data for Mn, Fe, Co, and Ni

Atomic number and element	$\lambda_{K\alpha}$, Å	λ_K edge, Å	$(\mu/\rho)_i$ Ni $K\alpha$, g/cm² [a]
25 Mn	2.102	1.896	344
26 Fe	1.936	1.743	380
27 Co	1.789	1.608	53
28 Ni	1.658	1.488	59

[a] From Heinrich.[33]

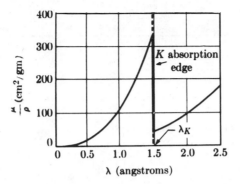

FIGURE 16. Variation of the mass absorption coefficient of nickel with x-ray wavelength (from Cullity,[26] Figure 1-8).

the photons do not have enough energy to excite K radiation. As λ decreases, the photons are more energetic, and the mass absorption coefficient decreases. At the λ_K edge the photons have exactly the right energy to cause secondary fluorescence and K lines are excited. As λ becomes progressively shorter, the photons are so energetic that they become less absorbed, μ/ρ decreases, and the amount of K emission decreases. This same argument can be made for L or M excitation.

Although the characteristic x-ray lines are very intense with respect to the continuum, they may not be the primary cause of secondary fluorescence. The most efficient production of secondary x-rays occurs at a wavelength λ just less than the λ_K edge. The continuum radiation varies in wavelength and some part of the spectrum may lie in a very favorable wavelength range to cause fluorescence. The parameter S can be used to denote the ratio of indirect (secondary) x-ray production by continuum x-rays to the total x-ray production from the primary electron beam as well as from secondary production by continuum x-rays. Experimental values of S as compiled by Green[35] are plotted versus Z, the atomic number, for the K and L_{III} ($L\alpha$) series in Figure 17. The dashed lines marked K and L_{III} (a) as calculated by Green and Cosslett[36] are also plotted in the figure. The data show that for light elements the amount of secondary fluorescence by continuum is small. However, for elements such as Cu the value of S is 10% and S exceeds 10% for L lines of many of the high atomic number elements.

C. Depth of X-Ray Production

In order to predict the depth of x-ray production (range) and the x-ray source size (spatial resolution), the amount of electron penetration must

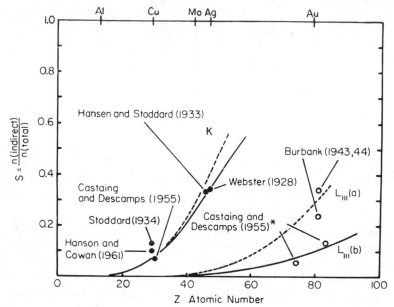

FIGURE 17.   Variation of the ratio of indirect to total x-ray production(s) with atomic number for the K and $L\alpha$ lines (from Green[35]).

be known. The problem of defining the electron penetration has been discussed earlier (Section II). From experimental results the dependence of mass range ρR on acceleration voltage E_0 was shown to be given by an equation of the form $\rho R = K E_0{}^n$, where K is a constant and n varies from 1.2 to 1.7. The mass range ρR is assumed to include all electrons of energies between E_0 and $E \simeq 0$. The mass range for continuum x-ray production will be similar to that of electrons since x-ray radiation can be produced even at low electron energies. The mass range for characteristic x-ray radiation will usually be smaller than that of electrons since characteristic x-rays can only be produced at energies above the critical excitation potential E_c. Therefore the dependence of mass range for characteristic x-ray production of a particular x-ray line in a given matrix $\rho R(x)$ will be given by an equation of the form

$$\rho R(x) = K(E_0{}^n - E_c{}^n) \tag{15}$$

Using a simplified energy loss expression, Castaing[37] obtained an equation of the form

$$\rho R(x) = 0.033(E_0{}^{1.7} - E_c{}^{1.7})A/Z \tag{16}$$

where E_0 and E_c are in keV, ρ is in g/cm³, $R(x)$ is in μm, and A and Z are the mean atomic weight and atomic number, respectively, of the bombarded material. Using an average value of A/Z equal to 2.3, since to a first approximation ρR is independent of target material, we obtain $\rho R(x) = 0.076(E_0^{1.7} - E_c^{1.7})$. Duncumb,[38] using the simplified Thomson-Whidding-ton energy loss expression, proposed that

$$\rho R(x) = 0.04(E_0^{2.0} - E_c^{2.0}) \tag{17}$$

for x-ray production. Cosslett and Thomas,[6] however, showed that the electron range obeys a power law with an index lower than 2.0. Therefore a square law for range cannot be used. Based on experimental measurements of x-ray production and electron range, Andersen and Hasler[39] developed a mass range equation for x-ray production as

$$\rho R(x) = 0.064(E_0^{1.68} - E_c^{1.68}) \tag{18}$$

In addition Reed,[40] when considering spatial resolution in x-ray micro-analysis, obtained an expression for $\rho R(x)$ based on the experimental voltage dependence found by Cosslett and Thomas[6] as discussed in Section II of this chapter. Reed defined the mass range equation as

$$\rho R(x) = 0.077(E_0^{1.5} - E_c^{1.5}) \tag{19}$$

Colby *et al.*,[15] studying thin films, derived a mass range equation

$$\rho R(x) = 0.033(E_0)^{1.5}A/Z \tag{20}$$

for high overvoltages. This relationship has been shown experimentally to be satisfactory when the overvoltage $U = E_0/E_c$ is equal to or greater than ten.[15]

To define the x-ray source size or spatial resolution, one can assume that the lateral x-ray production range is the same as $R(x)$, the x-ray range. This relation, however, assumes that the size of the electron beam impinging on the sample is vanishingly small. The total x-ray spatial resolution is therefore equal to the sum of the x-ray range $R(x)$ and the size of the electron beam d_p hitting the sample. In the electron microprobe and scanning electron microscope the electron beam size d_p is usually much smaller $[d_p/R(x) \leq 0.2]$ than the x-ray source size.

The x-ray range for Cu $K\alpha$ radiation in Cu is plotted in Figure 18 as a function of operating potential E_0 using the mass range relations [equations (16), (18), and (19)] as discussed above. The x-ray ranges differ from each other by as much as a factor of two. The lack of agreement may be due in part to the various approximations to the energy loss equations used. The mass range equation obtained by Andersen and Hasler[39] was based on experimental data and is perhaps the more correct. In further

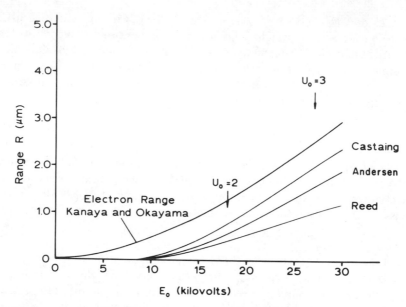

FIGURE 18. X-ray range for Cu $K\alpha$ radiation in Cu as a function of operating potential E_0.

discussion we will use the Andersen and Hasler x-ray range [equation (18)] for predictions. We assume that an uncertainty of $\pm 20\%$ is associated with the values obtained. The electron range according to Kanaya and Oka-yama[10] is also plotted on the figure. The maximum electron range R is greater than the x-ray range $R(x)$, as expected, resulting in the smaller x-ray source size obtained.

Figure 19 shows the x-ray range for various x-ray lines Al $K\alpha$, Cu $K\alpha$, Cu $L\alpha$, and Au $L\alpha$ generated within the element targets Al, Cu, and Au as a function of voltage.[41] The three matrix elements shown in Figure 19 were chosen to represent the range of specimen densities which are likely to be analyzed; the x-ray lines were chosen to represent the common x-ray wavelength ranges measured. The x-ray ranges $R(x)$ are given by equation (19) due to Andersen and Hasler.[39] The electron range R for Al is shown for comparison purposes. The x-ray ranges depend not only on the density of the matrix (Al = 2.7 g/cm³, Cu = 8.93 g/cm³, Au = 19.3 g/cm³) but also on the value of the critical excitation energy of the x-ray line produced. For example, the x-ray ranges for Cu $L\alpha$ radiation (13.3 Å, $E_c = 0.933$ keV) in Cu are quite different from the x-ray ranges for Cu $K\alpha$ radiation (1.541 Å, $E_c = 8.98$ keV) in Cu at the same energy, E_0. Therefore, one can see from Figure 19 that x-ray ranges vary with the element x-ray line as well.

For practical x-ray analysis, the Al $K\alpha$ in Al curve is suitable for estimating the x-ray ranges for $K\alpha$ characteristic x-rays from elements such as Na, Mg, Al, Si, P, and S (5–20 Å) or $L\alpha$ x-rays from Fe, Co, or Ni in a matrix of density around 3 g/cm³. This curve then represents x-ray ranges expected from aluminum alloys or silicate minerals. The Cu $K\alpha$ in Al curve is useful for estimating the $K\alpha$ x-ray ranges for elements such as Fe, Co, or Ni (1–2 Å) or the $L\alpha$ x-ray ranges for elements such as Au, W, and Ta in similar low-density materials. The Cu $K\alpha$ in Cu curve would represent $K\alpha$ characteristic x-rays from such elements as Fe, Co, Ni, Cu, and Zn (1–2 Å) in a matrix of density around 9 g/cm³. In other words, the Cu curve would represent the x-ray range expected for metallurgical samples such as ferrous alloys, Ni- and Co-base superalloys, and Cu-base alloys. The Cu $L\alpha$ in Cu curve would represent light element K lines (5–20 Å) or Fe, Co, Ni, Cu, or Zn L lines (5–20 Å) in a matrix of density around 9 g/cm³. The Au $L\alpha$ curve shows the mass range for a heavy element matrix of density around 19 g/cm³. X-ray lines in the wavelength range 1–2 Å would be considered in this case.

To obtain an x-ray source size or spatial resolution ≤ 1 μm, the voltage of the electron beam may have to be adjusted depending on the matrix material and line measured. For samples of geological interest such as silicates (Al K in Al curve), the operating voltage must be less than 10 kV.

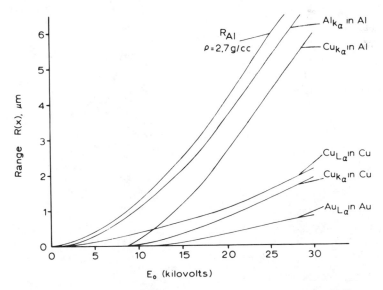

FIGURE 19. X-ray range for the Al $K\alpha$, Cu $K\alpha$, Cu $L\alpha$, and Au $L\alpha$ lines generated within the element targets Al, Cu, and Au as a function of operating potential E_0 (from Goldstein[41]).

With 10 kV, however, the K lines of elements such as Fe, Ni, and Cu are just barely excited. To ensure a minimum overvoltage of about two, a 15-kV operating potential is used and a source size of $\sim$2.5 μm (Cu $K\alpha$ in Al curve) is obtained. On the other hand, if the L lines of elements such as Fe, Ni, and Cu are used, 10 kV is quite adequate. For metallurgical samples of density $\sim$9 g/cm³, a 1-μm x-ray source size can be obtained by operating at $\sim$20 kV; however, if the L lines are used, operation at 10 kV will allow for a source size of less than 0.5 μm. For heavy element matrices and for x-ray wavelengths of 1–2 Å, beam sizes $\leq$1 μm can be obtained with operating voltages below 30 kV. An operating voltage of 20 kV is often necessary to ensure an adequate overvoltage ratio. For light elements measured in heavy element matrices, the beam size would be larger. In order to minimize the x-ray source size and maximize the peak to background ratio, it is often best to set the voltage of the electron beam E_0 so that the overvoltage $U = E_0/E_c$ is a factor of 2–3 for the shortest wavelength x-ray that is to be measured.

The discussion of x-ray ranges can be summarized with the aid of Figure 20. This figure shows a comparison of some of the x-ray production regions from two specimens, one with a density of $\sim$3 g/cm³ (Al) and the other with a density of $\sim$10 g/cm³ (Cu), which are bombarded with a 20-kV focused electron beam. The x-ray range for Al $K\alpha$ in Al is much greater

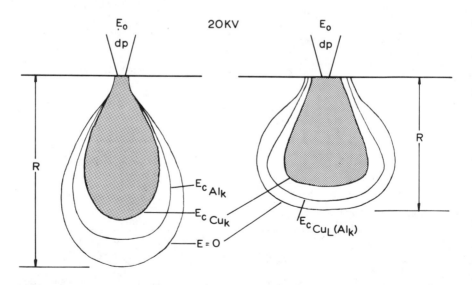

FIGURE 20.   Comparison of x-ray production regions from specimens with densities of (left) 3 and (right) 10 g/cm³.

than that for Cu K in Cu. In addition the Cu $K\alpha$ radiation in Al comes from a depth closer to the surface in Al than does the Al $K\alpha$ radiation. However, Cu $K\alpha$ radiation in Al comes from much deeper in the sample than does Cu $K\alpha$ radiation in Cu. The Al $K\alpha$ radiation is produced much closer to the surface in Cu ($\rho \sim 10$ g/cm³) than in Al ($\rho \sim 3$ g/cm³) itself. If the energy E_0 decreases below E_c for Cu $K\alpha$ radiation, the L lines will still be excited and they will be closer to the sample surface.

Up to this point it has been assumed that the only contribution to the x-ray source size is the electron range through which characteristic x-rays for a given element are produced. As noted earlier in this chapter, however, x-rays can also be produced by secondary fluorescence in which characteristic x-rays of another element present in the sample or continuum x-rays of $\lambda < \lambda_K$ edge are absorbed and produce additional characteristic x-rays. Characteristic x-rays obtained by this process are produced from a volume which is much larger than that encompassed by the electrons producing the primary intensity. Therefore, the x-ray spatial resolution is decreased. Fortunately, the contribution of fluorescence is usually a fairly small fraction of the total intensity, and the effect is usually small.

The effect of fluorescence on x-ray resolution or spot size can be determined by a simple experiment in which an artificial boundary between two pure elements is constructed. The method used is to measure the intensity of the excited element at a series of points along a line normal to the boundary. If any x-ray intensity is measured at distances sufficiently far from the boundary so that there is no excitation by electrons, the measured intensity is due to fluorescence. Such experiments have been carried out,[42–44] and fluorescence effects have been measured as far as 25 μm or more from the boundary. The amount of fluorescent radiation when compared to the electron-excited radiation is usually less than 5% at a distance 5 μm from the interface and is usually less than 1% at a distance 25 μm from the interface.

It is difficult to predict the x-ray source size when x-ray fluorescence is a factor. In most cases the x-ray source sizes given in Figure 19 indicate the practical values that are most useful. The effect of secondary fluorescence is, however, very important when accurate measurements of composition must be made at the interfaces between two phases where a discontinuity in composition occurs (see Chapter XII).

VI. AUGER ELECTRONS

When electrons or x-rays bombard a sample and ionize the atom, vacancies are formed in the inner electron shells. The atom rearranges or deexcites by filling the vacancy with an electron from a shell of lower energy and emits energy as a result of the transition. However, not all the energy is released as x-ray photons; some of the energy may be released by emitting an electron of a specific energy (Auger effect) (note Figure 13a). In either case the energy of the x-ray photon or of the Auger electron is characteristic of the emitting element.

One can describe the relative amounts of x-ray and Auger electron production through a term called the fluorescence yield ω. The K fluorescence yield ω_K can be defined as the ratio of the number of photons of all x-ray lines in the K series emitted per unit time divided by the number of K-shell vacancies formed during the same time,[31]

$$\omega_K = \sum_i \frac{(n_K)_i}{N_K} = \frac{n_{K\alpha} + n_{K\beta} + \cdots}{N_K} \tag{21}$$

where N_K is the rate at which K-shell vacancies are produced and $(n_K)_i$ is the rate at which photons of a spectral line i are emitted. The L and M fluorescence yields ω_L and ω_M are defined similarly. The Auger yield a is defined as $1 - \omega$ for each shell. If it were not for the Auger effect, ω would always be one.

The variation of the Auger yield with atomic number and series is shown in Figure 21.[45] One should note that the Auger yield is very high for light elements, $Z < 15$, in the K series and for almost all elements in the L and M series. The energies necessary to create vacancies in the K shells of the light elements, $Z < 15$, and in the L shells of elements $Z < 40$,

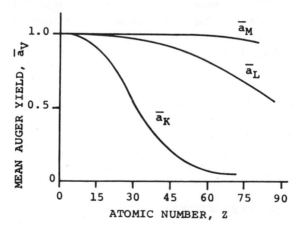

FIGURE 21.   Variation of the mean Auger yield as a function of atomic number for initial vacancies in the K, L, and M atomic levels (from Mac-Donald[45]).

where a_K approaches one, are less than 2 keV. It is apparent then that the x-ray fluorescence yield ω_K is greatly decreased for both the light elements and the long-wavelength L and M radiation. This decrease results in a corresponding loss of sensitivity. Therefore, Auger analysis may prove more useful than x-ray analysis for the light elements.[6]

The electronic transitions involved in the production of Auger electrons are illustrated in Figure 22.[46] A vacancy is created by electrons or fluorescent x-rays in an inner level such as the K level shown. The atom can return to an equilibrium state by means of an Auger transition. The electron from the M level fills the vacancy in the K level. The energy released from the electron transition M to K is absorbed by emission of an

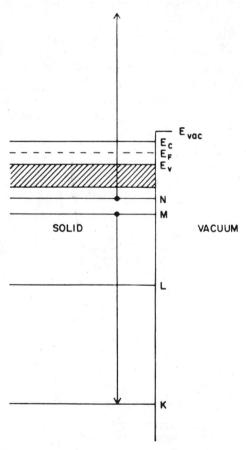

FIGURE 22.   Electron energy diagram illustrating a transition giving rise to an emission of an Auger electron (from Stein *et al.*[46]).

N electron. The electron in the N level is ejected with energy approximately equal to $E_K - E_M - E_N$. In this process, a doubly ionized atom having two vacancies is created by filling the initial vacancy and by ejecting one electron by the Auger process. The Auger yield a_K is higher in elements of low atomic number because the electrons are more loosely bound and E_K is relatively low. The same argument holds for the L and M lines.

There is a large number of possible electronic transitions which can produce Auger electrons of specific energies. In practice, the most common Auger energies correspond to the most probable electron transitions, those giving rise to the strongest x-ray spectral lines. Most elements have Auger transitions which result in ejected electrons with energies generally between 50 and 2000 eV. These energies lie just above the energy range of the low-energy secondary electrons discussed earlier. The observed Auger electron spectrum consists of broad peaks which are relatively weak. These peaks are superimposed on the secondary and scattered electron spectra and it is very difficult to measure the position of these peaks because of their relative weakness. When the derivative $dN(E)/dE$ of the electron distribution curve is obtained, the signal to noise ratio is greatly increased and the peak position can be accurately measured.[47] A differentiated spectrum obtained from a 304 stainless steel is shown in Figure 23. The

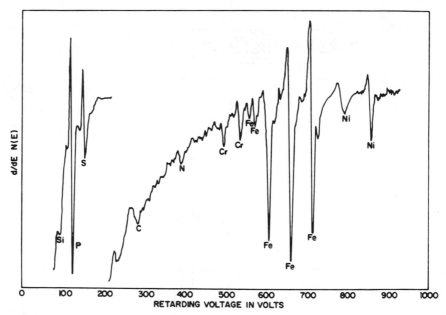

FIGURE 23.   A differentiated spectrum obtained from a 304 stainless steel (from Stein *et al.*[45]).

Auger peaks for Fe, Ni, Cr, P, S, and Si (L vacancies) as well as C and N (K vacancies) are clearly seen.

It is of interest to determine the effective volume from which Auger electrons are collected. This volume is determined by the range and beam width from which the Auger electrons are produced. Experimental evidence obtained by Palmberg and Rhodin[48] showed that the mean escape depth (range) for Auger electrons in Ag (without significant loss of energy) varies between 4 and 8 Å for energies of 72 and 362 eV, respectively. When Auger electrons are produced at greater than the escape depth their characteristic energy is partially changed or entirely absorbed by production of secondary electrons, etc. before reaching the surface. Higher energy electrons will have correspondingly larger escape depths. As discussed by MacDonald,[45] the average depth analyzed by Auger electron spectroscopy is a function of the energy of the Auger peak and is typically 1–10 Å for Auger electrons in the energy range of 10–1000 eV. The spatial resolution of the Auger signal is determined by the same factors which influence the low-energy secondary electron resolution. For Auger electrons of energies less than 1 keV, the spatial resolution is determined by the diameter of the electron beam d_p. The Auger electron range is therefore independent of the spatial resolution of the electrons.

Since Auger analysis is obtained from less than 10 Å of material in depth, it is sensitive to even a few monolayers of surface contaminants. The composition of the contaminants will be measured rather than the bulk material at the surface. Therefore Auger analysis must be performed in high-vacuum environment of at least 10^{-7} torr. Even at pressures of 5 × 10^{-8} torr, only semiquantitative surface studies can be performed since the electron beam breaks down hydrocarbons, which are then deposited onto the surface of the sample. For high-sensitivity Auger spectroscopy, a vacuum of better than 10^{-9} torr and a metal-sealed system are required. A few scanning electron microscopes with this vacuum capability have been produced.

VII. SUMMARY—RANGE AND SPATIAL RESOLUTION

Electron beam–specimen interactions are responsible for the signals which can be used to reveal the topography and local chemistry of the sample. Figure 24 summarizes the range and spatial resolutions of the various signals (backscattered electrons, secondary electrons, x-rays, and

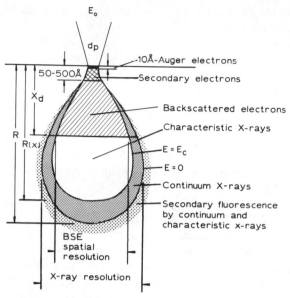

FIGURE 24.   Summary of range and spatial resolutions of backscattered electrons, secondary electrons, x-rays, and Auger electrons produced from an electron microprobe–scanning electron microscope (from Goldstein[41]).

Auger electrons) available from the electron microprobe–scanning electron microscope for a pure element of low to medium atomic number. The spatial resolution for x-rays and backscattered electrons is approximately the same as the respective ranges $[R(x), x_d]$. On the other hand, the spatial resolution of the secondary electron and Auger electrons is independent of the range of these signals. The spatial resolution of secondary and Auger electrons is approximately the same as the probe diameter, d_p, whereas the spatial resolution of the backscattered electron and x-ray signals is usually much larger than d_p. As illustrated in Figure 19, the x-ray spatial resolutions of several x-ray lines measured in the same material are different.

REFERENCES

1. H. A. Bethe, *Ann. Phys. (Leipzig)*, **5**, 325 (1930).
2. R. R. Wilson, *Phys. Rev.*, **60**, 749 (1941).
3. P. Duncumb and S. J. B. Reed, in *Quantitative Electron Probe Microanalysis* (K. F. J. Heinrich, ed.), National Bureau of Standards, Special Publication 298, Washington, D.C. (1968), p. 133.

4. V. E. Cosslett and R. N. Thomas, in *The Electron Microprobe* (T. D. McKinley, K. F. J. Heinrich, and D. B. Wittry, eds.), Wiley, New York (1966), p. 248.

5. V. E. Cosslett and R. N. Thomas, *Brit. J. Appl. Phys.*, **15**, 883 (1964).

6. V. E. Cosslett and R. N. Thomas, *Brit. J. Appl. Phys.*, **15**, 1283 (1964).

7. V. E. Cosslett and R. N. Thomas, *Brit. J. Appl. Phys.*, **16**, 779 (1965).

8. K. Murata, T. Matsukawa and R. Shimizu. *Jap. J. Appl. Phys.*, **10**, 678 (1971).

9. P. Duncumb and P. K Shields, *Brit. J. Appl. Phys.*, **14**, 617 (1963).

10. K. Kanaya and S. Okayama, *J. Phys. D. Appl. Phys.*, **5**, 43 (1972).

11. G. Shinoda, K. Murata and R. Shimizu, in *Quantitative Electron Probe Microanalysis* (K. F. J. Heinrich, ed.), National Bureau of Standards, Special Publication 298, Washington, D.C. (1968), p. 155.

12. K. F. J. Heinrich, in *X-ray Optics and Microanalysis, IV International Congress on X-Ray Optics and Microanalysis, Orsay, 1965* (R. Castaing, P. Deschamps, and J. Philibert, eds.), Hermann, Paris (1966), p. 159.

13. H. E. Bishop, in *"X-ray Optics and Microanalysis, IV International Congress on X-Ray Optics and Microanalysis, Orsay, 1965"* (R. Castaing, P. Deschamps, and J. Philibert, eds.), Hermann, Paris (1966), p. 153.

14. M. P. Palluel, *C. R. Acad. Sci., Paris*, **224**, 1492 (1947).

15. J. W. Colby, W. N. Wise, and D. K. Conley, in *Advances in X-Ray Analysis* (J. B. Newkirk and G. R. Mallett, eds.), **10**, 447, Plenum Press, New York (1967).

16. E. F. H. St.G. Darlington, in *V International Congress on X-Ray Optics and Microanalysis, Tubingen, 1968* (G. Möllenstedt and K. H. Gaukler, eds.), Springer Verlag, Berlin (1969), p. 76.

17. R. Shimizu and K. Murata, *J. Appl. Phys.* **42**, 387 (1971).

18. O. C. Wells, *Appl. Phys. Letters*, **16**, 151 (1970).

19. O. Hachenberg and W. Brauer, in *Advances in Electronics and Electron Physics* (L. Marton, ed.), Vol. 11, Academic Press, New York (1959), p. 413.

20. D. B. Wittry, in *X-Ray Optics and Microanalysis, IV International Congress on X-Ray Optics and Microanalysis, Orsay, 1965* (R. Castaing, P. Deschamps, and J. Philibert, eds.), Hermann, Paris (1966), p. 168.

21. L. Reimer, in *Scanning Electron Microscopy: Systems and Applications 1973*, Conference Series 18, Institute of Physics, London (1973).

22. H. Seiler, *Z. Angew. Phys.*, **22**, 249 (1967).

23. T. E. Everhart, O. C. Wells, and C. W. Oatley, *J. Electron. and Control*, 7, 97 (1959).

24. K. Murata, in *Scanning Electron Microscopy/1973*, IITRI, Chicago, Illinois (1973), p. 267.

25. A. N. Broers, in *Scanning Electron Microscopy/1970*, IITRI, Chicago, Illinois (1970), p. 3.

26. B. D. Cullity, *Elements of X-Ray Diffraction* Addison-Wesley, Reading, Massachusetts (1956).

27. S. T. Stephenson, in *Handbuch der Physik* (S. Flügge, ed.), Springer-Verlag, Berlin (1957), p. 337.

28. H. A. Kramers, *Phil. Mag.* **46**, 836 (1923).

29. E. Lifshin, in *"Summer Course Notes—Electron Probe Microanalysis and Scanning Electron Microscopy, June 19–23, 1972 at Lehigh University"* (1972).

30. C. Barrett and T. B. Massalski, *Structure of Metals*, 3rd ed., McGraw-Hill, New York (1966).

31. E. P. Bertin, *Principles and Practice of X-Ray Spectrometric Analysis*, Plenum Press. New York (1970).

32. J. A. Bearden, NYO-10586, U.S. Atomic Energy Commission, Oak Ridge, Tennessee (1964).

33. K. F. J. Heinrich, in *The Electron Microprobe* (T. D. McKinley, K. F. J. Heinrich, and D. B. Wittry, eds.), Wiley, New York (1966), p. 296.

34. B. L. Henke and E. S. Ebisu, in *Advances in X-Ray Analysis*, Vol. 17, Plenum Press, New York (1974), p. 150.

35. M. Green, in *X-Ray Optics and X-Ray Microanalysis, III International Symposium, Stanford University, Stanford, Calif., 1962* (H. A. Pattee, V. E. Cosslett, and A. Engström, eds.), Academic Press, New York (1962), p. 361.

36. M. Green and V. E. Cosslett, *Proc. Phys. Soc. (London)*, 78, 1206 (1961).

37. R. Castaing, in *Advances in Electronics and Electron Physics* (L. Marton, ed.), Vol. 13, Academic Press, New York (1960), p. 317.

38. P. Duncumb, in *Proceedings of the Second International Symposium on X-Ray Microscopy and X-Ray Microanalysis, Stockholm, 1960* (A. Engström, V. E. Cosslett, and H. Pattee, eds.), Elsevier, Amsterdam (1960), p. 365.

39. C. A. Andersen and M. F. Hasler, in *X-Ray Optics and Microanalysis, IV International Congress on X-Ray Optics and Microanalysis, Orsay, 1965* (R. Castaing, P. Deschamps, and J. Philibert, eds.), Hermann, Paris (1966), p. 310.

40. S. J. B. Reed, in *X-Ray Optics and Microanalysis, IV International Congress on X-Ray Optics and Microanalysis, Orsay, 1965* (R. Castaing, P. Deschamps, and J. Philibert, eds.), Hermann, Paris (1966), p. 339.

41. J. I. Goldstein, in *Metallography—A Practical Tool for Correlating the Structure and Properties of Materials*, ASTM Special Technical Publication 557, ASTM (1974), p. 86.

42. S. J. B. Reed and J. V. P. Long, in *X-Ray Optics and X-Ray Microanalysis, III International Symposium, Stanford University, Stanford, Calif. 1962* (H. A. Pattee, V. E. Cosslett, and A. Engström, eds.), Academic Press, New York (1963), p. 317.

43. J. I. Goldstein and R. E. Ogilvie, in *X-Ray Optics and Microanalysis, IV International Congress on X-Ray Optics and Microanalysis, Orsay, 1965* (R. Castaing, P. Deschamps, and J. Philibert, eds.), Herman, Paris (1966), p. 594.

44. F. Maurice, R. Seguin, and J. Henoc, in *X-Ray Optics and Microanalysis, IV International Congress on X-Ray Optics and Microanalysis, Orsay, 1965* (R. Castaing, P. Deschamps, and J. Philibert, eds.), Hermann, Paris (1966), p. 357.

45. N. C. MacDonald, in *Scanning Electron Microscopy/1971*, IITRI, Chicago, Illinois (1971), p. 91.

46. D. F. Stein, R. E. Weber, and P. W. Palmberg, *J. Metals*, 23, 39 (1971).

47. L. A. Harris, *J. Appl. Phys.*, 39, 1419 (1968).

48. P. W. Palmberg and T. N. Rhodin, *J. Appl. Phys.* 39, 2425 (1968).

IV

IMAGE FORMATION IN THE SCANNING ELECTRON MICROSCOPE

D. E. Newbury

I. THE SEM IMAGING PROCESS

To understand the process of image formation in the scanning electron microscope, we must examine the component parts of the image formation system: the scanning system, the signal detectors, the amplifiers, and the display. For this discussion, the upper portion of the electron probe forming system, i.e., the electron gun and the upper condenser lenses, is relatively unimportant. Details of these components have been given in the discussion of electron optics in Chapter II. The entire instrument is discussed in detail by Oatley.[1] We shall assume that a focused electron probe of diameter d_p is produced by the electron gun and lenses. The components of the image formation system are illustrated in Figure 1, and with this diagram we shall follow the sequence of events in the formation of the image. The electron beam travels down the optical axis of the electron optical column from the gun through the first two condenser lenses with each lens progressively demagnifying the beam. When the beam encounters the first set of scan coils, it is deflected off the optical axis. The second (lower) set of scan coils acts on the deflected beam and produces a second deflection, causing the beam to again cross through the optical axis. The scan coils rest in the bore of the final lens, and during passage

D. E. NEWBURY—National Bureau of Standards, Washington, D.C.

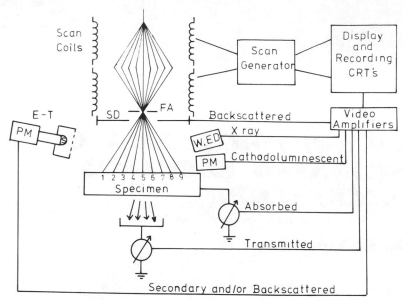

FIGURE 1.   Image formation system in the scanning electron microscope. FA, final aperture; SD, solid state electron detector; ET, Everhart–Thornley electron detector; PM, photomultiplier; S, scintillator; W, ED, wavelength- and/or energy-dispersive x-ray detectors; CRT, cathode ray tube.

of the scan coils, the beam is further demagnified by the action of this lens, which is normally the strongest lens in the microscope. The effect of the "double-deflection scanning" is to cause the beam to travel laterally as a function of time in a plane below the scan coils and perpendicular to the optical axis. The important point to note is that the beam movement is time resolved. The rays numbered 1–9 in Figure 1 are produced sequentially; only one ray path exists through the scan coils at any point in time. All of the rays pass through the same point on the optical axis after the second deflection. At this final crossover, an aperture is placed which serves to define the beam divergence α given by

$$\alpha = r_a/D \tag{1}$$

where r_a is the radius of this final aperture and D is the distance from the aperture to the specimen.

If we place a planar specimen at the level indicated in Figure 1, the rays 1–9 will strike the specimen at a series of discrete points, 1–9. A cathode ray tube (CRT) is scanned in synchronism with the specimen. The electromagnetic scan coils of the microscope and of the CRT are driven by the same scan generator, so that for each beam position on the specimen there is a unique position on the CRT. A one-to-one correspondence is thus established between each point on the specimen and each point on the CRT. The synchronized scan is contrived so that the geometrical arrangement of a group of points scanned on the specimen is maintained for the corresponding points on the CRT. Thus, if a square raster of points is scanned on the specimen, a square raster is scanned on the CRT. Similarly, when a single line of points is scanned on the specimen, a line is scanned on the CRT.

At each point on the specimen, the beam will dwell for some fixed time τ which is determined by the rate at which scanning takes place. During this dwell time τ the electrons of the beam interact with the specimen. The total interaction time for a given electron is very much less than τ, so that as soon as the electron beam leaves a particular spot, the interaction for that spot is complete. The details of the interaction of high-energy electrons with the specimen are explained in Chapter III. A variety of effects of the interaction is produced: high-energy backscattered electrons, low-energy secondary electrons, x-rays, and radiation in the ultraviolet, visible, and infrared regions. Each of these products of the interaction carries information about the nature of the specimen. All of the interaction products can be monitored, simultaneously if desired, through the use of appropriate detectors. The signals formed in the detectors are suitably amplified and used to control the brightness of the CRT (intensity modulation). For each point on the specimen, a point is established on the CRT, and the brightness of this point is related to a detector signal derived from the characteristics of the beam–specimen interaction. If this interaction were constant at every point on the specimen, then a uniform brightness B would be produced on the CRT. Actually, in a real specimen the beam–specimen interaction varies from place to place around the specimen due to the local character of the specimen. The interaction may be affected by topography, composition, crystallinity, magnetic or electric character, or by other properties of the specimen. Since the interaction varies from point to point on the specimen, the signals produced by the detectors will vary, and hence different values of brightness will be generated at points in the CRT. The combined system of the scan coils, scan generator, signal detectors, amplifiers, and display cathode ray tubes can thus map an area scanned on the specimen to an area scanned on the display. The geometrical relationship of a group of points on the specimen is reproduced on the CRT, and at each point, an intensity is produced which is related in some way to the specimen. Magnification results from the mapping process according to the ratio of the area scanned on the CRT to the area scanned on the specimen. Thus with a fixed scan size on the CRT, the effect of reducing the size of the area scanned on the specimen is to increase the magnification of the map. The information present in the small area is stretched by the mapping process to cover a larger area. Typically, the CRT's in the SEM have an area of 100 mm square; at a magnification of 10×, an area 10 mm square is scanned on the specimen, while at 10,000× the area on the specimen is 0.01 mm square.

An important concept in scanning electron microscopy is that of the "picture element" or "picture point." The SEM image is written on a high-quality CRT by a focused electron beam; on a typical CRT of 100 mm square, it is possible to focus the writing beam so that 1000 distinct and separate lines can be recognized and along a given line 1000 distinct beam spots can be written without overlapping. Such an image will appear to be of very high quality to the human eye, which, unaided, cannot resolve lines packed 10 per mm. The CRT screen can be imagined to be divided into 10^6 elements each 0.1 mm square, the maximum number which can be written on such a screen by the beam. The brightness of each element is approximately constant over the element size. This element is the fundamental unit of the image; the image cannot be subdivided further. This basic element is known as the picture element. From the one-to-one correspondence established between the areas scanned on the specimen and on the CRT, the picture elements of the CRT image can be related to an equivalent area on the specimen, which is found by dividing the area of the scan by 10^6 for a 1000-line CRT. Thus, at 1000×, the area scanned on the specimen is 100 μm square, and each picture element related to the specimen is 0.1 μm square. The concept of picture element size at the specimen is important when determining the optimum probe size for a given magnification. To operate at optimum probe current, it is not useful to have the probe size smaller than the picture element size, since a portion of the scanned area of the specimen is not being illuminated by the beam and beam current is lower than it need be. If the probe size is larger than the picture element size, then the number of valid picture elements in the final image is decreased, since the beam position at the specimen is not uniquely confined to one element on the CRT. The information is integrated over several elements on the CRT and the quality of the final image is decreased. When the probe size is equal to the picture element size, the maximum spatial quality of the final image is realized and the maximum useful current is obtained in the probe.

The picture element related to the specimen is also a useful concept in understanding the depth of focus of the SEM. If the beam is focused so that at a given working distance and magnification it is smaller than the picture element size, there exists a range of working distance over which the image is in effectively perfect focus. As long as the beam is not larger than a picture element, no overlap will occur. This criterion can be relaxed somewhat, since the unaided human eye cannot really resolve ten elements per millimeter. The image quality only begins to deteriorate noticeably when the beam covers about three or more picture elements. The depth of focus can thus be calculated as twice the distance over which the beam diameter increases from an initial value d_p to a value equal to (arbitrarily) the diameter of three picture elements. For a magnification of $1000\times$, a beam diameter of 500 Å, and a divergence of 5×10^{-3} rad, the depth of focus is at least 50 μm. (See Chapter II for a description of depth of focus.)

The image formation process in the SEM, the mapping of an area of the specimen onto a CRT, is very unlike the image formation process in an optical or transmission electron microscope. In those systems, rays emitted from the object pass through the lens and are formed into an image; the points in the image plane and in the object plane are connected by ray paths. In the SEM, no image in this conventional sense ever exists; no ray paths exist between the SEM image and the object. The SEM image is an abstract construction; it is simply a map. An image is a likeness of an object. Does the map of the specimen produced by the SEM convey information which we can interpret as an image? The answer to this question is affirmative. SEM images of rough specimens with extensive topography can be interpreted according to our prior experience with objects illuminated with light, despite the considerable difference between the processes of light and electron interaction with solids. However, images may be affected by every aspect of the electron–specimen interaction. To interpret images, we must understand the mechanisms by which the signal differs from point to point on the image. These differences in signal are referred to as contrast C which is defined as

$$C = \Delta S / S_{\text{av}} \qquad (2)$$

where ΔS is the change in the signal between any two points in the image, and S_{av} is the average signal. Considering a particular interaction product, e.g., backscattered electrons, the number of backscattered electrons may vary at different points. In this case, we are dealing with a *number effect*; such contrast is known as *number contrast*. Electrons which have left the specimen may be perturbed in flight along their trajectories to detectors which are external to the sample, e.g., by magnetic or electric fields. As will be shown, such *trajectory effects* can lead to contrast formation, with such contrast known as *trajectory contrast*. Since detector characteristics are important in this discussion, we shall consider signal detectors before proceeding with the topic of contrast formation.

II. SIGNAL DETECTORS

As explained in Chapter II, the act of focusing a probe of the order of 10 nm (100 Å) in diameter from a source of 50 μm in diameter results in a loss of most of the current emitted at the filament. With an emission current of 150 μA, a focused probe of 10 nm diameter will only contain about 10 pA (10^{-11} A). Considering a total coefficient of emission of electrons, both backscattered and secondary, equal to unity, a maximum of only 10 pA of signal is available. Despite some directionality of emission for the back-scattered and secondary electrons, the emissive signal is effectively radiated into a solid angle of a unit hemisphere. Only in special cases will we be able to collect all of the emitted current; hence the collected signal may range as low as 1 pA (10^{-12} A). To make use of such a low signal in a video display, a very high amplification must be employed. As we shall see, the noise or randomness in the signal is our chief limit in obtaining high resolution. More noise is introduced in the amplification process, described in detail by Oatley.[1] Conventional amplifiers can introduce an unacceptable amount of noise when amplification of currents of the order of a few pA is required. Moreover, bandwidth requirements (frequency response) must be considered. With the probe size equal to the picture element, the maximum usable frequency which we require in our display occurs in the case where alternate elements are white and black. During the frame time t_f, $N \times N$ picture points are scanned; the highest frequency component for alternate black–white points would be $N^2/2t_f$. The lowest frequency component would be the case where a signal change from white to black occurs across the frame. This component would have a frequency of $1/t_f$. The bandwidth Δf required is thus

$$f_{\text{high}} - f_{\text{low}} \simeq N^2/2t_f \tag{3}$$

For $N = 500$ lines and $t_f = 4$ sec, a raster speed usable for focusing the image, Δf is about 30 kHz. For amplifying small currents, the large gain which is necessary requires a large value of resistance in the conventional amplifier. The combination of a required large bandwidth and a small current is a severe challenge for the conventional amplifier. Both the resistance and the bandwidth appear in the equation for the noise generated by the amplifier:

$$V_{\text{noise}} = 4kTR \, \Delta f \tag{4}$$

where V_{noise} is the voltage signal due to noise, k is Boltzmann's constant, T is the absolute temperature, R is the amplifier resistance, and Δf is the bandwidth. At low signal currents, conventional amplification results in severe signal degradation. For these reasons, detectors which use direct electron multiplication have been developed.

A. Everhart–Thornley Detector

By far the most widely used detector in scanning electron microscopy is the scintillator–photomultiplier detector invented by Zworykin, Hillier, and Snyder[2] and further developed by Everhart and Thornley.[3] The detector is popularly known by the names of the latter workers. The Everhart–Thornley (ET) detector is shown in Figure 2. The basic component of the detector is a scintillation material which emits light when struck by high-energy electrons. The light created in the scintillator passes down the light guide (or light pipe) by total internal reflection and is led to the window of a photomultiplier. The light creates a cascade of electrons in the photomultiplier. Such a system creates a very large gain in amplification at a cost of only a small noise contribution, providing the scintillator is efficient.[4] The light guide can also be bent through an angle to allow optimum positioning of the scintillator relative to the specimen. Most scintillators require electron energies of 10–15 keV for efficient light production. With an incident beam energy of 20 keV or more, the majority of the primary backscattered electrons can activate the scintillator. However, the low-energy secondary electrons are not sufficiently energetic to activate the scintillator. To make use of the secondary electrons, the scintillator is maintained at a potential of about +12 kV to accelerate low-energy electrons to a sufficient energy for scintillator excitation. To protect the incident beam from the deflection it might suffer due to the presence of an unshielded 12-kV surface in the specimen chamber, the scintillator is surrounded by a Faraday cage, which is maintained at a low voltage relative to the specimen, usually −50 to +250 V, called the "collector

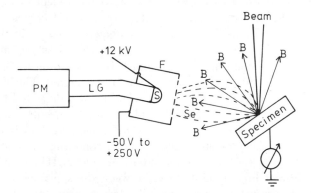

FIGURE 2.   Everhart–Thornley emitted-electron detector. B, backscattered electrons (solid trajectories); Se, secondary electrons (dashed trajectories); F, Faraday cage; S, scintillator; LG, light guide; PM, photomultiplier.

voltage." The low positive voltage serves to collect the secondary electrons from the specimen. Once attracted to the vicinity of the cage, the secondaries can then respond to the 12-kV potential on the scintillator.

The choice of scintillation materials must consider scintillator efficiency, time of fluorescent decay, and durability. High efficiency of light production is an obvious requirement. The time of decay of the electron-induced scintillation must also be considered. When the electron beam leaves one picture element, the luminescence of the scintillator must decay to a low level before the next picture element is reached or else integration of picture element information will occur, leading to decreased image quality. This is especially important at TV scanning rates, where the picture point time is extremely short. Finally, most scintillation materials show degradation with exposure to energetic electrons and must be replaced periodically to maintain peak efficiency.[4] A recent review of a range of scintillation materials provides a guide for the proper choice for use under various operating conditions.[5]

The ET detector offers the microscopist a very flexible system for studies of both the secondary and the backscattered electrons. With the Faraday cage maintained at +250 V, secondary electrons are collected with high efficiency from the specimen; secondary electron trajectories are bent through large angles by the +250 V potential. Many secondary electrons which are not emitted in the exact direction of the ET detector are collected nevertheless. The ET detector thus has a large solid angle of collection for secondary electrons. By lowering the voltage on the Faraday cage, the microscopist can continuously decrease the fraction of secondary electrons which are collected. When the potential of the Faraday cage is set at −50 V relative to the specimen, secondary electrons are almost completely excluded from collection. The primary electrons which are emitted with much higher energies are unaffected by the −50 V potential and continue to be collected. Thus, contrast effects which are due to secondary electrons alone can be determined by examining the image with and without the secondary electrons included in the signal. The high-energy primary electrons, B in Figure 2, are not significantly deflected toward the collector by the +250 V potential. Only those backscattered electrons emitted along the line of sight between the specimen and the scintillator are collected. The ET detector has a much smaller solid angle of collection for backscattered electrons than for secondary electrons. It should be noted that when a direct line of sight exists between the specimen and the ET detector, the backscattered electrons will always form part of the signal. Although the secondary electrons can be excluded by a negative voltage on the Faraday cage, the backscattered electrons cannot. The backscattered electrons can be excluded while the secondary electrons are collected by interrupting the direct line of sight with a shield which is opaque to the primary electrons. A positive voltage on the Faraday cage will still provide a reasonably efficient collection of secondaries around the shield.

The ET detector offers flexibility in placement relative to the specimen. The light pipe can be bent through a considerable angle without much loss of transmitted light. For example, the ET detector can be placed under a thin specimen for collection of transmitted electrons. Backscattered electrons show a pronounced directionality of emission from tilted surfaces. It is occasionally advantageous to establish an unusual detector–specimen configuration, such as placing the detector below the normal of a highly tilted specimen. Such a position can be obtained with a suitably bent light pipe. The ET detector is not without disadvantage. High beam currents can significantly degrade the scintillator; this results in poor performance at low currents. Under such conditions, a radiation-resistant scintillator, such as europium-doped calcium fluoride, should be used.[5]

The success of the ET detector is dependent on the fact that the magnetic field in the specimen chamber due to the final condenser lens is small. The lens pole piece is specially designed to achieve this condition (see Chapter II). Otherwise, a large magnetic field would deflect the secondary electrons and reduce the efficiency of collection.

B. Solid State Detector

The solid state or semiconductor detector operates on the following principle. When energetic electrons enter such a semiconductor, electron–hole pairs are created. If electrodes are placed on opposite faces of the semiconductor and a potential is applied to the electrodes with an external circuit, then the free electrons will be attracted to the positive electrode while the holes flow in the opposite direction; a current will be observed to flow in the external circuit. This current, when suitably amplified, can be used as a video input. In this scheme, a current will flow in the circuit even when no electrons strike the detector, producing a background or "dark" current which does not carry information and thus degrades a real signal.[1] Alternatively, if a *p–n* junction is formed in the semiconductor, an electric field naturally exists across the junction which can serve to separate the generated electron–hole pairs without an external field. In this configuration (Figure 3), the dark current is minimized and signal integrity is maintained. Generally, the higher the energy of incident electrons, the greater the production of electron–hole pairs. The detector is thus energy sensitive. Secondary electrons have very little effect unless

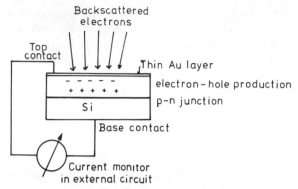

FIGURE 3.   Solid state or semiconductor detector for emitted electrons. High-energy electrons create electron–hole pairs which are separated by a p–n junction, producing a current in an external circuit.

accelerated by a suitable potential as in the ET detector. Such a system has been used by Crewe, Isaacson, and Johnson.[6]

Usually, the solid state detector is used to detect backscattered electrons. The semiconductor chips can be made quite large, 25 mm in diameter or more, and placed close to the specimen. With the detector immediately above the specimen, the solid angle of collection for backscattered electrons is very nearly a hemisphere. For backscattered electrons, the collection efficiency of a solid state detector is normally an order of magnitude better than that of the ET detector. An array of solid state detectors can be placed around the specimen to examine directional effects in the backscattered fraction. The cost of the solid state detector is a small fraction of that of an ET detector, since a simple planar p–n junction makes an adequate detector. However, the solid state detector is only superior at high beam currents. The amplification of the current in the external circuit can only be accomplished by conventional amplifiers, which do not yet give high-quality performance with small beam currents and fast scan rates. Improvement in amplification techniques would greatly increase the usefulness of the solid state detector.

C. Specimen Current Detection

When a beam carrying current i strikes a specimen, beam electrons are backscattered, producing a current $i_{BS} = \eta i$, where η is the coefficient of backscatter, and secondary electrons are emitted, producing a current $i_S = \delta i$, where δ is the coefficient of secondary electron production. If the specimen is connected to ground, a current i_{SC} will flow in the conducting path. This current is known as the "specimen current," "absorbed current," or "target current." The magnitude of the current is found from a balance of all currents entering and leaving the specimen:

$$i_{SC} = i - i_{BS} - i_S \tag{5}$$

Taking differentials of these terms, we find

$$\Delta i_{SC} = -\Delta i_{BS} - \Delta i_S \tag{6}$$

since $\Delta i = 0$ for a constant incident beam current. If we take the backscattered and secondary electron currents together as an "emissive current" i_E, then

$$\Delta i_{SC} = -\Delta i_E \tag{7}$$

The differential is the change in current between two points in the image. Equation (7) indicates that the change observed in an image formed with the specimen current signal will be opposite to the change observed with the emissive signal. Contrast in the specimen current signal is dependent on contrast in the emissive signal. Every electron leaving the specimen affects the specimen current. Thus, although strong directionality exists in the backscattering of electrons, it is only the act of leaving the specimen which affects specimen current. The effects of backscattering alone on the contrast can be studied by floating the specimen at a potential of $+50$ V; low-energy secondary electrons are prevented from escaping the specimen and only backscattered electrons contribute to Δi_E. Since the directionality of the backscattered electrons is not important in the specimen current, signal contrast measurements can be made accurately in specimen current and related to the backscattered signal number contrast, providing secondaries are eliminated by biasing, with the following equations:

$$i_{SC} = i - i_{BS} = i - \eta i = (1 - \eta)i \tag{8}$$

From equations (6) and (8),

$$\frac{\Delta i_{SC}}{i_{SC}} = \frac{-\Delta i_{BS}}{(1 - \eta)i} \tag{9}$$

$$\frac{\Delta i_{BS}}{i} = -(1 - \eta)\frac{\Delta i_{SC}}{i_{SC}} \tag{10}$$

$$\frac{\Delta i_{BS}}{i_{BS}} = \frac{-(1 - \eta)}{\eta}\frac{\Delta i_{SC}}{i_{SC}} = \left(1 - \frac{1}{\eta}\right)\frac{\Delta i_{SC}}{i_{SC}} \tag{11}$$

The main advantage of the specimen current signal for imaging is the independence of the signal with regard to the placement of an external detector. Since the specimen acts as its own detector, a signal is recovered in all circumstances, even when the specimen is close to the final lens or in the lens bore. The major disadvantage is the necessity for conventional amplification of the signal, which suffers the limitations described above. In recent developments, use of direct specimen current input to an operational amplifier has resulted in satisfactory imaging with 10 pA of current and a 100 sec frame time.[7] Finally, since effects on electrons after they have left the specimen do not influence the specimen current, it is not possible to observe contrast due to trajectory effects with specimen current.

D. Detection of Electromagnetic Radiation

The detection of x-rays produced during the energy loss which the incident electrons suffer is covered in detail in Chapter VII and will not be discussed here. It is sufficient to be aware that the x-ray signal can be used to form an image through suitable detection and signal amplification.

Some specimens emit long-wavelength electromagnetic radiation in the infrared, visible, and ultraviolet regions when bombarded by energetic electrons; this is the phenomenon of cathodoluminescence. The radiation so produced can be detected directly by a photomultiplier directed at the specimen, providing absorption in the window covering the photomultiplier is not too severe. Cathodoluminescence in the visible region can usually be studied with the same photomultiplier used in the ET detector. The radiation at all wavelengths can be monitored continuously, or a wavelength-dispersive device can be placed in the optical path between the specimen and the photomultiplier.[8] The intensity of cathodoluminescence is often low, and a detector has recently been described which maximizes the efficiency of light collection.[9,10] This device is shown in Figure 4. A first surface mirror in the form of an ellipse of revolution is placed above the specimen. A hole drilled through the mirror admits the electron beam to the specimen which is placed at one of the two foci of the mirror. At the second focus, a fiber optics light guide collects most of the light emitted by the specimen. The light guide then conducts the light to the photomultiplier, through a wavelength-dispersion device if required.

III. CONTRAST FORMATION

With an understanding of the process of image formation in scanning electron microscopy, and particularly the signal detectors, we are prepared

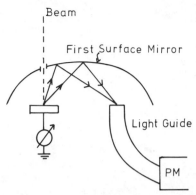

FIGURE 4.   High-efficiency cathodoluminescence detector. The specimen is placed at one focus of the elliptical mirror, and light emitted from the specimen is collected at the second focus by a light pipe. For electron imaging simultaneously, the specimen current signal must be used, since the mirror surrounds the specimen. PM, photomultiplier. (Adapted from Horl and Mugschl[9] and van Essen.[10])

to study contrast formation, that is, the reasons images show the intensity variations they do. If two points P_1 and P_2 differ in signal S, we say that contrast exists between these points, and the measure of this contrast C is

$$C = (S_1 - S_2)/S_{av} = \Delta S/S_{av} \qquad (12)$$

where S is the signal; S_1 and S_2 are the signals at points 1 and 2; and S_{av} is the general or average signal level for all points. In a measurement, S_{av} is taken as $(S_1 + S_2)/2$.

A. Atomic Number Contrast

In general, the signal between two points will be different because of some physical difference in the interaction of the beam electrons with the specimen or in the subsequent behavior of the interaction products after they have left the specimen. As a first example of a contrast mechanism, consider a flat specimen normal to the beam. The specimen has regions which are chemically different from one another; a multiphase alloy or rock would be an example of such a specimen. As explained in Chapter III, the backscattering of electrons increases with increasing atomic number (Figure 5). Thus, in an image of a multiphase specimen formed with backscattered electrons, we expect to see regions of high signal corresponding to the highest atomic number and low signal corresponding to the lowest atomic number, with regions of intermediate atomic numbers producing

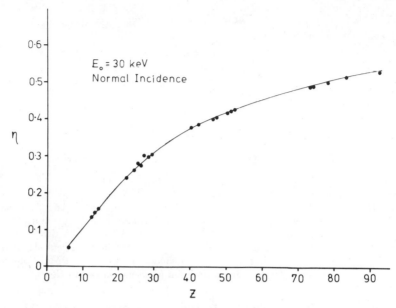

FIGURE 5.   The physical origin of atomic number contrast. The backscattering coefficient η increases with increasing atomic number Z. (Plotted from data of Heinrich.[11])

intermediate signal levels. The difference in the backscattered signal between regions of different atomic number is dependent on the exact values of atomic numbers which are involved. For a fixed ΔZ, the difference in η decreases as Z increases. If the region in question is a solution of different elements, then the effective backscatter coefficient is the weighted average of the backscatter coefficients of the pure elements.[11] When the signal is displayed on the CRT, we normally adjust the amplification and CRT controls so that the highest atomic number appears white, the lowest is black, and the intermediate values are various shades of gray. In this way we make use of the full dynamic range of the CRT, i.e., all shades of the gray scale from white to black which the CRT can display are possible in the image, providing an input signal exists to give a particular value. With only three phases, we would have only white, black, and one shade of gray. An image of a multiphase alloy is shown in Figure 6. Since the backscattered signal affects the specimen current signal, atomic number contrast can also be obtained in an image formed with specimen current. However, by equation (7), the phase with the lowest atomic number will appear white. The secondary electron coefficient δ is not a strong, regular function of atomic number as is the backscatter coefficient. Hence, if we formed an image of the specimen of Figure 6 with the secondary electron signal only, no appreciable contrast would be obtained.

We tend to interpret images based on our previous experience. The closest analogy to atomic number contrast in our experience is color. Species of different chemical composition frequently appear to us in different colors when examined with light. Thus, when the image of Figure 6 is examined, it is not unfamiliar to us, despite the fact that the electron interaction with the solid is not the same as the light interaction, because the shading has regularity like color.

As mentioned above, contrast can arise from effects due to different numbers of electrons leaving the specimen. This is pure number contrast; atomic number contrast is clearly an example of pure number contrast. If the interaction process results in directional emission of products, as is the case for backscattered electrons, and if the electron detector is

FIGURE 6.   Atomic number contrast. Specimen: composite of tungsten and alumina in an aluminum matrix. The specimen is held in a block of mounting plastic. Signal: specimen current (increasing atomic number causes decreasing brightness). The black phase is tungsten, dark gray is aluminum, light gray is alumina, and the white region is the plastic (mostly carbon).

directional, as is the case for the ET detector, then contrast can arise due to the trajectories of the electrons. Such contrast is termed pure trajectory contrast. Some contrast mechanisms combine aspects of both number and trajectory contrast. In the ensuing discussions, we shall characterize contrast mechanisms in terms of number and/or trajectory contrast.

B. *Topographic Contrast*

In the SEM we frequently examine rough specimens. The contrast which we obtain from such specimens is related to the topography present, and this is referred to as topographic contrast. This contrast is obtained from effects of topography both on the backscattered and the secondary electron fractions. Since most images formed with the ET detector contain contrast contributions from both signals, we must consider the interactions separately to interpret the combined image.

1. Topographic Contrast with Backscattered Electrons

It is observed that when a planar specimen is tilted away from normal beam incidence, the backscattering coefficient increases gradually, approaching unity at grazing incidence, e.g., Figure 7. Thus, if a specimen consists of a faceted surface with planar surfaces inclined at various angles to the beam, we would expect those surfaces tilted toward grazing

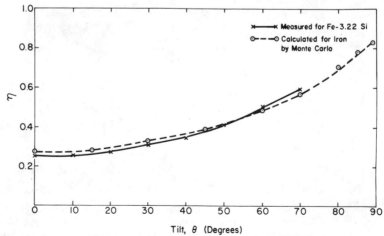

FIGURE 7. Backscattering coefficient η as a function of tilt angle θ (normal incidence is 0° tilt). Specimen: Fe-3.22% Si; beam energy 30 keV; specimen biased to +50 V to eliminate secondary electron emission.

incidence to show the highest signal, and hence the brightest level in the image. This contrast would be a pure number effect. However, a second effect dominates this backscatter vs. incidence angle effect. The backscattering of electrons is highly directional; the peak in backscattering tends to lie in the plane containing the normal to the surface and the incident beam direction. Moreover, when the ET detector is used to form the image, the high directionality of this detector for backscattered electrons ensures that only those electrons that are scattered toward the detector are collected. With a faceted specimen, then, only those facets that direct their scattering toward the detector will produce a signal. In a sample with randomly oriented facets, many will appear nearly black, since the high directionality limits acceptance to a small range of orientations. Such an image is shown in Figure 8a, where facets on an intergranular fracture surface are shown. Topographic contrast with the backscattered electron signal and the ET detector is thus a combination of number and trajectory effects, with the trajectory effects tending to dominate. If the specimen current signal is used instead to form the image, the trajectory effects are lost and the topographic contrast is due to the number effect alone (Figure 8b). The specimen current signal has been inverted so that facets inclined at a high angle appear bright. Comparing facets in the circled regions in Figure 8a,b, the facets show even illumination in the specimen current image (Figure 8b), indicating that each is tilted at approximately the same angle relative to the beam. In the backscattered image with the ET detector (Figure 8a), these facets show a great range in brightness, with only the facet facing the detector producing a high signal. The importance of trajectory effects with backscattered electrons in the ET detector signal must always be considered in interpreting an image.

2. Topographic Contrast with Secondary Electrons

It is observed that the secondary electron coefficient δ increases markedly as a function of increasing angle of tilt θ. This observation is explained by the following argument. As described in Chapter III, secondary electrons can only escape from depths less than about 10 nm (100 Å) for most materials. At such shallow depths the amount of elastic scattering which an incident primary beam has suffered is minimal, so most beam electrons are traveling nearly parallel to the incident direction. The incident beam is thus approximated as a straight line near the surface. As the specimen is tilted away from normal incidence ($\theta = 0°$), the length of beam path R which lies within $R_0' = 10$ nm of the surface increases (Figure 9) according to the relation

$$R = R_0 \sec \theta \tag{13}$$

FIGURE 8. Fractured polycrystalline iron. Intercrystalline fracture produces a facetted surface. The appearance of the image formed with various signals is illustrated: (a) Backscattered electron signal, ET detector. Note strong shadowing; the detector is at the top of the image. (b) Specimen current signal, inverted. No strong shadowing exists. (c) Secondary and backscattered electron signal, ET detector; note the softening of the shadows compared to (a). Information is obtained from all surfaces. (d) Specimen current signal, direct. Compared to (a), (b), and (c), the image appears inverted.

The secondary electron production rate is assumed to be constant along this path, so when the path length near the surface increases, more secondary electrons escape from the specimen. If we form an SEM image of an area containing facets tilted at various angles to the beam, more secondary electrons will be found to escape from the surfaces at high tilt. The magnitude of the contrast effect has been calculated by Everhart, Wells, and Oatley by the following argument.[12] The secondary electron signal S

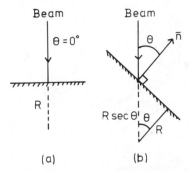

FIGURE 9.   Secondary electron emission as influenced by specimen tilt. R is the maximum escape depth for secondary electrons. When the specimen is tilted, the path length within a distance R from the surface increases.

from inclined surfaces is proportional to sec θ by equation (13). Taking the derivative of equation (13) by θ we obtain

$$dS = S_0 \sec \theta \tan \theta \, d\theta \qquad (14)$$

The contrast dS/S is found from

$$\frac{dS}{S} = \frac{S_0 \sec \theta \tan \theta \, d\theta}{S_0 \sec \theta} = \tan \theta \, d\theta \qquad (15)$$

Thus, for a given angle θ, the contrast change which results from a small change in angle, as in comparing the contrast between two surfaces tilted at slightly different angles, is given by equation (15). At $\theta = 45°$, a change in angle of 1° produces a contrast of $dS/S = 0.0175 = 1.75\%$, while at $\theta = 60°$, $dS/S = 0.03 = 3\%$ for $d\theta = 1°$. Topographic contrast carried by secondary electrons is thus quite sensitive to the geometrical configuration of the specimen surface and small differences in local tilt should be detectable.

When a beam of circular cross section A is intercepted by a plane at an angle other than a right angle to the beam axis, an ellipsoidal intersection is formed. For a constant beam current of n electrons in unit time, the current density J_A striking the specimen decreases as

$$n/A = J_A = J_{A_0} \cos \theta \qquad (16)$$

where J_{A_0} is the beam current density at $\theta = 0$. This effect does not cause contrast formation because the production of secondary electrons is proportional to the total current striking the specimen and not to the current density.

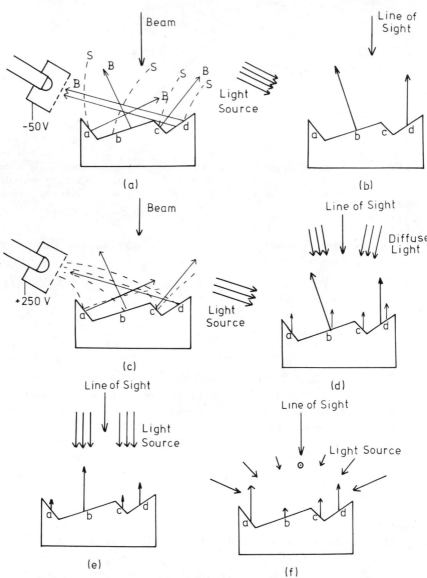

FIGURE 10.   Interpretation of topographic contrast by light analogy. (a) Backscattered electrons only, ET detector. A signal is obtained only from
those facets that scatter electrons toward the detector. Facet d will
appear bright, b less bright, and a and c totally dark. (b) Light
illumination analogy to (a). With the line of sight indicated and
oblique illumination from the source, facets b and d reflect light,

C. Interpretation of Topographic Images

The previous discussion of the characteristics of secondary and back-scattered signal images with the ET detector forms the basis for inter-pretation of topographic images. One of the most interesting aspects of SEM images of rough samples is their striking resemblance to images of rough objects illuminated with light in a certain way and viewed by eye. SEM images of small areas at high magnification can be interpreted from our previous experience with light illumination of large, rough objects despite the differences between the interaction of electrons and of light with the surfaces of solids. The analogy between light and electron illumination is illustrated in Figure 10. The rough surface is seen in cross section, and the electron beam direction and ET detector position are indicated (Figure 10a). Lines labeled B represent backscattered electrons while lines labeled S represent secondary electrons. Backscattered electrons travel in the direc-tions indicated; only those that travel toward the ET detector contribute to the signal. The appearance of the shadowing seen in such a backscattered electron image, e.g., Figure 8a, is duplicated with the light illumination shown in Figure 10b. The light source is placed at the position of the elec-tron detector and the viewing direction is from above, along the electron beam direction. Such an arrangement of lighting is referred to as oblique illumination; the illumination is strongly asymmetric, coming entirely from one side of the specimen, and strikes at a shallow angle to the general specimen plane. When the secondary electron signal is added to the back-scattered electron signal in the ET detector, e.g., Figure 10c, the efficient collection of secondaries from all surfaces, including those not facing the detector, serves to soften the harsh shadowing of the backscattered image (Figure 8a, c). Information is obtained from all surfaces of the speci-men that are illuminated by the incident beam. In terms of light illumina-tion, the addition of the secondary electron signal is equivalent to the addition of a second light source illuminating the rough object parallel to the line of sight (Figure 10d). Illumination parallel to the line of sight is

---

while a and c are shadowed. (c) Backscattered and secondary elec-tron collection, ET detector. A signal is collected from all facets, hence shadowing is greatly diminished. (d) Light analogy to (c). A combination of oblique illumination and diffuse vertical illumina-tion gives reflection from all facets. b and d appear brightest because of the oblique source. (e) Pure vertical illumination. Surfaces nearly perpendicular to the vertical line appear brightest. This is analogous to the specimen current image, e.g., Figure 8d. (f) Oblique illumina-tion encircling the specimen. Surfaces inclined to the line of sight appear brightest. This is analogous to the inverted specimen current image, e.g., Figure 8b.

known as vertical illumination. The image formed with the backscattered and secondary electron signals combined thus has an equivalent light illumination composed of both vertical and oblique components. Such a situation is approximated by placing a second light source at a higher angle (Figure 10d).

The illumination in the specimen current image of a rough surface (Figure 8d) is equivalent to pure vertical light illumination (Figure 10e). The brightness of each surface is determined only by the inclination of the surface to the beam because both η and δ are smoothly increasing functions of increasing tilt. Since the direct specimen current image (Figure 8d) shows inverse effects with respect to the emissive mode image according to equation (7), those surfaces that are nearly normal to the beam appear to be brightest. Thus, the analogy with vertical light illumination occurs, since surfaces which are normal to the line of sight reflect more light when vertically illuminated and appear brighter. In the inverted specimen current image (Figure 8b), surfaces at high tilts to the beam are bright. This image is the greatest departure from our experience with light. The closest light analogy is oblique illumination from a light source surrounding the specimen on all sides, i.e., a ring of light slightly above the plane of the specimen (Figure 10f). Such a situation is extremely uncommon in the real world, and hence an image formed with this signal is difficult to interpret. We must rely on our knowledge of the electron interaction and the detector characteristics for interpretation.

In our experience with visual images, the illumination is usually diffuse and somewhat asymmetric. A scene is usually lighted from one side, producing shadowing. The reflection of light is directional and the detector, the human eye, is capable of only detecting light reflected toward it. In SEM images made with the ET detector, the electron illumination is highly directional, i.e., only along the beam, but the signal detection for secondary electrons is diffuse; electrons traveling away from the detector may still be collected. Thus, interpretation of SEM emissive mode images from the light illumination analogy is carried out by matching the directional components and the diffuse components of each system. The light source is put at the position of the ET detector and the light detector, the eye, is placed at the position of the directional component in the electron system, the electron beam. For contrast mechanisms other than topographic contrast and atomic number contrast, the light analogy breaks down. We must interpret images formed with these other contrast mechanisms, e.g., electron channeling, magnetic contrast, etc., on the basis of our knowledge of the physics of the electron–specimen interaction. These mechanisms are discussed in Chapter V.

IV. SIGNAL CHARACTERISTICS AND IMAGE QUALITY

Assuming that a contrast mechanism operates to produce different signal levels at various points in the image, what are the conditions on the visibility of this contrast to the human eye? The first condition which must be satisfied concerns characteristics of the signal. We can examine signal characteristics by plotting the signal as a function of scan position on the oscilloscope (*Y*-modulation). Such a plot is shown schematically in Figure 11a. If we superimpose a large number of line scans across the same features of the specimen (Figure 11b), the following signal characteristics are observed. The plot of the signal shows regular changes from point to point which are associated with characteristics of the specimen. At any particular point, the signal plotted on successive scans is not precisely the same. A small random variation in the signal near the same average value occurs with successive scans. This variation occurs because the SEM is basically a counting device for discrete events: electrons or photons are counted with the various detectors for a time τ, the picture point time, to form the signal which is amplified and displayed. The counting of discrete events is subject to statistical fluctuations in the count. Thus, if

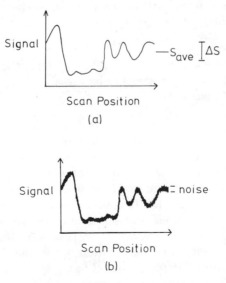

FIGURE 11.   (a) Signal versus scan position, single scan. The average signal and differential signal change across a feature are indicated. (b) Multiple scans superimposed showing the noise component.

the beam is held on a particular point, the number of electrons arriving at a detector and counted in a fixed time τ will vary in a series of nominally identical counts. If $\bar{n}$ is the mean of the counts, we can expect a variation of $\bar{n}^{1/2}$ about this mean for a series of determinations. This random variation in the signal is known as noise, and such randomness decreases the information which is available to us in the signal S. A measure of the signal quality is the signal-to-noise ratio S/N. The signal at the point has an average value of $\bar{n}$, while the noise is equal to $\bar{n}^{1/2}$. Thus

$$S/N = \bar{n}/\bar{n}^{1/2} = \bar{n}^{1/2} \tag{17}$$

From this relationship, it is apparent that the signal-to-noise ratio improves as the total number of electrons counted per picture point increases. Random variations form a less significant portion of a large number than a small number. For a fixed beam current, the S/N ratio decreases as the scanning speed increases, because of a decreased dwell time τ at each picture point. For a fixed dwell time at each picture point, the S/N ratio decreases as the beam current decreases and hence fewer electrons impinge on the specimen.

The S/N ratio is related to the image quality in the following way. From a study of the characteristics of human visual perception, Rose has determined that the true signal change dS must be greater than five times the noise for the human eye to be able to distinguish the true signal change, due to the specimen characteristics, from the signal change due to the noise in an image formed with the signal[13]:

$$dS > 5\bar{n}^{1/2} \tag{18}$$

We can use this criterion to develop a relation giving the minimum acceptable S/N ratio which is necessary to observe a given contrast $C = dS/S$. Given that $S \sim \bar{n}$ and $N \sim \bar{n}^{1/2}$,

$$C = dS/S > 5\bar{n}^{1/2}/\bar{n} = 5/\bar{n}^{1/2} \tag{19}$$

to satisfy the criterion of equation (18). Noise can be introduced at all stages in the signal path from the specimen to the display CRT. However, in a system based on the ET scintillator–photomultiplier system, the number of information carriers, electrons or photons, is a minimum at the stage of beam–specimen interaction and electron collection. At all later stages in a system with an efficient scintillator, the signal is progressively increased and the S/N ratio increases.[4] However, this increase in the S/N ratio cannot recover any information not in the signal at an earlier stage. This basic limitation on the signal has aptly been referred to as the "noise bottleneck" by Wells[14] and Broers.[15]

Let us consider the electron counting process at this noise bottleneck. Assuming that the secondary electron coefficient is near unity and nearly all secondary electrons are collected, then for each beam electron, one signal-carrying electron is created. In this situation, equation (19) indicates that to observe a given contrast C, the number of electrons which must be incident on each picture element on the specimen is

$$\bar{n} > (5/C)^2 \tag{20}$$

in order to satisfy the Rose criterion. Another random process operates on the information carriers at this production–collection stage. The process of secondary electron production is subject to random fluctuations. With a truly constant primary beam current striking the specimen at a fixed position, a random fluctuation in secondary production still leads to noise in the signal arriving at the detector. This noise process enters as a multiplicative factor to the value of five in the Rose visibility criterion. Various estimates of the factor have been given, but we shall follow the example of Oatley *et al.*[16] and take the value as two. Thus equation (20) becomes, considering fluctuations in the beam and in secondary production,

$$\bar{n} > (5 \times 2/C)^2 = 100/C^2 \tag{21}$$

This $\bar{n}$ is referred to as the threshold value of the mean number of electrons which must be incident on each picture element for a given contrast to be visible. The number is only an estimate: the Rose criterion is somewhat subjective since the sensitivity of the human eye is not constant from person to person. The assumptions made above on the secondary electron coefficient and collection may vary with operating conditions.

Given that the S/N ratio is not degraded in the subsequent stages of amplification and display, equation (21) can be transformed into an equation for the beam current i by considering the charge flux into a picture element in the picture element time τ. The charge flux is given by the mean number of electrons per picture element multiplied by the electronic charge $e = 1.6 \times 10^{-19}$ coul. The value of τ is found by dividing the frame time t_f, i.e., the time to scan a whole raster, by the number of picture elements. For a high-quality image with 10^6 picture elements, the threshold beam current i_{th} to detect a given contrast level is given by

$$i_{th} > \frac{\bar{n}e}{\tau} \tag{22a}$$

$$i_{th} > \frac{(100/C^2) \times 1.6 \times 10^{-19} \text{ coul}}{(t_f/10^6) \text{ sec}} \tag{22b}$$

$$i_{th} > \frac{1.6 \times 10^{-11} \text{ A}}{C^2 t_f} \tag{22c}$$

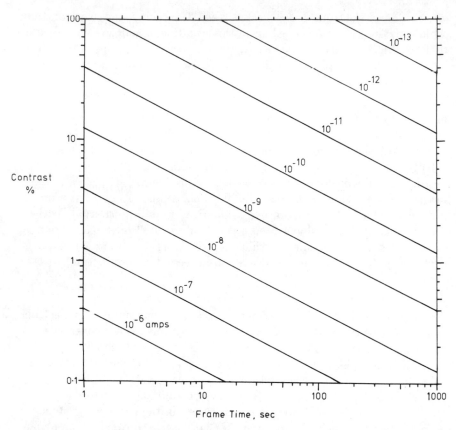

FIGURE 12.   Threshold beam current as a function of natural contrast level (contrast in the signal at the specimen) and frame time, assuming 10^6 picture elements.

This equation has great importance in scanning electron microscopy, for it sets the fundamental limitation on the beam characteristics for successful operation. In the next section it will be demonstrated that our ability to resolve fine detail is frequently limited by the noise in the signal and not by a basic inability of the electron lenses to form a sufficiently small beam diameter. Following Joy,[17] equation (22) can be conveniently described in graphical form (Figure 12); for any given contrast and frame time, the threshold current is easily found. The threshold current calculated by equation (22) is an estimate of the minimum acceptable beam current. For satisfactory images, the threshold current must be exceeded by at least a factor of two. When the beam current is equal to the threshold current, the noise forms a significant portion of the signal changes in the image, producing an unsatisfactory "grainy" appearance which tends to mask specimen-related signal changes. An example of this effect is shown in Figure 13.

V. RESOLUTION LIMITATIONS IN THE SEM

The limit of SEM resolution, defined as the minimum distance between the centers of two distinctly separated objects, is a function both of the

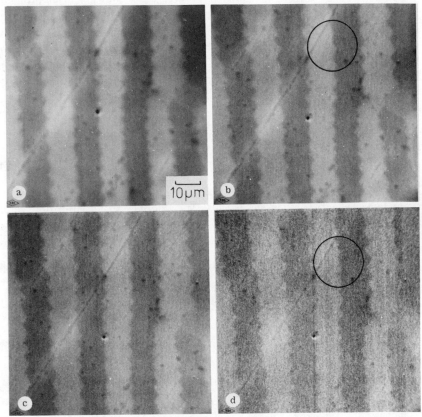

FIGURE 13.   Image of magnetic domains in Fe–3.22% Si observed with type II magnetic contrast. The effects of operating near the threshold current are illustrated. Beam current: (a) 1×10^{-7}, (b) 5×10^{-8}, (c) 2×10^{-8}, (d) 5×10^{-9} A. Note the loss of short-range details, the edge cusps, in the circled region while the long-range details, the stripe domains, are still visible. For the conditions, the threshold current is about 2×10^{-8} A.

properties of the signal and of characteristics of the electron–specimen interaction. We shall first consider limitations due to the signal properties described above.

A. Signal Limitations

As an example of how the signal properties affect the resolution, we shall consider the case illustrated in Figure 14. Figure 14a shows the signal intensity as a function of scan position across a small particle. When two such objects are placed close to one another, the intensity profiles begin to overlap (Figure 14b). We can use the Rayleigh criterion to set a limit on how close the points can be placed and still resolved. By the Rayleigh criterion, the points are resolved if the intensity minimum between them is approximately $\frac{1}{4}$ of the peak height.[18] The contrast for the Rayleigh criterion, assuming a signal of unity for the peak, is given by

$$C = dS/S = \tfrac{1}{4}/1 = \tfrac{1}{4} \tag{23}$$

The threshold current for this contrast for a 100-sec frame time is

$$i_{th} = 1.6 \times 10^{-11}/(\tfrac{1}{4})^2 100 = 2.6 \times 10^{-12} \text{ A} \tag{24}$$

Thus a threshold beam current of 2.6×10^{-12} A is needed, providing a secondary collection efficiency of unity, which is the assumption made previously, is achieved. The probe size carrying such a current is determined from the brightness equation (see Chapter II)

$$\beta = 4i/\pi^2 d^2 \alpha^2 \tag{25}$$

where β is the brightness in A/cm² ster, i is the beam current, d is the beam diameter, and α is the beam divergence. For a conventional tungsten

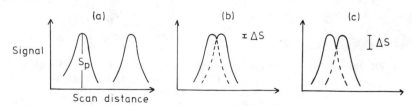

FIGURE 14. Criterion for resolution of two adjacent similar objects. (a) Signal profiles, well separated. (b) Overlapping profiles, not resolved. (c) Overlapping profiles, resolved, $S = \tfrac{1}{4}S_p$.

haiprin filament emitter in an SEM gun, $\beta \cong 5 \times 10^4$ A/cm² ster.[19] From the earlier discussion of depth of focus, a beam divergence of 5×10^{-3} rad (100 μm aperture and 10 mm working distance) is assumed. Substituting i_{th} for i and solving for d, we find

$$d \cong 9 \text{ nm } (90 \text{ Å}) \qquad (26)$$

Actually the beam diameter is limited by lens aberrations to some value which is larger than that given in equation (26). A beam size approximately 25% larger than that given by equation (26) is necessary to contain the desired current. In general, we will not be able to resolve detail which is smaller than the effective beam diameter. Since the distribution of current is Gaussian across the probe, it is usual to define the beam diameter as that portion of the beam containing 80% of the total current. The beam diameter by this arbitrary definition is about $\frac{7}{8}d$. Thus, the point-to-point resolution is limited to about 80 90 Å for a conventional SEM system. If the points are closer together than 90 Å, we cannot obtain sufficient current in a beam of smaller size to satisfy the threshold current for the Rayleigh criterion contrast. If a beam of 90 Å diameter is used with objects separated by less than 90 Å, the effect is to cause the intensity profiles to overlap more, thus failing to satisfy the Rayleigh criterion. A resolution limit of 90 Å is in good agreement with the guaranteed performance of most modern high-quality SEM's. It should be noted that this performance will *not* be achieved with all specimens or even with all objects in the same field of view of a particular specimen. Two important assumptions were made in deriving the resolution limit: (1) that the secondary electron emission collection ratio was approximately unity and (2) that the specimen was capable of producing the contrast of $C = 25\%$ necessary to satisfy the Rayleigh criterion. If either of these assumptions is not fulfilled, then the threshold current will not be reached and the feature of interest will not be visible. Many specimens do not produce strong secondary topographic contrast and for these specimens a limiting resolution of 20 nm (200 Å) or more is to be expected.[1] In a given field of view on a specimen some objects will not be in an optimum position relative to the detector for secondary electron collection or the secondary topographic contrast may be lower than for other objects in the field. Thus, while we may be able to resolve certain objects in a field, at other positions in the same field, the separation between equally spaced objects will be unresolvable. Optimum resolution performance is only achieved on optimum specimens. Considering other contrast mechanisms, such as electron channeling contrast or magnetic contrast, the level of the contrast is frequently much weaker than secondary topographic contrast, which leads to resolution limits of 100 nm (1000 Å) or more. These contrast mechanisms will be discussed in Chapter V.

A relaxation of the threshold current criterion for visibility occurs if a geometrical arrangement other than point-to-point is involved. If we are predicting the visibility of a line-type object, then the value of i_{th} predicted by equation (22) is an overestimate of the actual value; examples of objects in this class include dislocations in crystals revealed by electron channeling contrast[20] and sharp edges observed in secondary electron topographic contrast. The reason for this relaxation is the ability of human visual perception to make use of information across the entire image at once. In Figure 15, an image is presented of a line-type object viewed under conditions in which the beam current is close to the value of the threshold current. The noise component in the image is quite strong; nevertheless, the presence of the linear object is apparent because the observer's visual process recognizes a long-range pattern in the image. In this same image, it would be extremely difficult to discern a point-to-point separation of small objects because of the noise.

Examination of equation (25) indicates that improved resolution performance can be realized through the use of electron sources with a higher value of brightness β. Broers has described a high-brightness ($\beta = 4 \times 10^6$ A/cm^2 ster) electron source using a lanthanum hexaboride (LaB_6) emitter on an electron optical column which is optimized for high-resolution performance.[15,21] With this system he has obtained a resolution of surface details of 2.5 nm (25 Å) point-to-point. Crewe and his co-workers have described an SEM system which uses a pointed cold field emission source ($\beta \cong 10^7$ A/cm^2 ster).[22] This system has been used to resolve heavy atoms in long-chain molecules in the scanning transmission mode of operation.[23] Currently, extensive work is being carried out to make both of these electron sources, as well as other possible high-brightness sources,[24] available to the general user. The field emission source currently offers the highest available brightness but instabilities in the source, which have been recognized for some time,[2] require ultrahigh vacuum (UHV) in the source and techniques of rapid scanning with superimposed scans for useful micrograph preparation. Moreover, if high current capability is also desired, the LaB_6 source has been shown to be clearly superior to the cold field emitter.[25] Continued improvements in electron sources offer the most hope for improved SEM performance.

B. Resolution Limitation due to Beam–Specimen Interactions

The ultimate resolution may also be limited by the characteristics of the electron beam–specimen interaction. For example, atomic number

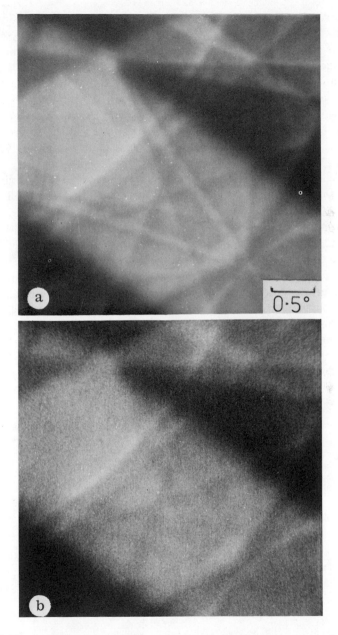

FIGURE 15. Example of visibility near the threshold current due to long-range order in the image. Despite a poor signal-to-noise ratio in (b), the long lines of the pattern are still visible. Electron channeling pattern from silicon. (a) 5×10^{-8}, (b) 5×10^{-10} A.

contrast results from backscattering of the primary beam electrons. As shown by Monte Carlo calculations (Chapter III), the primary electrons can diffuse to a considerable distance from the beam impact point due to scattering processes before escaping the specimen (backscattering). Thus, if a beam is focused to 20 nm (200 Å) and is scanned across a phase boundary, i.e., a boundary separating regions which are different in chemical composition (Figure 16), the beam electrons will begin to interact with the atoms of region 2 while the beam center is still 0.5 μm or more from the boundary. The backscatter coefficient thus changes gradually over a distance of about 0.5 μm as the beam approaches the boundary, whereas the true chemical change is abrupt, occurring over a distance of 1 nm or less. The position of the boundary is therefore poorly defined, i.e., the boundary appears "fuzzy." An example of this effect is shown in Figure 17. The beam size is considerably smaller than the image width of the phase boundary, as indicated by the sharp edges on the small particle lying on the surface which appears in secondary topographic contrast.

The resolution limit set by the electron interaction may not even be the same in two different directions in the same image. The mechanism of type II magnetic contrast, which is discussed at length in Chapter V, involves primary electrons. The specimen must be tilted at about 50–60° and at such high tilts, the interaction volume of the primary electrons

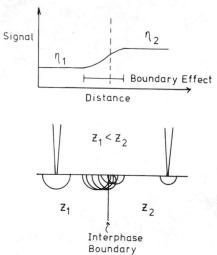

FIGURE 16.   Resolution limited by electron scattering effects. As the beam is scanned across the sharp interphase boundary, the signal changes over a wide lateral range due to the large interaction volume of the primary electrons.

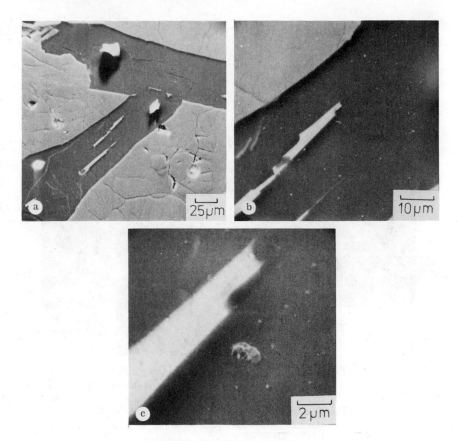

FIGURE 17. Atomic number contrast images showing a loss of definition at an interphase boundary due to the effect illustrated in Figure 16. In the high-magnification image (c), small surface particles seen in secondary electron topographic contrast appear sharper than the interphase boundary. Specimen: basalt.

becomes elongated in a direction "downwind" from the impact point while the interaction volume has a smaller dimension parallel to the tilt axis than it does at normal incidence. Thus, if a magnetic domain boundary is either parallel or perpendicular to the tilt axis, the resolution will differ considerably. Such an effect is observed in Figure 18.

Wells has analyzed the mechanisms of topographic contrast with primary and secondary electrons to determine those aspects of the beam–specimen interaction which may ultimately limit resolution.[14] Secondary electrons are produced by the impacting beam electrons; these secondaries are confined by their extremely short range within the specimen to the area

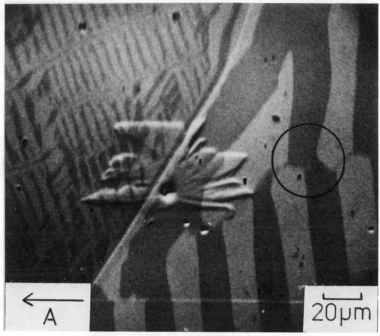

FIGURE 18.   Type II magnetic contrast image showing different edge resolutions
in different directions. The specimen is tilted 55° to the beam,
producing increased beam spreading perpendicular to the tilt axis Ā.
The images of domain walls perpendicular to the tilt axis are sharper
than domain wall images parallel to the tilt axis (circled region).

of impact of the beam electrons (Figure 19a). These secondaries carry
information which is related only to the beam impact area. Secondary
electrons are also produced by the backscattered electrons as they leave the
specimen. These carry information from regions of the surface which are as
much as 1 μm from the beam impact area; this portion of the signal is
effectively a noise component relative to the first portion, since we are only
interested in information from the area of the beam impact. The ratio
of the secondaries formed outside the impact area to those formed inside
has been estimated to be about a factor of four for most elements.[4] The
relative area density of information carriers is higher for beam-excited
secondaries than for backscattered excited secondaries, since the latter are
distributed over a much larger area. If a sharp change in topography occurs
in the beam impact area, then the signal change from secondaries in the
impact area will be much larger than the signal change from the secondaries
formed in the backscattered emission region, since topographic effects are
averaged over such a large area. Thus small particles and sharp edges

produce strong, distinct contrast with the secondary electron signal [Figure (19b)]. If an object is less sharply defined and has a gradually changing slope, then the secondary contrast is badly degraded by the noise component due to long-range emission, since the signal in the beam impact area is not much different than the average signal produced by integrating the secondary topographic effects over the emission region. The secondary topographic contrast mechanism is thus limited with regard to the resolution of certain features because of the weak contrast which is formed, and a significant portion of the secondary electrons which are collected is effectively a noise component. Since the secondaries formed by the incoming beam electrons have approximately the same energy spectrum as those formed by backscattered primaries, it is impossible to separate the two components experimentally.

Wells has shown that significant improvements in the resolution limit can be achieved by using the high-energy fraction of the backscattered electrons.[14] The diffusion of the beam electrons in the specimen away from the beam impact area is accompanied by a loss of energy due to inelastic scattering processes which occur. Those backscattered electrons that have not lost much of their initial energy, the "low-loss electrons," have left the specimen without diffusing over a great distance from the impact area. Since they have energies which are different from the electrons that have undergone a large number of scattering acts, they can be separated by

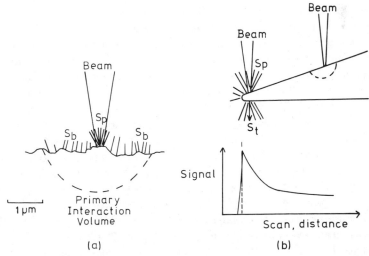

FIGURE 19.   (a) Production of secondary electrons by the incident primary electrons S_p and by the backscattered electrons S_b. (b) Enhanced escape of secondary electrons at a thin edge due to production by exiting transmitted electrons. The signal is high near the edge and decreases rapidly as the material becomes thicker.

energy filtering techniques.[26] Wells calculates that a point-to-point resolution of 1 nm (10 Å) should be possible, and a practical resolution of 3 nm (30 Å) has been achieved using a combination of a high-brightness source and an energy-filtered backscattered signal.[14,21]

VI. SIGNAL PROCESSING

Given that the threshold current to observe a certain level of contrast is exceeded, will an observer be able to detect this contrast information in an image displayed on a CRT? An affirmative answer to this question frequently depends on the correct application of signal processing techniques to the image. To understand the need for signal processing, we must again examine characteristics of the human visual process. Considering now a "black and white" photographic presentation of an SEM image (or any other, for that matter), made on a high-quality, continuous tone medium, the human eye can detect shades of gray, ranging from black to white. Although the film may have an effectively "continuous" range of shades of gray (on a macroscopic scale), the sensitivity of the human eye only allows a limited number of distinct shades, called "gray levels," to be discerned (Figure 20). These gray levels on the film can be related to the video signal of the SEM as shown in Figure 21. For a given range of the video signal, the eye is only able to detect a single shade of gray in the CRT image and photograph. The signal intensity will vary as a function of position in the image due to the characteristics of the specimen. If the signal intensity varies greatly due to the natural mechanism of the contrast formation, then all of the discernible gray levels in the film will be exposed (Figure 8c) and the image will appear to the observer to be of high quality with regard to the intensity range. This situation is referred to as employing the full "dynamic range" of the film. Such a condition is often obtained with the direct, i.e., unmodified, video signal produced by topographic contrast when formed with the combined secondary/backscattered electron signal of the Everhart–Thornley detector.

When the contrast is weaker in nature, such as the atomic number contrast formed between regions in a specimen with a small difference in atomic number, then the range of signal values will only result in a limited range of gray levels in the final image (Figure 22a). Such an image will appear to be of low contrast to the human eye and will not be very satisfactory (e.g., Figure 23a). Depending upon the individual, the ultimate sensitivity of the human eye is in the range of 3–5% brightness contrast, i.e., $\Delta B/B$, where B is the brightness[13]; hence, any contrast mechanism

FIGURE 20.   Gray scale formed by sequential time exposure of film. Approximately
12 levels can be discerned.

which does not result in a contrast of at least 5% in the final image will
not be visible to the observer, no matter how much the threshold current
for the contrast is exceeded.

Fortunately, the video signal of the SEM is time-resolved, i.e., the
signals for the various elements of the image arrive in a sequence in time
while they are being displayed as a sequence in space on the CRT. The

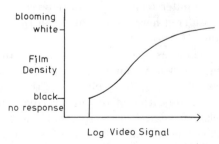

FIGURE 21.   Schematic illustration of film response as a function of video signal.
The response is a convolution of the CRT phosphor characteristics
and the film characteristics.

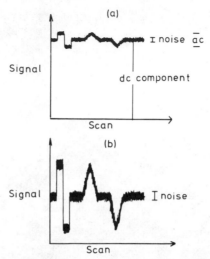

FIGURE 22.   Schematic illustration of the effect of black level suppression. The contrast of features is increased, but the noise component is increased as well.

signal for each point can therefore be manipulated electronically and changed to some new value to be displayed. The various techniques for manipulation of the signal are grouped under the term signal processing. These techniques are applied in real time, i.e., while the image is being formed. Another class of image manipulations exists in which the signal is recorded digitally giving an image function $I = f(X, Y)$, where I_n is the intensity at the point X_n, Y_n, and the function is transformed by digital techniques with a computer to a modified function $I' = f'(X, Y)$. The modified function can be used to reconstitute an image. Information can also be extracted on signal changes from point-to-point to give quantitative information on the number of edges or the size and frequency distributions of objects. These computer techniques are referred to as image processing and will not be considered here. Information on these techniques is available in the literature.[27]

Signal processing has as its objective to increase the information which is available to an observer in an image. It does this by preferentially treating the portion of the signal which carries the information of interest so that the resulting intensities in the image cover the full range of discernible gray levels. The signal processing can thus be viewed as a transformation T which acts on an image function $f(X, Y)$ to produce a modified image function $f'(X, Y)$:

$$f'(X, Y) = T(f(X, Y)) \tag{27}$$

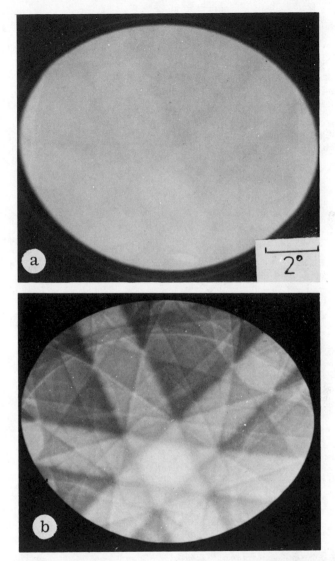

FIGURE 23. Example of an application of black level suppression to enhance the displayed contrast. The natural contrast of the electron channeling pattern is about 5% and is only barely visible in an image formed with the direct signal (a). Application of black level suppression greatly enhances the visibility of details in the ECP (b).

We shall consider the following signal processing techniques: (1) black level suppression; (2) gamma or nonlinear amplification; (3) signal differentiation; and (4) *Y*-modulation.

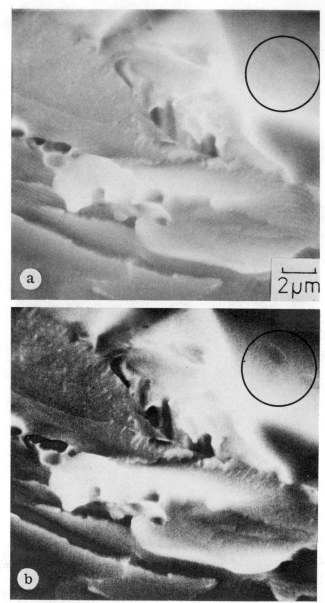

FIGURE 24.   Example of noise visibility enhancement due to black level suppres-
                   sion. The visibility of the domelike objects (circled region) is im-
                   proved through the use of black level suppression, but noise (grain-
                   iness) is more apparent in (b). (a) Direct signal. (b) Black level
                   suppression used. Specimen: fractured iron.

A. Black Level Suppression

We can analyze a signal like that of Figure 22a as having a time-varying or ac component superimposed on a static or dc component. The information content of the signal is mostly contained in the ac component, and it is this portion of the signal which we would like to distribute over the dynamic range of the film. This can be accomplished by applying the signal processing technique known as "black level suppression" or differential amplification. In black level suppression, most of the dc component is subtracted from the signal, and the differential signal is then amplified to cover the full signal display range of the CRT; the process is illustrated schematically in Figure 22b, and an example of the application of black level suppression to a real image to increase the contrast in the display is shown in Figure 23. The natural contrast which arises from the electron channeling effect (see Chapter V) is only about 5%; and hence, it is only barely visible in an image with the signal displayed in an unmodified form (Figure 23a). Black level suppression distributes the channeling contrast information over the full dynamic range of the photograph, increasing the information in the image enormously (Figure 23b). Black level suppression cannot make a contrast level visible unless the threshold current for that contrast level is exceeded for the recording conditions. The desired information must form a statistically valid portion of the ac component of the signal. Since the noise in the signal is also an ac component, the noise is also amplified by the differential amplification process. In the unmodified signal (Figure 22a), the noise at any particular signal level may not change the level sufficiently to cause that signal to be displayed in an adjacent (higher or lower), incorrect gray level. However, after differential amplification, the noise is expanded (Figure 22b) and it may cause the signal at a point to be displayed at an incorrect gray level in the final image. The random variation appears as "graininess" or a speckled appearance in the image (Figure 24) and it tends to suppress the visibility of fine scale brightness changes due to true specimen-related signal changes. Noise can be reduced by causing the beam to dwell for a longer time at each picture element so that more signal carriers are counted. The time constant of the amplifier can also be increased. Together these actions render the noise less statistically significant in the final micrograph.

B. Nonlinear Amplification (Gamma)

Often the information of interest resides in a narrow range of the signal whereas the signal does in fact cover the full dynamic range (Figure 25). Such a situation is found in the case where we are interested in details at the bottom of a depression where signal collection is poor. A useful signal processing technique in this situation is nonlinear amplification, which is called the gamma transformation. This transformation follows a simple power law in which the signal S is altered in the following manner[1]:

$$S_{\text{out}} = S_{\text{in}}^{1/\gamma} \tag{28}$$

The choice of gamma is usually an integer in the range 1–4, although a continuous function is often more useful.[28] The effect of gamma transformation is illustrated schematically in Figure 26. The gamma operator acts to spread a range of input signals over a larger range of output signals, thus causing preferential contrast expansion at either end of the gray scale at the expense of the other. A practical example of the application of gamma transformation to an image in which detail in a depression is of interest is given in Figure 27. It has been determined that it is preferential to expand the lower range of the signal rather than the upper range.[28]

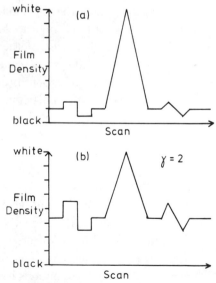

FIGURE 25.   Schematic illustration of the application of nonlinear (gamma) amplification. In (a) the details in the low-signal region are of poor visibility. In (b) nonlinear amplification with $\gamma = 2$ results in increased contrast of the low-signal details.

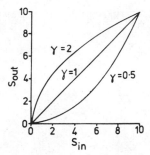

FIGURE 26. Gamma amplification curves. Integer values of gamma result in enhancement of low-signal details, and fractional gamma values result in enhancement of high-signal details. The output signal has been gamma-processed and subsequently expanded to a 0-10 V range.

Signal saturation can occur at either the black or white end of the scale, but saturation at the white end can produce the undesirable effect of "blooming" on the film, in which adjacent points to a written point may be exposed due to the high signal intensity.

C. Signal Differentiation

The SEM operator can choose between (1) low contrast-sensitivity over the entire range of signal intensity by displaying the unmodified signal or (2) using the black level or gamma transformations to give high contrast-sensitivity within a restricted range of signal intensity. By using a signal derived from a time derivative of the original signal to modulate the brightness of the CRT, we can obtain enhanced sensitivity to certain contrast changes at all signal levels.[29] The time derivatives of the signal, e.g., the first time derivative dS/dt, tend to enhance the high-frequency components at the expense of the low-frequency components. All areas in the image that produce a constant signal are displayed at the same gray level regardless of the constant signal intensity. Changes of the signal produce a change in brightness and hence gray level. The action of time derivative transformations is illustrated schematically in Figure 28. Note that the derivative signal is proportional to the slope in time of the original signal. If the average level is set at the middle of the gray scale, signals resulting from positive slopes are displayed at the white end of the scale, while signals from negative slopes appear near the black end. The derivative signal is thus related to the direction in which the beam crosses a feature. If the beam is scanning nearly parallel to a line object,

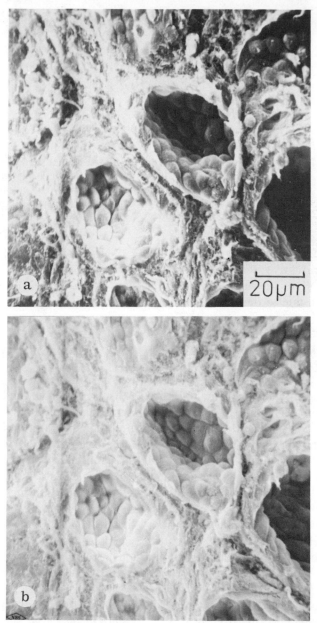

FIGURE 27.　　　Example of application of gamma processing to a real image. In (a) the domelike structures in the pits are of low visibility. In (b) the use of nonlinear amplification with $\gamma = 3$ improves the visibility of these details. Specimen: mouse thyroid.

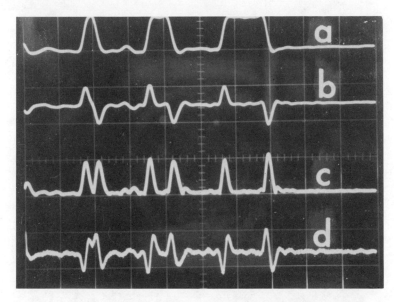

FIGURE 28. Effects of time derivatives on waveforms: (a) direct signal; (b) first time derivative; (c) absolute value of the first time derivative; (d) second time derivative, inverted. Note derivative enhancement wherever a signal change occurs in (a).

the signal changes very slowly across it and the derivative is therefore a small value. If ξ is the angle between the line object and the scan line, then the derivative signal S' is given by

$$S' = S'_{90} \sin \xi \qquad (29)$$

where S'_{90} is the value of the derivative signal when the scan is at right angles to a line object. When the scan is exactly parallel to the line object S' is zero. Thus, information parallel to the scan line is totally lost. These effects are illustrated in Figure 29, which shows the direct and time derivative images of a simple object, a hole in a flat plate. The direct image shows even illumination of the edge (Figure 29b). The first time derivative (Figure 29c) shows opposite edges white and black, and uneven signal intensity around the edge, with information loss where the scan line is tangent to the hole. The absolute value of the first derivative forces signal values near the black end toward the white end (Figure 29e). The second time derivative shows effects similar to the absolute value of the first time derivative (Figure 29g). If bidirectional (orthogonal) scanning is employed,

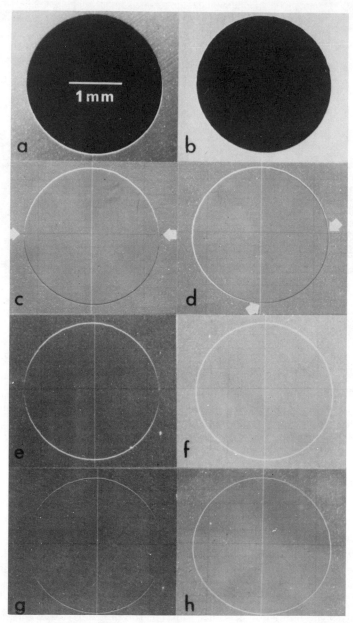

FIGURE 29.   Effects of time derivatives on a simple image, a hole bored in an
aluminum plate and set normal to the beam. (a) Emissive mode
image; (b) specimen current image, inverted. All derivative images

in which two successive scans at right angles are superimposed on the same film, the following effects are noted. The first time derivative shows only one dark quadrant, but the information loss still exists, rotated to 45° with respect to the scan axes (Figure 29d). With the absolute value of the first derivative and the second derivative, a full outline of the object is obtained (Figure 29f, h). The information loss line is eliminated. It has been shown that the second time derivative with bidirectional scans is an isotropic transformation.[28] That is, the image is independent of rotation of the scan axes. An isotropic transformation will not, however, make an anisotropic input function isotropic; it simply does not introduce any anisotropy.

The derivative operators enhance edges of objects and tend to "crispen" the image.[30] Such crispening can produce a more pleasing image, but it must be applied with caution, since the edge may be made sharper, i.e., effectively cover fewer picture elements, than the beam does, In this case, the crispening is a false and misleading effect.

The operation of the derivative operators on an image of a fracture surface is illustrated in Figure 30. The derivative transformations crispen the edges of the facets, but the impression of depth is lost because the static levels (low frequency) responsible for shadowing are lost (Figure 30b, c, d).

Signal mixing can be useful to produce a modified image in which an additive combination of the direct signal with a time derivative signal is displayed. The static levels are partially recovered, regaining some impression of shadowing, while edge sharpening is retained. These effects are shown in Figure 30 e, f, g.

D. Y-Modulation

In the Y-modulation transformation, the beam on the CRT is vertically (Y) deflected from a zero position; the deflection is proportional to the signal intensity. The image is thus built up from a series of line traces

---

are formed from the specimen current signal. (c) First derivative, undirectionally scanned (vertical scan lines). The signal varies around the circle. Note the loss of information along vertical lines, e.g., where the circle is tangent to a vertical line (arrows). (d) First time derivative, bidirectionally scanned. Information loss occurs along a diagonal (arrows). Note nonuniform signal around circle. (e) Absolute value of first derivative, undirectionally scanned. Note vertical loss line. (f) As (e), bidirectionally scanned. No information loss occurs, and edge intensity is more uniform. (g) Second derivative, undirectionally scanned. (h) As (g), bidirectionally scanned. No information loss occurs, and the intensity around the edge is uniform.

with Y-modulation rather than intensity modulation. This transformation produces a serious distortion of the image because the vertical deflection is a function of both the location in the scanning raster and the signal intensity. An example of this distortion is apparent in Figure 30h. Y-

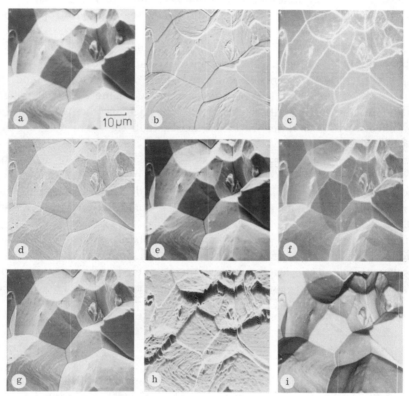

FIGURE 30.   Effects of various signal processing operations on a complex image. Specimen: fractured iron; beam: 20 keV. (a) Secondary and backscattered signal, ET detector, no signal processing applied. (b) First time derivative of (a), bidirectionally scanned. Note apparent flattening. (c) Absolute value of first time derivative, bidirectionally scanned. Note strong outlining of edges. (d) Second time derivative, bidirectionally scanned. Note crispening of edges and less apparent flattening. (e) Same as (b) but mixed in equal proportions with the direct signal. Note crispening and the appearance of depth. (f) Same as (c) but mixed in equal proportions with the direct signal. (g) Same as (d) but mixed in equal proportions with the direct signal. Note excellent enhancement of fine structure on grain surfaces and appearance of depth. (h) Y-modulation image. Note distortion of fracture facets. (i) Inverted, gamma-processed image, $\gamma = 4$.

modulation is useful in detecting low contrast levels which would not be apparent to the eye in an intensity-modulated image.[31]

Each of the signal transformations described above has limitations and deficiencies which must be recognized for successful application. Otherwise, artifacts may be introduced in the final image which may provide erroneous information about the specimen.

VII. IMAGE DEFECTS

It must be recognized that an SEM image has a built-in periodicity due to the regular spacing of the scan lines. We are always looking at objects through a grid of points and/or lines. If the object we are examining also has periodicity, moiré fringe effects may be produced. This effect is shown in Figure 31, where a grid object is examined at different magnifications. When the periodicity of the object approaches that of the CRT display, moiré fringes are formed, thus introducing an artifact to the image which must not be interpreted as a feature of the specimen.

In some situations, competing contrast mechanisms may be regarded as an image defect. Basically, if the signal change between two points may occur for a number of reasons, the mechanism producing the stronger contrast will tend to mask the weaker contrast mechanisms. The information carried by the weaker contrast is effectively rendered invisible by being forced into just one or two gray levels. Competing contrast mechanisms can sometimes be separated providing their characteristics are distinctly different. For example, if a trajectory contrast mechanism, such as topographic contrast formed with the backscattered signal in an ET detector, is dominating a number contrast mechanism, such as atomic number contrast, the trajectory topographic contrast can be eliminated by use of the specimen current signal, which is insensitive to trajectory effects of the electrons. However, as pointed out previously, topographic contrast with backscattered electrons also has a number component. This number component cannot be separated from the number effect due to Z differences, since the electrons backscattered due to tilt and Z differences are indistinguishable. However, the trajectory topographic effect is stronger than the number effect, and the elimination of trajectory effects is very helpful, as shown in Figure 32, in interpreting the atomic number contrast in this example.

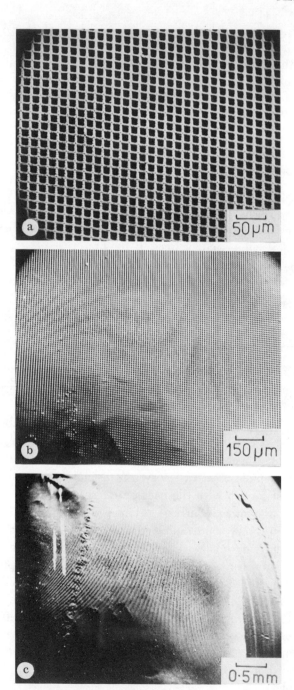

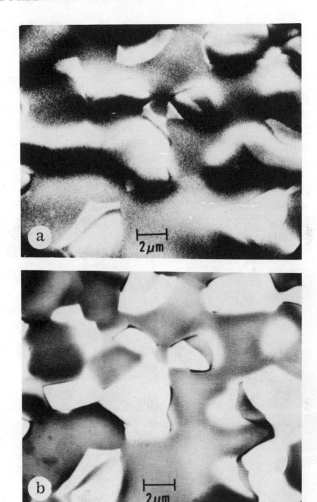

FIGURE 32. Elimination of trajectory component of backscattered topographic contrast. In (a) the ET detector image shows strong shadowing effects due to surface roughness which overwhelms atomic number contrast. In (b) the specimen current image shows good visibility for the atomic number contrast (bright: lead-rich phase; dark: tin-rich phase) due to elimination of trajectory effects which do not affect the specimen current signal. Note that tilt effects which produce number topographic contrast are not eliminated.

FIGURE 31. Moiré effect in images of a grid. (a) Regular grid pattern, high magnification. At lower magnifications (b) and (c) the scan periodicity and object periodicity interact, producing the fringe effects.

VIII. ELECTRON PENETRATION EFFECTS IN IMAGES

The distance of penetration of high-energy primary electrons into a specimen and the escape depth of secondary electrons in special geometrical situations can determine contrast effects in images. These effects, when properly understood, can be used to increase our knowledge of a specimen.

When specimens are examined at very high magnification, e.g., as in Figure 1 of Chapter 1, the edges of objects in an image often appear bright relative to the interior of the object. This phenomenon can be understood in

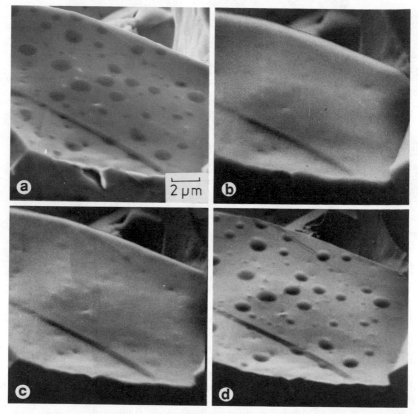

FIGURE 33. Polycrystalline iron, brittle fracture. The fracture surface has been exposed to atmospheric attack for six months. (a) Backscattered and secondary electrons, 30 keV; (b) backscattered electrons only, 30 keV; (c) backscattered electrons only, 20 keV; (d) backscattered electrons only, 5 keV.

terms of a beam penetration effect (Figure 19b). As beam electrons enter a specimen, secondary electrons are created near the surface. If the specimen is thin relative to the range of the primary electrons, then some primary electrons may be transmitted. As these electrons escape the specimen, more secondary electrons are created and are collected efficiently by the ET detector. Lateral spreading can also produce enhanced backscatter and secondary electron production near an edge. This results in enhanced contrast of edges. From the discussion of resolution limitations in Section V, it is apparent why such edges are the main features of high-resolution images.

When the object under examination is a surface film, comparison of the images obtained from the backscattered plus secondary electron signal with the images obtained from the backscattered electron signal alone at various beam energies can reveal striking differences. Figure 33 is a series of images of an iron fracture surface exposed to atmospheric attack. The backscattered plus secondary electron image (Figure 33a) obtained with a beam energy of 20 keV contains numerous spot-like objects which were observed to develop over a period of months after fracture. Observation of the same region with the backscattered electron signal alone at 30 keV and 20 keV (Figures 33b, c) does not reveal the spots. Evidently, these regions are so thin that the high-energy electrons penetrate the films without interacting significantly. However, a backscattered electron image at 5-keV beam energy does show the spots. At the lower beam energy, interaction near the surface results in contrast formation. From a knowledge of the mean beam penetration as a function of energy, a rough estimate of the film thickness could be made.

REFERENCES

1. C. W. Oatley, *The Scanning Electron Microscope*, Part 1, "The Instrument," Cambridge, The University Press (1972).
2. V. K. Zworykin, J. Hillier, and R. L. Snyder, *ASTM Bull.*, **117**, 15 (1942).
3. T. E. Everhart and R. F. M. Thornley, *J. Sci. Instr.*, **37**, 246 (1960).
4. T. E. Everhart, in *Proceedings of the 3rd Stereoscan Colloquium*, Morton Grove, Illinois, Kent Cambridge Scientific, (1970), p. 1.
5. J. B. Pawley, in *SEM/1974, Proceedings of the 7th Annual SEM Symposium* (O. Johari, ed.), IITRI, Chicago, Illinois (1974), p. 27.
6. A. V. Crewe, M. Isaacson, and D. Johnson, *Rev. Sci. Instr.*, **41**, 20 (1970).
7. H. Yakowitz, C. E. Fiori, and D. E. Newbury, in *SEM/1973, Proceedings of the 6th Annual SEM Symposium* (O. Johari, ed.), IIRTI, Chicago, Illinois (1973), p. 173.
8. M. D. Muir and D. B. Holt, in *SEM/1974, Proceedings of the 7th Annual SEM Symposium* (O. Johari, ed.), IITRI, Chicago, Illinois (1974), p. 135.

9. E. M. Hörl and E. Mugschl, in *Proceedings of the 5th Congress on Electron Microscopy*, Conference Series No. 14, Institute of Physics, London (1972), p. 502.
10. C. G. van Essen, *J. Phys. E* 7(2), 98 (1974).
11. K. F. J. Heinrich, in *X-Ray Optics and Microanalysis* (R. Castaing, P. Deschamps, J. Philibert, eds.) (1966), p. 159.
12. T. E. Everhart, O. C. Wells, and C. W. Oatley, *J. Electron. Control*, 7, 97 (1959).
13. A. Rose, *Advances in Electronics* (L. Marton, ed.), Vol. 1, Academic Press, New York (1948), p. 131.
14. O. C. Wells, in *SEM/1974, Proceedings of the 7th Annual SEM Symposium* (O. Johari, ed.), IITRI, Chicago, Illinos (1974), p. 1.
15. A. N. Broers, in *Microprobe Analysis* (C. A. Andersen, ed.), Wiley, New York (1973), p. 83.
16. C. W. Oatley, W. C. Nixon, and R. F. W. Pease, in *Advances in Electronics and Electron Physics* (L. Marton, ed.), Academic Press, New York (1965), p. 181.
17. D. C. Joy, in *SEM/1973, Proceedings of the 6th Annual SEM Symposium* (O. Johari, ed.), IITRI, Chicago, Illinois (1973), p. 743.
18. R. S. Longhurst, *Geometrical and Physical Optics*, Longmans, London, (1957). p. 207.
19. R. J. Woolf and D. C. Joy, in *Proceedings of the 25th Anniversary Meeting of EMAG, Institute of Physics, 1971* (W. C. Nixon, ed.), Institute of Physics, London (1971), p. 34.
20. G. R. Booker, D. C. Joy, J. P. Spencer, and C. J. Humphreys, in *SEM/1973, Proceedings of the 6th Annual SEM Symposium* (O. Johari, ed.), IITRI, Chicago, Illinois (1973), p. 251.
21. A. N. Broers, in *SEM/1974, Proceedings of the 7th Annual SEM Symposium* (O. Johari, ed.), IITRI, Chicago, Illinois (1974), p. 9.
22. A. V. Crewe, D. N. Eggenberger, J. Wall, and L. M. Welter, *Rev. Sci. Instr.*, 39, 576 (1968).
23. M. Isaacson, J. Langmore, and J. Wall, in *SEM/1974, Proceedings of the 7th Annual SEM Symposium* (O. Johari, ed.), IITRI, Chicago, Illinois (1974), p. 19.
24. J. B. Le Poole, J. Kramer, and K. D. van der Mast, in *Proceedings of the 5th European Congress on Electron Microscopy, 1972*, Conference Series No. 14, Institute of Physics, London (1972), p. 8.
25. A. N. Broers, in *Microprobe Analysis* (C. A. Andersen, ed.), Wiley, New York, (1973), p. 100 and Figure 6.
26. O. C. Wells, *Appl. Phys. Lett.*, 19, 232 (1971).
27. J. Lebiedzik, K. G. Burke, S. Troutman, G. G. Johnson, Jr., and E. W. White, in *SEM/1973, Proceedings of the 6th Annual SEM Symposium* (O. Johari, ed.), IITRI, Chicago, Illinois (1973), p. 121.
28. C. E. Fiori, H. Yakowitz, and D. E. Newbury, in *SEM/1974, Proceedings of the 7th Annual SEM Symposium* (O. Johari, ed), IITRI, Chicago, Illinois (1974), p. 167.
29. K. F. J. Heinrich, C. E. Fiori, and H. Yakowitz, *Science*, 167, 1129 (1970).
30. P. C. Goldmark and J. J. Hollywood, *Proc. IRE*, 39, 1314 (1951).
31. M. Oron and V. Tamir, in *SEM/1974, Proceedings of the 7th Annual SEM Symposium* (O. Johari, ed.), IITRI, Chicago, Illinois (1974), p. 207.

V

CONTRAST MECHANISMS OF SPECIAL INTEREST IN MATERIALS SCIENCE

D. E. Newbury and H. Yakowitz

I. INTRODUCTION

The versatility of the scanning electron microscope is due in great measure to the rich variety of mechanisms of electron interaction with solids. These interaction mechanisms can be employed to produce contrast in SEM images, revealing differences in the physical nature of the specimen from point to point. This chapter will discuss six contrast formation mechanisms of particular interest: (1) electron channeling contrast, which is related to the crystallographic nature of a specimen; (2) type I and (3) type II magnetic contrast, which can reveal magnetic domains in both crystalline and artificial magnetic structures; (4) voltage contrast, which arises from differences in the electrostatic potential of the specimen; (5) electron-beam-induced current contrast, a phenomenon of electron–hole pair production in semiconductor devices; and (6) cathodoluminescence, which arises from the emission of long-wavelength electromagnetic radiation in the ultraviolet, visible, and/or infrared regions. Of these mechanisms, 1–5 are confined to applications in the physical sciences, while mechanism 6, cathodoluminescence, finds application in physical, mineralogical, and biological sciences.

D. E. NEWBURY and H. YAKOWITZ—Institute for Materials Research, National Bureau of Standards, Washington, D.C.

II. ELECTRON CHANNELING CONTRAST

A. Introduction

The study of crystalline materials in the scanning electron microscope has been greatly advanced by the discovery of electron channeling effects from bulk crystals.[1] The electron channeling pattern (ECP) can provide crystal orientation and crystal perfection information from a surface layer less than 50 nm (500 Å) thick. The development of special instrument-operating techniques has provided useful ECP's from selected areas as small as 1 μm in diameter (selected area electron channeling patterns, SACP).[2–4] Electron channeling contrast can be employed in images scanned in the conventional manner to provide images of grains in microstructures, twins, and other crystallographic features.[5] In addition to the mechanisms of electron channeling contrast, this section will consider the techniques for obtaining selected area channeling patterns.

B. The Mechanism of Electron Channeling Contrast

As explained in detail in Chapter III, the interaction of the primary electrons of the beam with the atoms of the solid results in deflections of the electron trajectories from the incident direction due to high-angle elastic scattering and a loss of energy due to inelastic scattering. Some primary electrons escape the crystal with reduced energy (backscattering) while the remainder are absorbed in the case of a bulk, effectively infinitely thick specimen. The mechanisms of energy loss include the production of low-energy secondary electrons (less than 50 eV), x-ray production, and the production of long-wavelength light (cathodoluminescence). The secondary electrons have a range in the solid of about 2 nm (20 Å), whereas the range of the high-energy primary electrons is typically 100–2000 nm (1000–20,000 Å). In a crystalline solid, the periodicity of the atom arrangement can affect the way in which the primary electrons interact, especially in the initial interaction near the surface. In particular, we shall consider the electron channeling effect, which is illustrated in a simple manner in Figure 1. The electron channeling effect arises because of the differing atomic packing density which is observed along different crystallographic directions. Thus, when the primary electron beam is incident as in Figure 1a, the atomic packing is high and the electrons tend to interact close to the surface. When the beam is incident at a slightly different angle relative to the crystal (Figure 1b), the electrons penetrate more deeply into

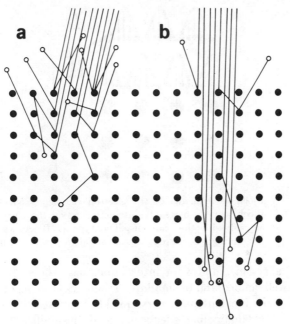

FIGURE 1.   Simple model of the electron channeling effect. The beam incidence angle at (a) causes strong interaction at the surface. At (b), the angle of incidence results in electron penetration in channels through many atom layers before significant interaction.

the crystal, passing between the rows of atoms along "channels." For the primary electrons, the probability of escaping from a solid decreases close to exponentially with depth below the surface. Thus, those electrons that initially channel into the crystal have a lower probability of escape than those that interact in the first atom layers.

Comparing the backscattered fraction η for Figures 1a and 1b, a difference will be observed, with more electrons escaping for the non-channeling condition (Figure 1a). The secondary electrons, which are produced continually along the paths of the primary electrons, are very strongly absorbed and only those secondaries that are produced near the surface have a large escape probability. Thus, if the secondary electron production coefficient δ is compared for the beam–crystal conditions of Figures 1a and 1b, a difference will again be observed, with a higher value of δ for strong interaction near the surface (Figure 1a). It should be noted that we are concerned here with channeling effects near the surface. The electron can be affected at any position in its trajectory by the periodic atomic arrangement. However, the channeling effect will only lead to

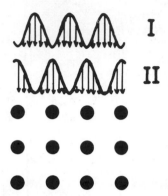

FIGURE 2.   Bloch wave model of the electron channeling effect. The type I
Bloch wave is a weakly interacting wave; the current vectors are
aligned with the channels. The type II Bloch wave is strongly
interacting; the current vectors are aligned with the atom positions.

significant changes in η and δ for interactions near the surface where the
beam is still well defined and of small divergence. Scattering of the primary
electrons from the incident path very quickly increases the divergence of
the beam, so that channeling effects are made insignificant since only a
very small fraction of the primary electrons are traveling parallel in any
given direction. The depth of average channeling penetration varies with
atomic number, but for most elements very little additional contribution
to the contrast results from depths below 50 nm (500 Å).

   The simple model of the electron channeling effect given in Figure 1
can be made more rigorous by the use of the Bloch wave model to describe
the properties of the high-energy primary electrons within the crystal.[6]
For any choice of primary beam–crystal orientation relationship, the
properties of the fast electrons can be described by a superposition of
Bloch waves, principally the type I wave, which interacts weakly with the
atoms, having nodes at the atomic positions, and the type II wave (strongly
interacting, with antinodes at the atomic positions) (Figure 2). For interaction with a given set of lattice planes, the proportions of the type I and
type II waves differ as a function of the beam–crystal orientation. For a
particular set of crystal planes of interplanar spacing d, the proportions of
the type I and type II waves are equal when the Bragg condition is fulfilled:

$$n\lambda = 2d \sin \theta \tag{1}$$

where n is the integer order of reflection, λ is the electron wavelength, and
θ is the angle between the beam and the planes. For particular values of λ,
which is fixed by the accelerating voltage, and d, the value of θ must have
a particular value, called the Bragg angle θ_B, for the condition of equation

(1) to be satisfied. For angles less than θ_B, the type II wave dominates, and the electrons tend to interact close to the surface, which produces larger values of η and δ. For angles greater than θ_B, the type I wave dominates and more electrons channel into the crystal; this effect results in lower values of η and δ. The intensity profile which results as a function of δ is illustrated in Figure 3. For $-\theta_B < \theta < \theta_B$, the intensity is high due to dominance of the type II wave. At $\theta = \theta_B$, a sharp intensity change results from increased contribution from the type I wave. For $\theta > \theta_B$, the influence of the first-order reflections upon the beam becomes negligible and the intensity reaches a background level. For $\theta = 2\theta_B$, $3\theta_B$, etc. a sharp transition again results. Note that the width of the transition decreases as the order of n in equation (1) increases.

In order for the electron channeling effect described above to produce contrast in an SEM image, the conditions $\theta < \theta_B$ and $\theta > \theta_B$ must be produced at sequential points in an SEM image. If these conditions are satisfied for a particular set of crystal planes, contrast will occur because of differences from point to point in the numbers of emitted electrons, both primaries and secondaries. Electron channeling contrast is thus a form of pure emission number contrast.

Electron channeling contrast can be obtained using the secondary and primary electron signals separately or combined. Since electron channeling contrast is number contrast, the requirement for a balance of currents leads to complementary changes in the specimen current. As explained in

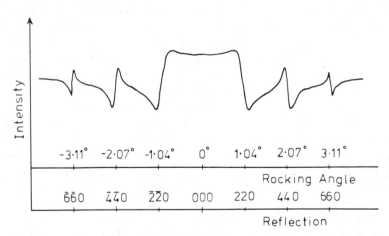

FIGURE 3. Plot of backscattered intensity versus scan angle relative to the crystal ("rocking curve"). The curve is plotted for reflections of the type (nn^0) in silicon with a beam incidence energy of 30 keV.

Chapter IV, the contrast with the specimen current signal is of the opposite sense from that of the emissive (secondaries plus primaries) signal.

C. The Electron Channeling Pattern

For electron channeling contrast to occur in the SEM image, a situation must be produced in which the beam–crystal orientation relationship changes so that regions with $\theta < \theta_B$ and $\theta > \theta_B$ are contained within the same field. One such situation is the case in which a large (approximately 5 mm square) single crystal is examined in the SEM at low magnification (approximately $20\times$). At low magnification, the normal scanning action results in angular beam deflections of approximately $\pm 8°$ from the optical axis. For 30-keV primary electrons, the Bragg angle for low-order planes, e.g., (200) in iron, is of the order of $2°$. Thus, in the normal scanning action the conditions $\theta < \theta_B$ and $\theta > \theta_B$ can be satisfied for those planes that make an angle of less than $8°$ with the optical axis, i.e., nearly parallel to the beam. Note that the surface plane does not influence the planes that interact with the beam to produce the phenomenon of electron channeling. Only those crystal planes that are nearly parallel to the incident beam can contribute to the channeling effect.

The specimen must be a large single crystal in this technique of low-magnification channeling pattern formation, since the large angular deflection shifts the beam impact several millimeters from the optical axis. The beam is scanned through the normal raster grid of points, resulting in conventional image formation (Chapter IV). However, the scanning action also produces a one-to-one correspondence between the angle that the beam makes relative to the crystal and the position on the final display cathode ray tube (CRT). Thus, the normal topographic image of the crystal surface is obtained along with the electron channeling effects that result from the changing beam–crystal relationship (Figure 4a). In Figure 4a, the silicon crystal has been deliberately chosen to be smaller than the area being scanned to emphasize the fact that a normal low-magnification image is obtained along with the electron channeling effects. The electron channeling effects result in the formation of a pattern of bright bands on a dark background (emissive mode) and fine lines; this pattern is called an electron channeling pattern or ECP ("Coates patterns" and "pseudo-Kikuchi patterns" have also been used in the literature). The pattern consists of contributions from many crystal planes, since a real crystal contains a periodic atom arrangement in three dimensions. Thus each point in the image actually consists of intensity contributions from many different planes ("many-beam conditions"), each of which has an intensity vs. θ

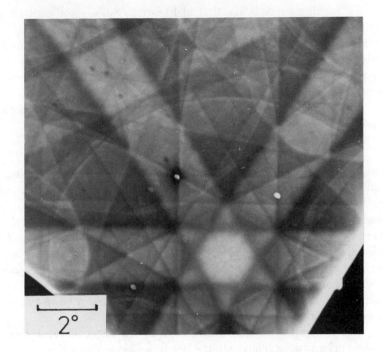

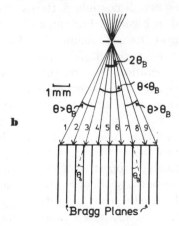

FIGURE 4.   (a) The electron channeling pattern generated by scanning at low magnification over a large single crystal. The material is silicon oriented near the [111] pole. Beam conditions: $E_0 = 30$ keV; $i = 1 \times 10^{-8}$ A; divergence $\sim 2 \times 10^{-3}$ rad. Emissive signal (backscattered and secondary electrons). (b) Scanning action used to generate the ECP of (a).

curve ("rocking curve") like that of Figure 3. Since significant channeling effects are only observed near Bragg conditions and sharp transitions are obtained, the effects of particular lattice reflections are distinct in the pattern. The bands are observed in regions where $\theta < \theta_B$, as shown in Figure 4b, and the fine lines represent the sharp change in contrast at the Bragg angles for higher order planes.

The projection of the ECP is gnomonic, and distance D in the ECP is related to the deflection angle θ by

$$D \propto \tan \theta \tag{2}$$

Distance in the field and angular deflection are linearly related at small angles, less than about 5°, where $\theta \simeq \tan \theta$. The trace of a lattice plane in the ECP bisects the band to which it is related. As shown in Figure 3, bands have a width of $2\theta_B$; the width is inversely proportional to the lattice plane spacing d and the accelerating voltage. If the total angle of scan is reduced by increasing the magnification, the bands in the ECP appear to increase in width, since a given distance in the field now corresponds to a smaller change in angle. The magnification can be increased to the point where the total scan angle is much less than $2\theta_B$, and a band can no longer be seen (Figure 5c).

The collection of ECP's obtained by tilting a crystal through all possible orientations relative to the beam can be assembled into a "channeling map," which represents all possible patterns. Such a channeling map for copper is illustrated in Figure 6. For cubic crystals, a complete map is obtained within the "unit triangle" defined by the [001], [011], and [111] poles, with all other possible orientations produced from this region by mirror and/or rotation operations. Several such maps have been presented in the literature.[3, 8] When an "unknown" pattern is obtained from an arbitrarily oriented crystal of the same material, the bands can be indexed by comparison with the map. The determination of orientation becomes a problem of pattern recognition. A solution of the indexing of the unknown pattern can be confirmed by comparing angles between bands, angular band widths, and symmetry properties of the unknown with that region of the map that represents the same orientation. Since the band spacings and angles vary with the crystal structure, lattice spacing, and accelerating voltage of the primary electrons, a new map would seem to be required for each crystal structure. Fortunately, in the cubic system the angles between bands and poles are independent of lattice spacing. Hence all cubic crystals have geometrically similar channeling maps, although band widths vary according to the lattice parameter. A copper (fcc) map can be used to index nickel (fcc) ECP's. Channeling patterns are subject to the rules for forbidden reflections (see, for example, Barrett and Massalski[9])

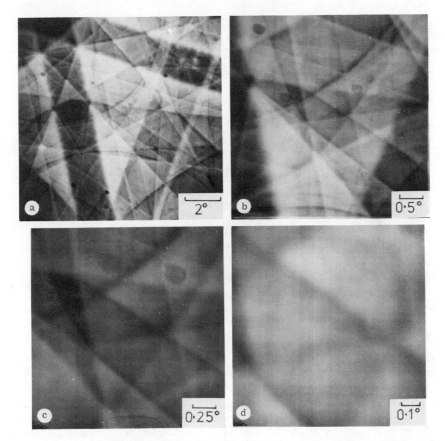

FIGURE 5.   Magnification series in the ECP mode of operation of Figure 4.
Magnification: (a) 20×; (b) 50×; (c) 100×; (d) 200×. Material:
silicon oriented near the [111] pole. Beam conditions: same as
Figure 4. Signal: specimen current. As the magnification increases,
the scan angle decreases and the portion of the ECP that is observed
decreases.

and the channeling maps for fcc, bcc, and diamond cubic (dc) crystals differ
because of the presence or absence of particular reflections. Thus the [111]
reflection is found in the fcc map and is absent in the bcc map, while
[222] is present in both. It is useful, therefore, in indexing ECP's from
cubic materials to have available separate maps for bcc, fcc, and dc crystals.
For noncubic crystals, the channeling maps are more complicated because of lower symmetry, and both band widths and interband angles
vary with the lattice parameters. These crystals can usually be oriented
by comparing the ECP's with Kikuchi maps, which are geometrically

FIGURE 6.   Electron channeling map for copper (face-centered cubic crystal
structure) covering the basic crystallographic repeating unit. Beam
conditions: $E_0 = 30$ keV. Signal: specimen current. (Courtesy of
C. G. van Essen.)

similar, obtained by transmission electron microscopy.[10] Computer-drawn maps can also be useful, although such maps present only lines and band edges without any intensity information.[10]

D. *Electron Optical and Signal Processing Conditions*

Electron channeling effects result in weak contrast from bulk speci-mens. Experimental measurements of the contrast $(S_{max} - S_{min})/S_{av}$, where S_{max} and S_{min} are the signals obtained at orientations corresponding to "white" and "black" in the channeling map (Figure 6), reveal a maxi-mum value of 8% or less.[6] This contrast is nearly independent of incident electron energy in the range that has been investigated, 5–30 keV. An atomic number effect exists, with contrast decreasing slightly with increasing atomic number. For almost all experimental conditions, the microscopist can expect to be working in the contrast range 1–5%. The weak nature of electron channeling contrast requires two important considerations for successful microscopy. First, the threshold beam current, i.e., the minimum beam current necessary to detect a given contrast in the presence of statistical fluctuations in the electron counting process, is high. As shown in Chapter IV, equation (22), the threshold current for 5% contrast with a 10-sec frame time is approximately 6×10^{-10} A. This threshold current must be considerably exceeded to produce an image of satisfactory quality. A beam current of at least 5×10^{-9} A should be set as the first step in establishing correct conditions for electron channeling contrast. The second condition that must be established as a consequence of the weak nature of electron channeling contrast is that adequate signal processing must be applied in displaying the image.

As discussed in Chapter IV, the human eye is relatively insensitive to contrast less than 5%. Figure 7a contains an ECP image obtained with no signal processing applied, i.e., the original signal was amplified linearly and displayed. The bands of the ECP are only barely visible. However, the SEM is eminently suited to process the signal so as to enhance the low contrast values. Various forms of signal processing are explained in detail in Chapter IV. Electron channeling contrast is successfully displayed using the black level operator (differential amplification). Figure 7b shows the same field as Figure 7a, but with the black level operator used to transform the signal. The fine details of the ECP are readily observed. Thus the second step in establishing correct conditions for channeling contrast is to set the black level to a high value and subsequently to increase the amplifier gain. This spreads the ECP information over the full range of available gray levels.

The third and final operating condition for channeling contrast concerns the beam divergence α at the specimen. Electron channeling contrast depends on the relative angle between the beam and the crystal. The sharp changes in contrast that occur at the Bragg angle for a particular set of planes frequently take place over a very small range of angle, typically $0.2°$ around θ_B. In a real beam, the electrons do not travel in perfectly parallel paths; the skewness in the paths of the electrons is known as the divergence. If the beam has a divergence which is greater than the fine angular details in the ECP, these details will be lost. For any beam–crystal orientation relationship, the contrast change is integrated over a range of angles equal to the divergence of the beam so that the intensity at each picture point is not related to a well-defined angle. The effect of divergence on the quality of the ECP is illustrated in Figure 8, where increasing beam divergence causes a loss of fine details in the ECP.

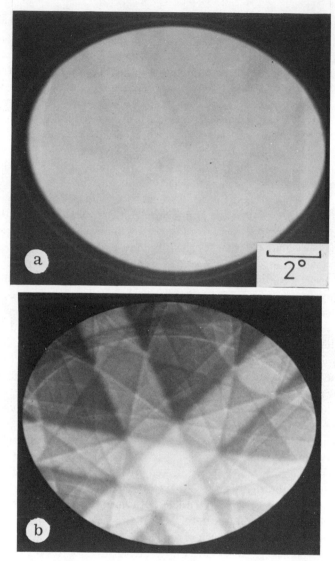

FIGURE 7.   Example of the application of black level suppression signal processing in order to enhance electron channeling contrast in the final display.
(a) Direct emissive mode signal; black level = 0. (b) Black level
suppression applied. Material: silicon; beam: 30 keV.

At very high divergence, only the low-order bands are observed
(Figure 8d). It is thus desirable to have as small a divergence as practical
in the beam. Ideally, we would like to have a perfectly collimated beam.

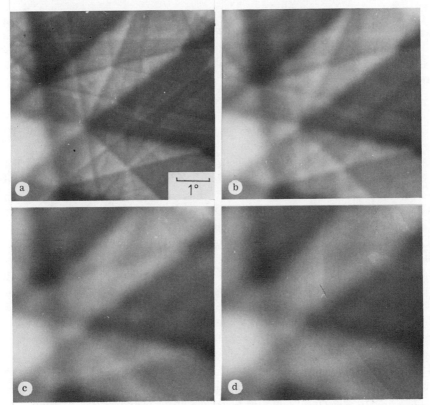

FIGURE 8.   Effect of increasing beam divergence (decollimation) on electron channeling pattern quality. The divergence changes from (estimated) 2×10^{-3} rad in (a) to 10^{-2} rad in (d).

The practical limitation on collimation is set by the brightness equation (Chapter II):

$$\beta = 4i/\pi^2 d^2 \alpha^2 \tag{3}$$

where β is the brightness, i is the beam current, and d is the size of the electron beam; β is a constant for a particular electron source. As α is decreased by one of several methods, the beam current decreases as the square of α. However, the beam current must not fall below the threshold current i_{th}. A typical set of instrument operating parameters for channeling contrast and a 2-sec frame time are: (1) a beam current of 5 nA and (2) a beam divergence of 3×10^{-3} rad. With a conventional tungsten hairpin filament source, which has a brightness of 5×10^4 A/cm^2 ster at 30 kV, the above conditions on i and α result in a probe diameter of approximately 400 nm (4000 Å).

In practice, the beam current can be set by monitoring the specimen (absorbed) current while adjusting the condenser lenses. The divergence is set by the diameter of the final aperture. For a given specimen to aperture distance D, the divergence α is related to the radius r_a of the final aperture by the equation

$$\alpha = r_a/D \tag{4}$$

For a working distance of 10 mm, a divergence of 2.5×10^{-3} is achieved with a 50-μm final aperture.

A second form of signal processing that can be very useful in working with electron channeling patterns is time differentiation of the signal (derivative operator, Chapter IV). This operator enhances the sharp transitions at edges and eliminates the constant signal level in low-order bands (Figures 9a, b). Time differentiation is especially useful when the ECP of a tilted specimen is required. The backscattering coefficient increases with increasing tilt; this effect is pronounced above 30° tilt. When the ECP is generated either by low-magnification imaging of a single crystal or by the rocking beam method, which will be subsequently described, the angle of the beam relative to the specimen changes by ±8° about the tilt angle, i.e., from 53° to 37° as the beam is scanned from the top of the field to the bottom. The backscattering coefficient, which produces the average level of the signal, changes by approximately 15% over this range of tilt. Since a high value of the black level operator must be used to observe electron channeling contrast, the change in average signal due to the change in the backscattering coefficient is amplified as well by the black level operator. The ECP image that results from a tilted specimen is illustrated in Figure 9c. The derivative operator recovers the information lost in Figure 9c, by enhancing the high-frequency detail while suppressing the low-frequency signal changes, such as the gradual change in average signal level (Figure 9d).

E. *Electron Channeling Contrast from Polycrystals*

When polycrystalline material consisting of very coarse grains ($\sim$1 mm) is imaged in the SEM under conditions appropriate for the detection of electron channeling contrast, an image such as that of Figure 10 is obtained. The scanning action of the beam again produces channeling bands, but the discontinuities in the crystal orientation at the grain boundaries cause the bands to terminate, since the crystal planes terminate on the boundaries. Only a small portion of a channeling pattern can be seen on each grain, since the change in scan angle is only a few degrees across each grain.

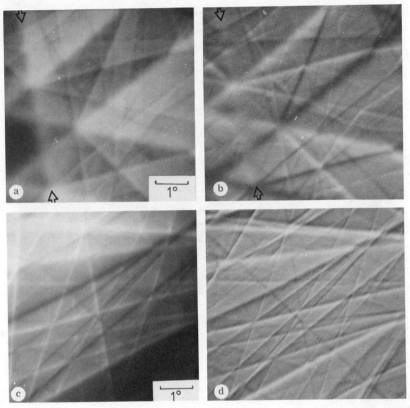

FIGURE 9.   (a,b) An example of the application of first time-derivative signal processing to an electron channeling pattern from silicon: (a) specimen current signal with black level suppression processing; (b) fully differentiated signal with unidirectional scanning, vertical scan lines. Note the near absence of the band edge indicated by the arrows. This line is nearly parallel to the scan line. (c,d) Example of the application of first-time-derivative signal processing to recover lost information. The specimen is tilted 35° from normal incidence. In (c), processed with black level suppression only, the effect of changing background level due to the η vs. θ relationship causes the signal to fall below the sensitivity of the film. In (d), the entire field is visible in the differentiated image since the slowly changing background is strongly attenuated.

At higher magnifications, the change in scan angle across the field is smaller since the beam is not deflected as far from the optical axis. The change in scan angle ϕ_M with magnification M is given by

$$\phi_M = \phi_{20}(20/M) \tag{5}$$

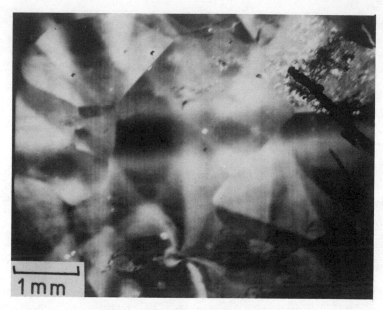

FIGURE 10.   Effect of reducing the grain size for the scanning situation of Figure 4b. A small portion of the ECP of each grain is observed. Material: Fe–$3\frac{1}{4}$% Si; beam: 30 keV; signal: specimen current.

where ϕ_{20} is the change in scan angle across the field at a particular magnification, in this case, 20×. Thus, at a magnification of 200×, ϕ_{200} is only about 1.5° compared to ϕ_{20} of 15°. When polycrystals with fine grains are examined at high magnifications, the change in scan angle across a grain is usually less than a Bragg angle, and bands can no longer be seen. In Figure 11, a nickel polycrystal with grains of approximately 300 μm is shown. The signal varies across most grains due to the change in scan angle, but distinct bands cannot be seen. For comparison, an optical micrograph of the same area is shown in Figure 11b (note the hardness indent for reference). The specimen is an electropolished mirror, with slight etching attack delineating certain boundaries. The SEM image (Figure 11a) shows strong contrast at all boundaries, since a change in crystal orientation produces a strong change in electron channeling effects. If the grain size is further reduced and a higher magnification is used to examine the specimen, the change in scan angle across a grain is very much less than a Bragg angle and a uniform signal is produced at all points in the grain. However, the grains produce different signal levels because the orientation varies from grain to grain, resulting in an image such as shown in Figure 12. Contrast is produced at any crystallographic dis-

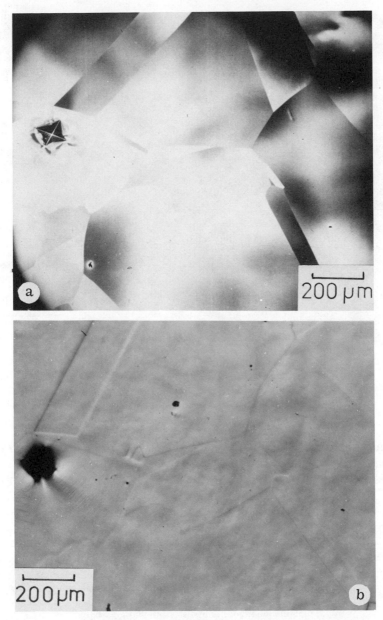

FIGURE 11. A further reduction of grain size from that of Figure 10 results in the necessity for high-magnification imaging and a smaller change in scan angle across the field. The scan angle on any grain is less than $2\theta_B$ and not sufficient to form recognizable ECP bands. The mottling is due to contrast changes at band edges. Material: nickel; beam: 30 keV: signal: specimen current. (a) SEM; (b) optical.

FIGURE 12. Crystallographic contrast from fine-grained material. The grain size is now so small that at the magnification required to observe grains, the change in scan angle across the field is only about 2°. Across any grain the change in scan angle is insufficient to go through a Bragg angle and the contrast is constant across each grain.

continuity, such as a twin boundary. The tone of a particular grain corresponds to the tone at that point in the ECP that represents the same angular relation of the beam to the lattice as the grain in the micrograph.

The channeling contrast observed from a small grain is usually uniform, but it can be nonuniform under special circumstances for two distinct situations. For certain crystal orientations relative to the beam,

---

FIGURE 13. Uniform and nonuniform channeling contrast from fine grains. Grains A and B show uniform contrast. The ECP's of these grains, (b) and (c), show constant contrast in the range of scan angle given by the circles corresponding to the micrograph. Grain C shows a smoothly changing contrast from dark to light. The ECP of this grain, (d), shows a band edge in the range of scan angle. Grain D shows a bandlike, irregular contrast; this is identified as bend contour channeling contrast. The ECP of grain D, (e), shows distorted bands indicative of bent crystal. Material: Fe-$3\frac{1}{4}$% Si; beam: 30 keV; signal: specimen current.

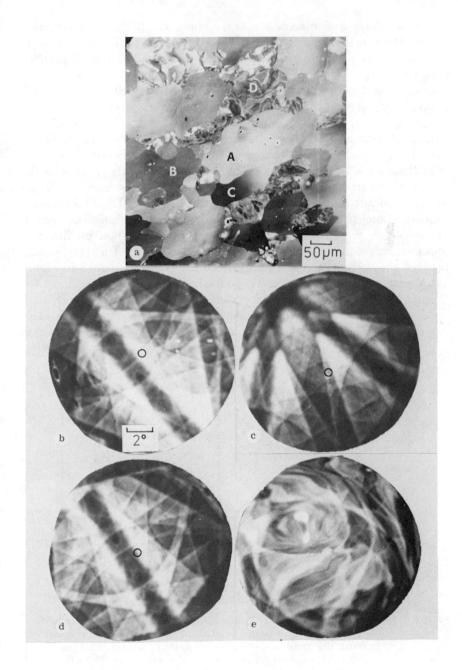

the small change in scan angle across the grain, e.g., 0.3°, can be sufficient to pass through the Bragg condition for a particular set of lattice planes, i.e., the edge of a low-order band or higher order fine line, and to produce a signal change (Figure 13). Grain C is oriented in such a way that the edge of a band is very near the center of the ECP (Figure 13d). Thus a small change in scan angle, indicated by the small circle in Figure 13d, is sufficient to cause a significant signal change. A second type of non-uniform channeling contrast results if the crystal is elastically or plastically bent. In this case, a relative change in the beam–lattice orientation occurs even though the beam does not change angle significantly relative to the surface of the specimen. If the crystal is bent through at least one Bragg angle, a contrast contour is observed (Figure 13a, grain D). This contrast contour is analogous to a bend contour in thin-foil transmission electron microscopy.[5] Such SEM bend contours give information on areas of imperfect crystalline material in a microstructure.[5] The relationship between the orientation and a simple bend contour is illustrated in Figure 14. A simple bend about an axis is equivalent to a line trace in the ECP (Figure 14a); if the line trace intersects any contrast change such as a band, then a contrast contour appears in the micrograph (Figure 14b). An example of a simple bend applied to polycrystalline gold sheet is

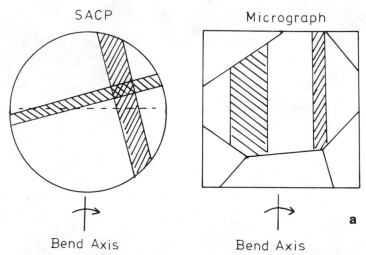

FIGURE 14. Relationship of the bend contour in the micrograph to orientation (a) Schematic: a simple bend about the axis indicated is equivalent to a line trace in the channeling pattern; the bands intersected by this trace appear on the grain in the micrograph. (b,c) Bend contours observed in polycrystalline gold after a simple bend about a vertical axis. Material: gold; beam: 30 keV; signal: specimen current.

illustrated in Figure 14 c. The grains show approximately parallel bend contours since all grains have been bent about a common axis. The shading of the bend contours varies from grain to grain because the orientation is different. The simple bending of each grain produces contrast contours corresponding to a line trace at some point in the channeling map, which depends on the exact orientation of each grain.

The resolution limit in conventional microscopy with channeling contrast is set by the beam diameter rather than the mechanism of contrast formation. The beam diameter is determined by the selection of beam current and divergence in the brightness equation, equation (3). For useful operating conditions with a tungsten thermionic emitter electron gun, the minimum beam diameter is approximately 0.4 μm. In special circumstances, these operating conditions can be modified and the resolution limit can be improved. After an area of interest has been located, a photographic image can be recorded with a long frame time, 100 sec or more. The threshold current falls as the time per picture point is increased, since more electrons are accumulated for each picture point and the random count fluctuations (noise) become less significant. Thus, for a longer frame time, a lower value of beam current is acceptable, and from equation (3), a smaller beam diameter can be obtained. In Figure 15, the limiting resolution of the spurlike structure on the twin boundary in polished, flat nickel is less than 0.2 μm. The area of Figure 15 a was first located with the instrument operating with a beam current of 5 nA and the contrast was maximized for the twin boundary of interest by tilting and observing the image. The beam current was then reduced to 0.5 nA and the image was recorded with a frame time of 100 sec, resulting in the image of Figure 15 b. Increasing the frame time even more is possible, but such increases are eventually limited by instrument stability.

F. Selected Area Channeling Patterns (SACP)

As explained previously, the minimum area from which information on orientation can be obtained with an ECP generated as in Figure 4b is about 4 mm square, since the magnification must be low. This limitation to large specimens has been overcome in the selected area channeling pattern (SACP) technique, in which the beam is rocked through the desired range of angles while it is confined to a very small area on the specimen, typically a 10-μm diameter circle or smaller.[2,3] The basic principle of the technique, illustrated in Figure 16, is to lower the crossover point of the scanned rays from the level of the final aperture so as to coincide with the specimen surface. This condition can be achieved in a

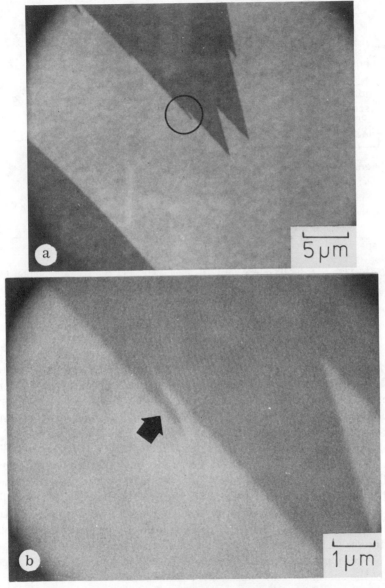

FIGURE 15.   Spatial resolution limit with electron channeling contrast. The spur
on the twin boundary is resolved to approximately 200 nm (2000 Å).
Electron source: conventional tungsten hairpin; beam: 30 keV;
signal: emissive mode; material: nickel. (a) $i = 5 \times 10^{-9}$ A, frame
time 40 sec; (b) $i = 5 \times 10^{-10}$ A, frame time 100 sec.

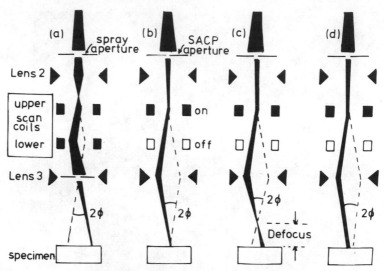

FIGURE 16.   Ray diagrams illustrating schematically the scanning action in the
selected-area channeling pattern (SACP) mode: (a) conventional
micrograph scanning; both scan coils active. (b) SACP scanning:
lower scan coil off; crossover at the specimen surface. (c) SACP
scanning, crossover above specimen surface. (d) SACP scanning,
crossover below specimen surface.

double-deflection scanning system by deactivating the lower set of scan
coils. The upper set of scan coils deflects the beam off-axis and the action
of the final lens brings the rays to a crossover point below the final aperture.
Typically, the beam is rocked through 12° for a 20× magnification setting
in this single-deflection technique. As the magnification is increased, the
off-axis deflection is decreased, and the angle of rock is decreased. Since
the rays are far off-axis at the position of the final aperture, this normally
small aperture must be replaced with an aperture at least 2 mm in diameter
to allow the rays to pass. In this case, the final aperture no longer defines
the beam divergence. To limit the beam divergence, a small aperture,
∼100 μm in diameter, must be used in place of one of the large splash
apertures higher in the column. This aperture is placed either at the position of the splash aperture of the second condenser lens or else at the position of the splash aperture above the scan coils.

A second way in which the crossover can be made coincident with the
specimen surface is to use a deflection coil located below the final lens. The
beam is then scanned in the normal manner through a crossover in the
final aperture (Figure 16a) and the post-lens scan coils bring the diverging

rays back to a crossover on the specimen. Since a crossover is formed at the position of the final aperture, this aperture serves its normal function to limit the beam divergence.

In both of the SACP techniques, the vertical position of final crossover can be changed. In the single deflection technique, the crossover position is a function of the strength of the final lens, and in the post-lens deflection technique, the final crossover depends on the excitation of the post-lens scan coil. Varying the vertical position of the crossover changes the area scanned in the SACP mode (Figures 16b–d). For crossover positions near a specimen at fixed elevation, the change in scan angle with lens strength is negligible. The change in scanned area as a function of crossover position can be used to produce a "through-focus series," in which the crossover point is sequentially placed above, at, and below the specimen surface (Figures 16 and 17).

In Figure 17a, a high-magnification (1000×) micrograph of a group of grains in Fe–$3\frac{1}{4}$% Si is shown; the beam is nearly normal to the specimen at all points. In the SACP mode of operation with the crossover point above the specimen surface, a large area is scanned, and across this area, the beam–specimen angle changes by about 12° due to the rocking action (Figure 16b). Thus both spatial and angular information is contained in the image, and the grains labeled A and B in the conventional image (Figure 17a) can be recognized in the SACP image (Figure 17b). As the crossover point is brought closer to the specimen surface, the area scanned decreases and the effective magnification increases.

More of the ECP of grain A, located at the center, is observed as grain A fills more of the field. When the crossover is coincident with the specimen surface, the magnification is a maximum and the area scanned lies entirely within grain A; the selected area electron channeling pattern of grain A is thus obtained, and the orientation of grain A relative to the beam is found to be near the [001] pole. When the crossover point is moved below the specimen surface, the area scanned increases and grains surrounding grain A are again observed. Note that the image is now reversed by 180° from the image of Figure 17 b, since the first scanned ray now strikes the left side of the field first, instead of the right (Figures 17b, d). The through-focus series can be carried out rapidly, since all that is involved in changing the image in the sequence illustrated in Figures 17(b–f) is to change the strength of the objective lens by approximately 0.1 A. The through-focus series thus provides a rapid means to identify precisely and rapidly the area from which the SACP is obtained.

The two major parameters of interest in the SACP technique are the size of the selected area, which we wish to minimize, and the angle of rock, which we wish to maximize to aid in the recognition of the ECP. The selected area is a minimum when the crossover is coincident with the specimen surface. If the focusing were perfect, this crossover would be practically a geometric point and the effective magnification would be nearly infinite. However, lens defects, principally spherical aberration, cause the crossover to have a finite lateral size. The minimum obtainable diameter d_{min} of this area is given by

$$d_{min} = \tfrac{1}{2}C_s\phi^3 \tag{6}$$

where C_s is the spherical aberration coefficient and ϕ is the semiangle of rock, i.e., the cone semiangle in Figure 16b. Equation (6) indicates that the selected area can only be made smaller by reducing the angle of rock, which is undesirable, or by decreasing C_s. The coefficient C_s is a decreasing function of lens excitation, and hence it is advantageous to operate with the specimen as close to the pole piece of the final lens as possible, so the excitation can be maximized. The interrelationship of the angle of rock, minimum selected-area size, and working distance from the final lens has been determined in detail for a particular system by Booker and Stickler.[11] At 2 mm working distance, a selected area of 8 μm diameter with $2\phi = 10°$ can be obtained.[11] Since d_{min} can be reduced by decreasing ϕ, in circumstances where a small scan angle is acceptable, e.g., when the SACP is used to assess crystal perfection rather than orientation, the ECP information can be obtained from areas as small as 1 μm diameter.

In specifying the performance of an SACP system, it is important to note both the minimum selected-area size and the corresponding angle of rock. It should also be noted that SACP performance is only a function of the lens–scan coil system and not a function of the brightness of the electron source. An increase in source brightness will allow a smaller beam size to be obtained for a specified divergence and current, an important

FIGURE 17.   "Through-focus series" in the SACP mode. (a) Conventional micrograph; the SACP of grain A is desired. (b) SACP mode scanning, crossover above specimen (compare with Figure 16), both spatial and angular information is present. Portions of the channeling patterns of grain A and neighboring grains are seen. (c,d) crossover brought closer to the specimen surface; the magnification increases, and the pattern of grain A increases in size. (e) Scanning entirely within grain A, the SACP of only this grain is obtained (compare with Figure 16b). The orientation is near the [100] pole. (f) Crossover below the surface (compare with Figure 16d). The magnification decreases and patterns of grain A and surrounding grains are observed. The image is reversed by 180° from (b) due to the reverse in the scanning sequence. Material: Fe–$3\tfrac{1}{4}$% Si; beam: 30 keV; signal: specimen current.

advantage in high-resolution micrographs with channeling contrast, but higher brightness will not improve SACP performance with regard to the minimum selected area. SACP performance can be improved by using techniques to correct for the spherical aberration effect, such as the dynamic focusing technique of van Essen, [12] which has been used to obtain SACP's with $2\phi = 10°$ from selected areas 1 μm in diameter. [13]

A comprehensive study of the accuracy of orientation determinations by means of SACP's has not been made. Joy *et al.* have compared the SACP technique with x-ray Kossel patterns and report that the accuracy of the SACP technique is "better than 1°" in determination of the crystal normal parallel to the optical axis. [14] In principle, points in the ECP can be located with a precision which is limited by the beam divergence, about 0.15°. However, the accuracy is limited by the distortions inherent in the SEM recording system, the scan generator, scan coils, and the cathode ray tube display. Such distortions can introduce uncertainties in the relative location of the points in the ECP of 5% or more. Considering that the usual angular diameter of ECP is 10°, this distortion factor limits the accuracy of orientation determination to a minimum of 0.5°. Improvements in the scan linearity and orthogonality can reduce this figure.

G. Crystal Perfection Effects on Electron Channeling Patterns

Two of the most important properties of electron channeling patterns are that (1) the quality of the ECP is strongly dependent on the perfection of the crystal and (2) the ECP information comes from a shallow surface layer, less than 500 Å. A considerable number of studies have been carried out relating the quality of ECP's to some measure of crystal perfection. [15–18] ECP quality has been expressed in terms of line width, minimum observable angular detail, and maximum contrast. All of these factors have been found to change markedly with crystal perfection. Crystal perfection has been assessed in terms of applied strain or defect accumulation, as measured by the dislocation density observed in the transmission electron microscope.

Two different kinds of ECP behavior as a function of strain are noted. Deformation at low relative temperatures. $T/T_m < 0.3$, where T_m is the absolute melting temperature, usually results in a loss of fine lines, line broadening, and decreased contrast of the bands relative to background; the bands remain straight. An example of this behavior is illustrated in Figure 18, where ECP's from an aluminum crystal are compared before and after deformation by rolling. A severe decrease in ECP quality is

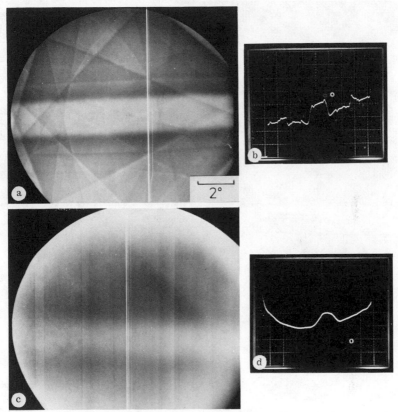

FIGURE 18. Effects of deformation on ECP quality; cold work. (a) ECP from an annealed aluminum single crystal. (b) Line trace, intensity vs. scan angle at the position indicated in (a). (c) Same crystal deformed at room temperature to a 5% reduction in thickness by rolling. (d) Line trace at position indicated. Note the loss of fine detail and decrease in contrast after deformation. Beam: 30 keV; signal: emissive.

observed after a 5% reduction in thickness by rolling. Deformation at high relative temperatures, $T/T_m > 0.5$, again results in a loss of fine lines, line broadening, and decreased band contrast; but the bands frequently distort, showing irregular band width and bending bands. This behavior is illustrated in Figure 19, where ECP's after successive tensile strains in Pb–1.5% Sn are compared. The distortion of bands becomes extremely severe after large deformations (Figure 19c). The band distortion has been shown to arise from polygonization and the development of subcells with successive rotation vectors from cell to cell.[5,15] The exact behavior of ECP's as a function of crystal perfection is a complicated interaction of

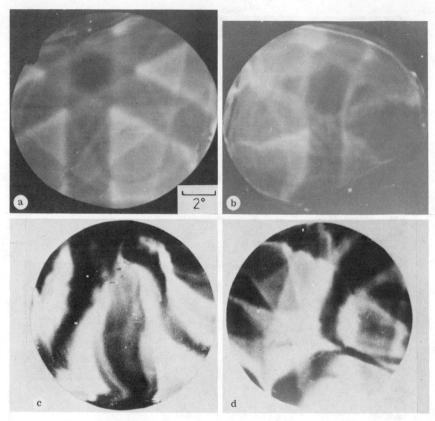

FIGURE 19. Effects of deformation on ECP quality; hot work. (a) Annealed Pb–1.5% Sn. (b) Same grain after a 2.5% total specimen strain in simple tension at room temperature. Note band bending and distortion. (c,d) Same specimen, SACP's obtained after 25% specimen tensile strain; repolished to eliminate surface topography. Note severe band distortion. Beam: 20 keV; signal: specimen current.

many factors, which are currently under intense study. Current applications to materials studies emphasize relative crystal perfection measurements rather than absolute measurements.

Finally, in making studies relating ECP quality to crystal perfection, the preparation of the surface can greatly affect ECP quality. In Figure 20, the effect of an amorphous surface film on the ECP's from annealed niobium is illustrated. It is thus important in relative measurements to ensure that the surfaces that are compared are of similar quality, and in absolute measurements, the surface contribution to ECP degradation must either be eliminated or else accounted for in the measurement.

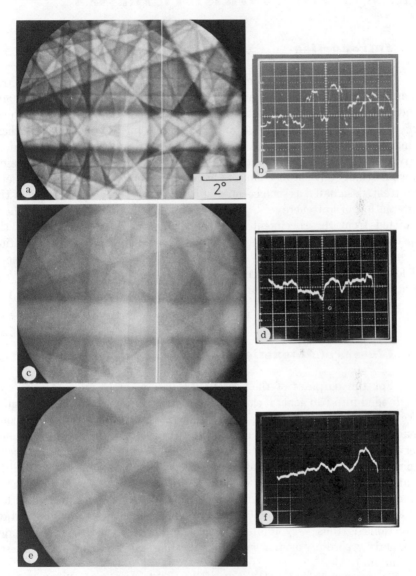

FIGURE 20. Effect of surface condition on ECP quality. Specimen: electro-polished niobium. (a,b) As-polished; (c,d) same crystal after deposition of 100 nm (1000 Å) of carbon by evaporation; (e,f) same crystal after anodizing to 75 nm (750 Å) of oxide. Note degradation of ECP quality due to the presence of a surface layer.

III. MAGNETIC CONTRAST IN THE SEM

A. Introduction

The magnetic field associated with certain materials, such as ferro-magnetic crystals and magnetic tapes, can affect the interaction of high-energy primary electrons with these materials or the resulting products of the interaction. In the scanning electron microscope (SEM), these magnetic effects can be used to produce image contrast between regions having differing magnetization directions known as magnetic domains. Two distinct contrast mechanisms have been recognized. Type I magnetic contrast, designated as such because of its order of discovery, arises from the interaction of low-energy (<50 eV) secondary electrons, which have been emitted from the specimen, with the *external* leakage fields above the specimen surface. Type II magnetic contrast results from the interaction of high-energy primary electrons from the beam with the *internal* magnetic field of the specimen. This section will discuss the characteristics of types I and II magnetic contrast and the types of magnetic materials that can be studied with each form of contrast.

B. Classes of Magnetic Materials

For the purposes of the present discussion, magnetic materials will be divided into two general classes, illustrated in Figure 21. Class I (Figure 21a) contains those materials that have strong leakage or demagnetizing magnetic fields outside the surfaces of the specimen. Such external fields serve to maintain the magnetic flux continuity between the bulk magnetic domains. Examples of materials in this class include the uniaxial crystalline materials cobalt and yttrium orthoferrite ($YFeO_3$). In such materials, the magnetization vector lies parallel or antiparallel to a low-index direction of low multiplicity, such as [0001] in cobalt. The choice of magnetization directions is thus very limited in such materials. Class I also includes magnetic recording media, such as recording tapes coated with iron or chromium oxide.

The materials in class II are characterized by having negligible leakage magnetic fields outside the specimen. Flux path closure can be realized entirely within the specimen due to the formation at the surface of domains whose magnetization is parallel to the surface (Figure 21b.) Materials in this class, which are said to show cubic anisotropy, include iron and nickel. The magnetization can lie along any $\langle 100 \rangle$ direction, and the availability

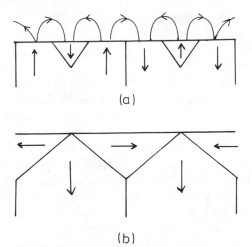

(a)

(b)

FIGURE 21.   Schematic illustration of two types of magnetic materials: (a) Uniaxial materials with a strong leakage (demagnetizing) field above the
the surface. Examples: cobalt, $YFeO_3$, magnetoplumbite. (b)
Materials of cubic anisotropy; surface closure domains prevent the
formation of significant leakage fields. Examples: iron, nickel.

of these directions due to high multiplicity allows the formation of closure
domains on most surfaces.

C. Type I Magnetic Contrast

The first observation of magnetic contrast in a scanning electron beam
instrument was made by Dorsey.[29] Dorsey examined magnetic tape upon
which a waveform had been recorded. When using the secondary electron
signal and with the specimen rotated properly relative to the detector, an
image with the periodic nature of the recorded wave was obtained. Dorsey's
work has been extended to the imaging of magnetic domains in crystalline
materials by more recent work.[20,21]

A simple illustration of the mechanism of this magnetic contrast,
which has become known as type I contrast, is shown in Figure 22a,
and a typical SEM image of type I contrast is shown in Figure 22b.
The conventional secondary electron detector, usually of the Everhart–
Thornley type, attracts secondary electrons with a positive ($\sim +250$ V)
potential on a Faraday cage (Figure 22a). The trajectories of the low-
energy secondary electrons are greatly perturbed by this collection voltage,
and the paths of flight can be extremely curved. The detector thus has a
large solid angle of collection at the specimen for secondary electrons. If

a magnetic field of magnetization **B** exists above the specimen, the electron passing through such a field has a force exerted on it, the Lorentz force **F**:

$$\mathbf{F} = -|e|\ \mathbf{v} \times \mathbf{B} \tag{7}$$

where **v** is the electron velocity and e is the charge. The acceleration from this force deflects the electron trajectory from the path it would take without the presence of a magnetic field. By proper arrangement of the specimen relative to the detector, the additional magnetic deflection can cause some secondary electrons that normally would be collected to be lost. Secondary electrons emerging from an adjacent domain of opposite magnetization are deflected in the opposite manner, i.e., toward the detector. A difference in the number of secondary electrons detected at the electron collector is thus observed between the two domains. The number of secondary electrons emitted from the sample is the same in the adjacent domains. The contrast arises because of effects on the electron trajectories after they have left the specimen. The usual cosine distribution of secondary electrons produced at normal beam incidence is tilted alternatively in opposite domains (Figure 22). Type I magnetic contrast is thus a form of *trajectory* contrast, in which the effects on the electrons are external to the specimen, as compared with *number* contrast, in which different numbers of electrons leave the specimen at different places.

The primary backscattered electrons, most of which have energies of the order of >20 keV (30 keV incident beam energy), are not sufficiently deflected by passing through the leakage fields to affect collection with the ET detector. Hence no significant contrast is obtained with primary electrons. Since equal numbers of electrons leave the specimen at all points, no contrast is obtained when the absorbed or specimen current signal is used.

Type I magnetic contrast is dependent on the directionality of collection of secondary electrons with the Everhart–Thornley detector. If all the secondary electrons were collected, no contrast could be obtained. The contrast mechanism depends on some of the secondary electrons escaping collection. The actual magnitude of the contrast is strongly dependent on this directionality. Several workers have succeeded in increasing the magnitude of the contrast by increasing the directionality of the detector[21-23]

FIGURE 22.   (a) Schematic illustration of the type I magnetic contrast mechanism. The leakage field above the specimen surface deflects the trajectories of secondary electrons. Proper rotation relative to the detector provides deflection toward the detector from domains of one magnetization and away from the detector in domains of opposite magnetization. (b) An example of domains observed in type I magnetic materials.

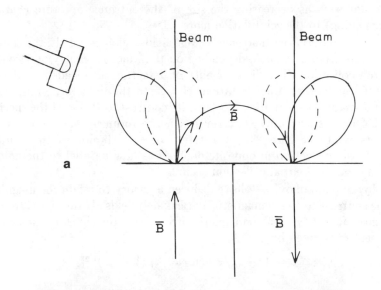

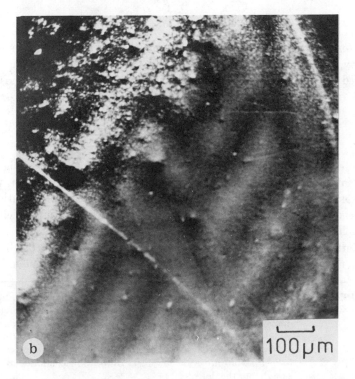

by means such as decreasing the size of the entrance aperture (Faraday cage opening) to the scintillation material.

For a material with strong leakage fields, such as cobalt, the contrast may range from 2 to 20%, depending on the exact geometry of secondary electron collection. The directionality of detection also leads to a dependence of the contrast upon rotation relative to the detector. Thus, with a set of domains in alternating black–white contrast, rotation of the specimen by 180° causes the contrast to reverse, i.e., a formerly black domain now appears white (Figure 23). A rotation of 90° eliminates the contrast, because the deflection in opposite domains is now parallel to the detector face and no differential collection occurs.

Joy and Jakubovics[24] have deduced a theory to relate the magnitude of the contrast to the demagnetizing (leakage) fields. If the signal from the specimen i is unity with no magnetic effects, then the signal in the presence of a field H is given by

$$i = 1 - (2/\pi)[\sin^{-1}\mu + \mu(1 - \mu^2)^{1/2}] \tag{8}$$

where

$$\mu = \frac{|e|}{mv} \int_0^\infty H_x \, dz \tag{9}$$

and e/m is the charge-to-mass ratio for the electron, v is the electron velocity, and H_x is the component of the field parallel to the surface and resolved along a line normal to the specimen–detector axis. For most cases, $\mu \ll 1$ and equation (8) reduces to

$$i = 1 - \frac{4\mu}{\pi} = 1 - \frac{4|e|}{\pi mv} \int_0^\infty H_x \, dz \tag{10}$$

The integral is evaluated along the trajectory of the electron flight from the point where it emerges to the detector. Note that the electron velocity v appears in the denominator of the multiplicative factor in equation (10), and hence a low velocity, such as that obtained with secondary electrons, results in a large change in i from unity.

Fathers, Joy, and Jakubovics have given an example of the use of equation (10) in the magnitude of contrast from bubble domains.[25] From

---

FIGURE 23.   Effect of specimen rotation on type I magnetic contrast. Specimen: magnetic recording tape with a recorded square wave. (a) Most domains are arranged for enhanced secondary electron collection, giving white (high signal) against the background. (b) A 180° rotation about the specimen normal results in contrast reversal; formerly white areas are now dark. In (b) the secondaries are deflected away from the detector.

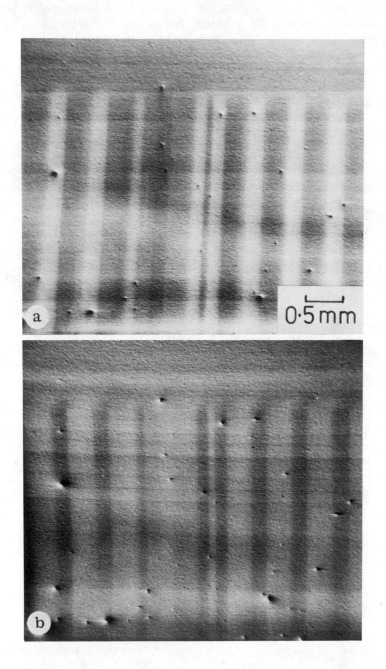

an expression for the magnetic scalar potential, the field integral, equation (10), can be determined. By examination of a plot of this function (Figure 24), the characteristics of the contrast can be predicted. As the domain is approached along the detector–specimen axis, the contrast rises to a maximum at the domain boundary. As the domain is crossed, the contrast falls to zero at the middle, reverses, and rises to a negative maximum at the opposite domain boundary, decreasing away from the domain. The contrast behavior predicted by this theory agrees well with that observed in practice (Figure 25).

Type I magnetic contrast arises solely from effects on secondary electrons. The numerical distribution of the energies of secondary electrons is virtually independent of the energy of the incident primary electrons, at least over the range of accelerating voltages usually used in SEM work, i.e., 10–50 kV. Hence there is virtually no dependence of the magnitude of type I contrast on the accelerating voltage. At very low accelerating voltages, i.e., less than 3 kV, the production of secondaries per incident primary electron increases (see Chapter III). This affects the signal i, i.e., the constant number of secondary electrons emitted, but not the contrast. An improved signal-to-noise ratio is achieved at low acceleration voltages despite the decrease in gun brightness (Chapter II), thus improving the image.

Examination of an image obtained with type I magnetic contrast (Figure 22b) reveals that the sharp change in magnetization which occurs at a domain wall (usually less than 100 nm wide) does not produce a sharp contrast change. The image of the wall is basically diffuse and poorly defined over a distance of several micrometers or more. This is a result of the lack of a sharply defined leakage field above the surface; the flux lines of the leakage field do not change abruptly at a domain wall. The limiting resolution in type I magnetic contrast is thus not a function of the size of the electron beam. The resolution limit is related to an inherent feature of the contrast mechanism, namely the diffuse nature of the demagnetizing fields.

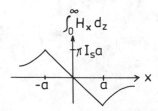

FIGURE 24.   Calculation of type I magnetic contrast for a bubble domain (example courtesy of Dr. D. C. Joy, Oxford University).

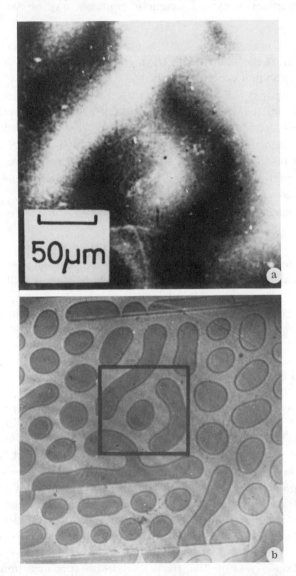

FIGURE 25.   (a) Bubble domain in yttrium orthoferrite observed with type I magnetic contrast. Note the contrast change from black to white across the domain diameter as predicted by the calculation in Figure 24. (b) The same region as (a) observed with a magneto-optical technique. (Example courtesy of Dr. D. C. Joy, Oxford University.)

Characteristics of type I magnetic contrast may be summarized as follows:

1. Type I magnetic contrast is obtained with secondary electrons only. No contrast is obtained with the high-energy primary (backscattered) electrons or specimen current signals.

2. The contrast is a form of pure trajectory contrast, arising from the deflection of emitted secondary electrons by leakage magnetic fields external to the specimen.

3. The magnitude of the contrast can be as high as $dS/S = 20\%$ and is dependent on the geometry of secondary electron collection.

4. The contrast is dependent on specimen (magnetization direction) rotation relative to the electron detector.

5. The magnetization vector must be out of the specimen (leakage fields).

6. The contrast is independent of accelerating voltage.

7. The resolution limit is dependent on the contrast mechanism; the limit is typically several micrometers due to the diffuseness of the leakage fields.

8. The contrast across the domain is a function of the field integral and is nonuniform with respect to position.

D. Type II Magnetic Contrast

The observation of contrast from magnetic domains in a material from class II, i.e., those materials characterized as having negligible leakage fields, was made by Philibert and Tixier,[26] who observed fir-tree domains in iron. The nature of the mechanism of this contrast, which arises from the interactions of the primary electrons with the *internal* magnetic fields, has recently been explained.[27-30] The mechanism of type II contrast is illustrated in Figure 26. As the primary electrons interact with the atoms of the solid, scattering elastically through large angles and losing energy (Chapter III), the presence of an internal magnetic field in the crystal causes the electrons to be deflected due to the action of the Lorentz force, equation (7). For the particular beam–specimen magnetization arrangement given in Figure 26, the magnetic deflection in alternate domains causes the electron backscatter coefficient to be alternately higher and lower than in the case of no magnetic effect. The domains thus appear in light–dark contrast due to differences in the backscatter coefficient. The specific geometric conditions that must be fulfilled are: (1) the specimen must be tilted relative to the beam, and (2) the magnetization vector

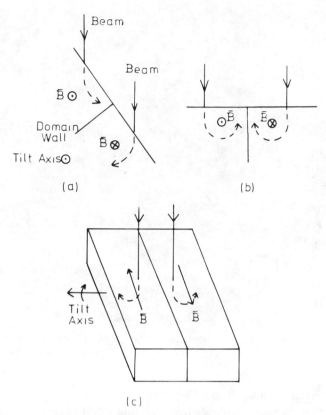

FIGURE 26. The mechanism of type II magnetic contrast: (a) Correct conditions: high tilt (55°) and magnetization parallel to the tilt axis. The Lorentz force acting on the beam electrons brings them closer to the surface in one domain and deeper into the specimen in a domain of opposite magnetization, producing a difference in the backscattering co-efficient. (b) Incorrect conditions: at normal incidence, the cyclotron effect of the Lorentz force does not cause a difference in depth of the beam electrons in domains of opposite magnetization. Clockwise or counterclockwise rotation occurs. (c) Incorrect conditions: high tilt but magnetization perpendicular to the tilt axis. The cyclotron action does not cause a difference in depth, only rotation in a clockwise or counterclockwise sense in domains of opposite magnetization.

must lie parallel to the tilt axis. Under these conditions (Figure 26a), the "cyclotron action" of the magnetic field tends to bring the electrons closer to the surface for one magnetization direction and farther from the surface for the opposite magnetization. The contrast that is obtained under these conditions is illustrated in Figure 27.

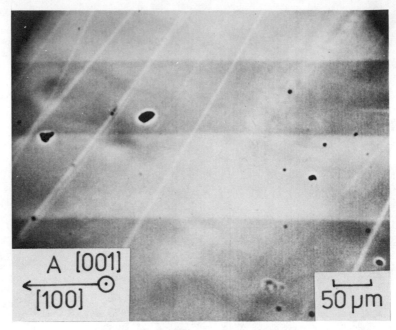

FIGURE 27.   An example of domains in Fe-3.22% Si observed with type II
magnetic contrast. Crystallographic directions and tilt axis indicated.

The contrast results from differences in the backscatter coefficients or
the numbers of escaping electrons from the oppositely magnetized domains.
Hence it is a form of pure *number* contrast. Since actual differences in the
number of electrons leaving the specimen occur, the requirement of current
conservation (Kirchhoff's current theorem) results in differences in the
absorbed or specimen current in the domains. Type II magnetic contrast is
thus obtained with the backscattered electron and absorbed current
signals. The secondary electrons, emitted very close to the surface, are
not significantly affected by the internal magnetic fields; hence, no significant
type II contrast is obtained with the secondary electron signal.

For normal beam incidence (i.e., 0° tilt), the cyclotron action of the
magnetic field does not cause a differential change in depth of electron
penetration in domains of opposite magnetization (Figure 26b). Hence,
there is no difference in the backscatter coefficient between domains and no
contrast results. The contrast is thus dependent on the tilt relative to the
beam. The form of this dependence is given in Figure 28, where experimental
measurements, Monte Carlo electron trajectory calculations, and analytical
calculations are compared.[28-30] The contrast increases with increasing tilt

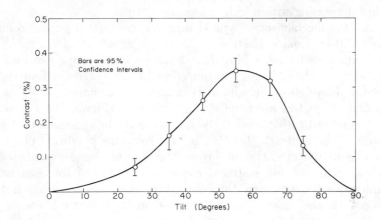

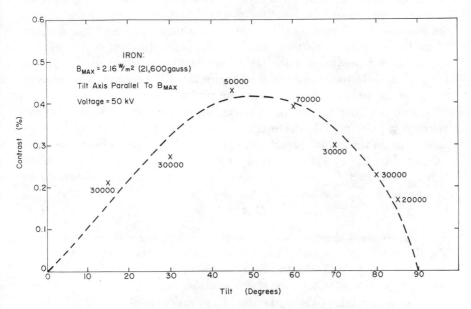

FIGURE 28. (a) Type II magnetic contrast measured for Fe–3.22% Si, $E_0 =$ 30 keV, as a function of tilt. The specimen current signal was used for the measurements. (b) Calculation of type II magnetic contrast as a function of tilt with the Monte Carlo electron trajectory modeling technique. The numbers at each point indicate the number of electron trajectories calculated.

away from normal incidence, reaching a maximum for a tilt of $\theta \simeq 55°$, and then decreasing with a further increase in tilt.

Since the mechanism of type II magnetic contrast depends on interactions of the primary electrons with the internal magnetic field, the magnetization vector must lie within, or have a large component resolved in, the surface plane. The cyclotron effect of the magnetic field causes differential depth effects in oppositely magnetized domains, providing the magnetization vectors are parallel to the tilt axis (Figure 26a). Rotation of the specimen by 180° around the normal from the condition of Figure 26a reverses the contrast. Rotation by 90° from the condition of Figure 26a results in the condition of Figure 26c. In this case, no differential change in depth of the scattered electrons is obtained for domains of opposite magnetization and the contrast is zero. The contrast is thus dependent on the rotation of the magnetization vector relative to the beam, the contrast being a maximum when the magnetization is parallel to the specimen tilt axis. The effects on the observed contrast from a fir-tree domain pattern as a function of rotation about the specimen normal are illustrated in Figure 29. The contrast is a maximum for those sections of the fir-tree domain pattern that have the magnetization vector parallel to the tilt axis and zero when the magnetization is perpendicular to the tilt axis. At intermediate rotations (Figure 29d), the components resolved along the tilt axis result in intermediate contrast levels.

Since the Lorentz force on the electrons is proportional to the electron velocity [equation (7)], the magnitude of type II contrast is strongly dependent on the electron accelerating potential. The dependence has been calculated by Monte Carlo electron trajectory techniques, and the relation between contrast C and accelerating voltage E_0 has been found to be (Figure 30)

$$C \propto E_0^{1.44} \tag{11}$$

Thus the contrast increases sharply as the accelerating voltage is increased. The Lorentz force is also linearly dependent on the magnitude of the magnetization B, and Monte Carlo calculations confirm a linear dependence of type II contrast on the magnetization.

The magnitude of the contrast with optimal available SEM conditions (30 kV accelerating potential—the maximum often available—specimen tilt 55°, and magnetization vector parallel to the tilt axis) is found to be only $C = 0.3\%$ for iron, which has a saturation magnetization of 21,000 G. This extremely low contrast value results in a very high threshold current, typically 5×10^{-8} A for a 10-sec frame speed, as explained in Chapter IV. Moreover, the level of contrast is far below the minimum detectable value for the human eye, and therefore a high value of differential amplification ("black level") must be used to enhance the contrast to an acceptable

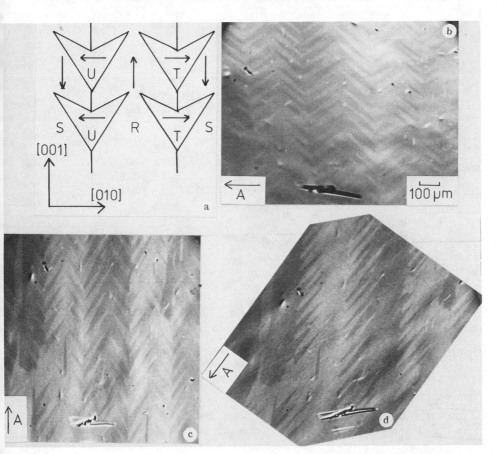

FIGURE 29. Effects of rotation about the specimen normal on type II magnetic contrast. Specimen: Fe single crystal; beam: 30 keV, 200 nA; signal: emissive with black level suppression. Tilt: 55°. (a) Schematic illustration of magnetization directions in the "fir tree" structure. (b–d) Rotation about the specimen normal. The tilt axis and crystal directions are indicated. Note that the domains are only in contrast when the magnetization is parallel to the tilt axis.

level. The lower level of contrast obtained with materials of lower magnetization, such as nickel, necessitates working at higher accelerating voltages to increase the contrast, via the voltage effect, to an acceptable level. In nickel, the contrast is not visible at 30 kV, but an increase to 50 kV accelerating potential provides satisfactory domain images (Figure 31).

One means to increase contrast is by energy filtering the collected primary signal. Monte Carlo calculations indicated that most of the electrons responsible for the contrast had lost only between 0 and 20% of their initial energy.[28] Experimental evidence substantiating this calcula-

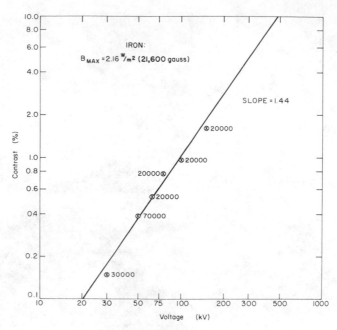

FIGURE 30.   Monte Carlo calculations of type II magnetic contrast as a function
of accelerating voltage. The magnetization is parallel to the tilt axis
and the specimen is tilted at 60°.

tion was obtained as well.[29] Therefore, if materials yielding low type II
contrast must be examined, raising the primary beam voltage and collecting only those scattered electrons having lost 0–20% of the original energy
should markedly increase the observed contrast.

Examination of a high-magnification image of domains obtained
with type II magnetic contrast (Figure 32) reveals that the domain
boundary is much more sharply defined than with type I contrast. The
magnetization vector is constant everywhere throughout a domain and
only changes as a function of position within the domain wall, which is of
the order of 100 nm or less. The resolution of a wall position is thus not
limited by the magnetic effects, at least above 100 nm, as in type I, but is
limited by the probe size or interaction volume. This can, however, be a
serious limitation because of the high threshold current necessary to discern
the weak type II contrast. A calculation of the probe size with the brightness
equation (Chapter IV) for type II contrast when a tungsten thermionic
electron gun is used results in a value of approximately 400 nm, which is
observed in rise time measurements across domain walls (Figure 33). This
value can be reduced by working at higher accelerating voltages, which
reduces the threshold current by increasing the contrast, or by energy

filtering to increase the contrast to achieve the same result. Beam spreading effects in the specimen can also limit the resolution. Energy filtering should, in principle, reduce the effects of beam spreading.

Characterization of type II magnetic contrast may be summarized as follows:

1. Type II magnetic contrast arises from the effect of the internal magnetic field on primary electrons. The contrast is pure number contrast.

2. Type II contrast is obtained with the primary (backscattered) electron and specimen current signals. No appreciable contrast is obtained with secondary electrons.

3. The magnetization vector must lie in the plane of the specimen or have a significant component resolved in that plane.

4. The contrast is strongly tilt dependent, with a maximum at 55° tilt relative to the beam.

5. The contrast is dependent on specimen magnetization rotation

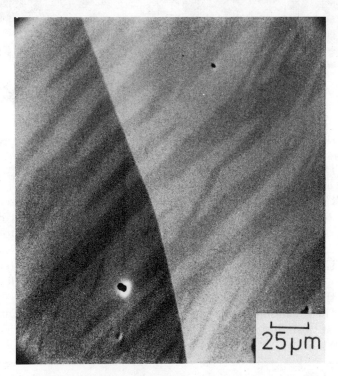

FIGURE 31.   Magnetic domains observed in nickel at 50 keV incident energy and a beam current of 200 nA. A subgrain boundary appears in electron channeling contrast; the domains cross the low-angle boundary without significant disruption.

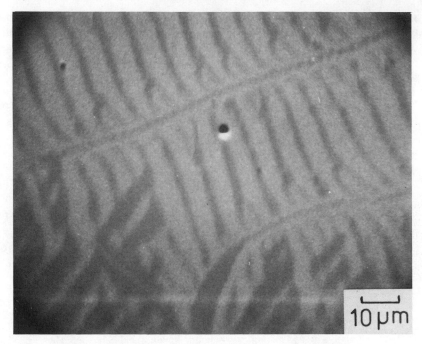

FIGURE 32.   High-resolution image of magnetic domains in Fe–3.22% Si. Domain
                       widths of 1 μm or less in width can be seen. Beam: 30 keV, 100 nA.

relative to the beam; maximum contrast occurs for the magnetization
vector parallel to the tilt axis.

6. The contrast is dependent on the electron accelerating voltage,
varying as $C \propto E_0^{1.44}$.

7. The contrast is "weak," e.g., iron at 30 kV, 55° tilt, magnetization
parallel to the tilt axis produces only 0.3% contrast between domains.
High threshold currents and high black level must be used to observe
type II contrast.

8. The contrast is carried by the highest-energy backscattered electrons. An improvement can be made in the contrast by energy filtering to
remove the low-energy backscattered electrons.

9. The resolution limit is determined by the probe size or interaction
volume, at least down to the size of the domain wall thickness, which is less
than 100 nm.

FIGURE 33.   (a) High-resolution image of a stripe domain in Fe–3.22% Si.
                       (b) Intensity profile across the domain boundary indicates a beam
                       size of approximately 0.5 μm. Beam. 30 keV, 100 nA.

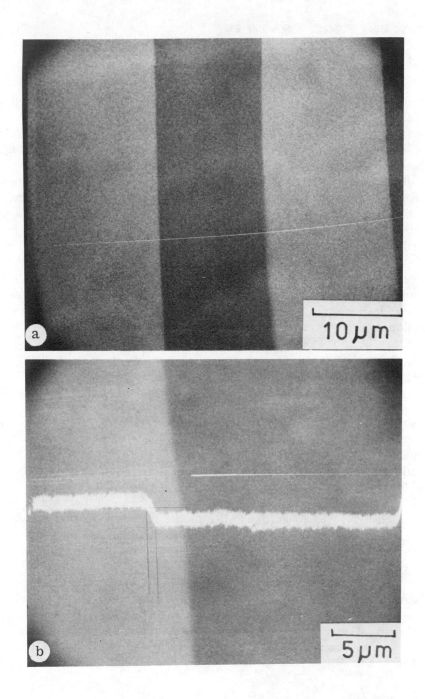

IV. VOLTAGE CONTRAST

The trajectories of secondary electrons having energies in the range $0 < E \leq 50$ eV are sensitive to the effects of surface potentials and electric field gradients near the specimen surface. For example, if a specimen is placed at a positive potential of 5 V with respect to ground, then there is a very high probability that secondary electrons with less than 5 eV energy will not arrive at the Everhart–Thornley detector; these electrons will not escape the specimen. On the other hand, a negative bias of a few volts will act as a booster for the escape of secondary electrons; hence more of these electrons are likely to arrive at the detector. For example, in an integrated circuit device, where portions of the specimen can be made more negative than the rest, we would expect to see the negative regions in bright contrast with respect to the regions at positive potential since more signal is collected from such regions. Such contrast is commonly termed voltage contrast. Clearly, the contrast arises from the variation in number of secondary electrons arriving at the detector. Equal numbers of secondary electrons leave the specimen at each point, but the number collected varies from different points due to trajectory effects. Voltage contrast is a type of pure trajectory contrast. Hence no contrast effects are observed with the specimen current signal when the usual voltages are applied, e.g., <10 V.

Voltage contrast is useful for direct examination of semiconductor device circuits since the effects of specimen biasing can be observed directly. Furthermore, the technique is useful for failure analysis; failed portions of integrated circuits can often be identified quickly and the source of the difficulty traced with SEM. This mode of examination is extremely useful, and therefore many attempts have been made to improve the sensitivity of the technique, i.e., to observe the effects of very small surface potential differences. Perhaps the earliest attempt was to place a slotted shield before the detector so as to maximize the probability of collecting secondary electrons whose trajectories are close to one another at the expense of other trajectories; secondary electrons having small differences in energy have similar trajectories.[32,33]

The absolute measurement of potential is, however, still in the speculative stage. The reason for this is that the trajectories of the secondary electrons of interest are also extremely sensitive to transverse electric field effects.[34] Therefore, while the presence of a uniform field may not affect the determination of relative potentials, it can seriously disturb attempts to make absolute measurements.[34] The effect of transverse fields can be reduced by applying a high accelerating field normal to the specimen surface, but this is not always practical.[35] A transverse field at the specimen surface causes secondary electrons to accelerate in a direction parallel to the specimen surface. The angular deflection of -4-eV secondaries is about 3.5 V-deg due to such transverse fields.[33] Now in an actual integrated circuit, peak fields of 10^4 V/cm are common, so that in certain specimen–detector configurations, secondary electrons from the specimen may miss the Everhart–Thornley detector altogether due to the presence of the transverse field.[33] Therefore attempts to carry relative potential detection to more sensitive limits involve special detectors designed to collect all secondary electrons[34,36,37] and energy-analyze.

Figure 34 shows an example of voltage contrast observed from a practical semiconductor device. The device is a metal–oxide–semiconductor–transistor (MOST) ladder. The electrodes beginning at the bottom

FIGURE 34. Voltage contrast image of a gate in a metal–oxide–semiconductor transistor (MOST) ladder. -5 V applied to gate with all other electrodes at ground potential.

are: source 1, gate 1, drain 1, source 2 (common), gate 2, and so forth for the four devices connected in series on the chip. Figure 34 also shows the effect of applying a −5-V potential to gate 1 (bottom) with all other electrodes grounded. Since negative potential enhances secondary electron collection from this regions, the electrode of gate 1 appears bright with respect to all other electrodes.[38]

A number of specialized detector systems meant to improve voltage contrast sensitivity have been constructed. The first of these was a concentric cylinder velocity analyzer positioned with the cylinder axis perpendicular to the symmetry plane of the specimen chamber.[34] This detector sorts the secondary electrons with respect to their velocities normal to the specimen surface. More important, their velocity distribution perpendicular to the symmetry plane, i.e., that affected by the transverse fields, can be determined. Voltages can be measured with an accuracy of ±500 mV or ±5%, whichever is larger, with this system.[34]

Other attempts to improve the measurement of small voltage differences were made by preparing special spherical screen or cylindrical detectors capable of collecting all secondaries, but also of sorting them with respect to energy at the collection point. This is done by biasing the electrodes that form the physical construction of the detector. Driver constructed a spherical system.[36] Fleming and Ward achieved sensitivities of 250 mV by merely providing a pair of biased screens between the specimen and the Everhart–Thornley detector.[39]

Banbury and Nixon described a cylindrical detector shown schematically in Figure 35.[37] This detector gave a 13% video signal intensity change per unit volt bias at the specimen near zero bias. Later investigations of cylindrical detectors showed that larger radius detectors provided better voltage sensitivity than smaller radius detectors. Measured sensitivity in the region −0.5 to +0.5 V at the specimen was 20% per volt with a 19 mm radius R.[40] A similar cylindrical detector constructed to maximize voltage sensitivity achieved a sensitivity of 33% per volt and had a log-linear response in the region −5 to +5 V at the specimen surface.[40]

The effect of surface contamination produced by the electron beam must be considered in any sensitive measurement of surface potentials. Contamination rasters produced by exposures of the order of 10^{-4} coul/cm² were observed to exhibit strong voltage contrast.[40] Hence the contamination material was a nonconductor, allowing the buildup of a surface charge. Such observations indicate that if the contamination layer and the specimen do not have the same surface properties, then voltage measurements may include effects having nothing to do with the surface of interest. Hence, for the most careful determinations of potential differences, some

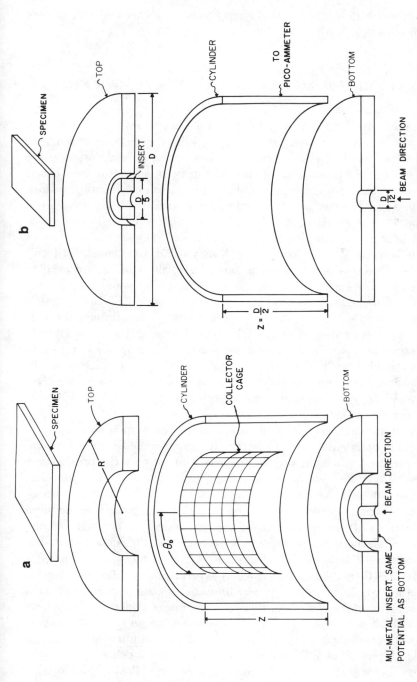

FIGURE 35. (a) Isometric cutaway view of cylindrical secondary electron detector of radius *R* and cylinder height *Z*. Each electrode is insulated from the others. (b) Isometric cutaway view of cyclindrical secondary electron detector designed for maximum voltage contrast sensitivity. Current changes on the cylinder of 33% per volt bias on the sample were observed in the range −1 to +1 V specimen bias.

means to eliminate contamination effects, such as a cold finger or oil-free vacuum system, is highly desirable.

V. ELECTRON-BEAM-INDUCED CURRENT (EBIC)

If a $p–n$ junction is placed in close proximity to a focused electron beam, the incident electrons create excess electron–hole pairs or carriers. The $p–n$ junction field collects these carriers during their diffusion in the specimen. Thus, a beam-induced current can be produced in an external circuit which contains the junction. This current can constitute the video signal in an SEM as well. If the carriers are generated further than a few diffusion lengths from the $p–n$ junction, recombination will take place with little or no contribution to the output signal. The zone of creation of electron–hole pairs was modeled by Kanaya and Okayama.[41] In this model, the beam is presumed to penetrate straight into the material to some depth. During penetration, the electrons suffer some mean energy loss and undergo elastic scattering. Then the electrons diffuse in all possible directions from this point. During diffusion, energy is lost through inelastic scattering, thus creating electron–hole pairs. Hence, in this simple model, the electron–hole pair creation zone is a sphere. While Monte Carlo simulation probably would give a theoretically more accurate picture of the creation zone, the Kanaya model can be used to predict, with good accuracy, such physical quantities as minority carrier lifetime and surface recombination velocity.[42]

The diffusion of p- or n-type impurities results in moderate to severe local plastic deformation due to lattice distortion caused by the high impurity concentration built up at the surface in forming the original $p–n$ junction. Therefore dislocations are often formed near the $p–n$ junction. The EBIC method can be used to image these dislocations; the contrast mechanism is probably due to enhanced recombination of the beam-induced excess carriers at the dislocation. Figure 36 shows such an image of dislocations. This method of examining lattice defects is, of course, limited to semiconductors due to the need for a field region within the specimen.[43]

The EBIC method can be used to experimentally infer considerable information about minority carrier diffusion lengths and lifetimes in a semiconductor device. To do this, one monitors emitter–base and/or collector–base junctions as the primary beam voltage in the SEM electron column is varied. Active regions in semiconductors are usually separated by oxide layers or may lie below aluminum interconnections. Hence the increased beam penetration with voltage serves to indicate which regions

FIGURE 36. Electron-beam-induced conductivity contrast (EBIC). Specimen: diffused square in silicon; crystallographic defects are visible. These are formed during boron diffusion. Enhanced recombination of electron–hole pairs takes place near the line defects, lowering the conductivity as compared to surrounding regions. (Courtesy of A. M. B. Shaw and G. R. Booker.)

are responsible for the observed EBIC effects; Figure 37* shows the emitter base in a transistor circuit (emitter grounded) examined with three different SEM primary voltages, 5, 10, and 20 kV, respectively. At 5 kV, the junction edges are bright since nearly all the excess carriers generated in the deflection region contribute to the electron-beam-induced current. As the SEM operating voltage is increased, penetration of the metallization layer becomes apparent and finally, at 20 kV, there is appreciable penetration into the base region. Quantitative evaluation of such information leads to determination of appropriate diffusion lengths and lifetimes in the device.[44]

---

*Figures 37 and 38 are color illustrations and will be found following p. 206.

The signal from an EBIC experiment can be derivative- or Y-modulation-processed. In the case of derivative processing, one is able to obtain more video information because the dc component of the $p–n$ junction can be eliminated. The Y-modulation processing can be useful since the displacement of the signal is proportional to the EBIC signal intensity. Hence regions of high induced current with respect to some datum are obvious. Furthermore, comparisons of micrographs taken of the same device under different primary SEM beam conditions may be facilitated by the Y-modulation processing.

VI. CATHODOLUMINESCENCE

In addition to electron scattering and x-ray production, electromagnetic radiation in the visible, infrared, and ultraviolet spectral regions may be emitted when an electron beam strikes a solid target. In certain types of specimens, the interaction of the high-energy electrons creates electron–hole pairs. If the electron–hole pairs recombine, energy is given off in the form of long-wavelength radiation. This radiation is commonly referred to as cathodoluminescence (CL). The CL signal can be obtained from a photomultiplier and used to modulate the brightness of the CRT so as to obtain an image of the emitting surface. Quantitative data can be obtained by recording the emitted spectral intensity as a function of wavelength or photon energy. The emission bandwidth for light emitted by electron transitions from excited states varies markedly with temperature, becoming sharper as the temperature is decreased.

A difficulty in studying CL excited by the electron beam in the scanning electron microscope is that of low intensity due to the relatively low beam power and small excited volume. Therefore a very efficient light collection system is needed to collect the maximum possible fraction of the emitted luminescence at the photomultiplier of the SEM in order to observe the relatively few photons emitted per unit time. One way of obtaining extremely efficient photon collection is to place half an ellipsoidal mirror so as to collect all light emitted from the specimen; a small hole permits the beam to pass so as to strike the specimen normally. The specimen is placed at one focus of the ellipsoidal mirror and the emitted light is collected by a lens located at the other focus. The light collected is collimated and enters a fiber optic bundle at normal incidence. These fiber optics conduct the light to the photomultiplier or entry slit of a monochromator[45–47] (see Fig. 4, Chapter IV). The collection efficiency of the ellipsoidal mirror system depends strongly on the vertical position of the specimen; hence intensity should be maximized by adjusting the specimen height before commencing experiments. Furthermore, when the beam is scanned over a raster of 200 μm or larger, there is a noticeable drop in intensity at the edges of the field of view.[46] In order to obtain enough intensity in any case, the current in the SEM beam must be approximately 1–10 nA, which results in the spot size ranging up to 0.5 μm; the CL spatial resolution is thus of this order.

Pfefferkorn and Blaschke have formulated a list of desirable features designed to allow optimal experimental exploitation of CL in the SEM[48]:

(1) a highly efficient photon collector system such as that just out-lined;

(2) an optical spectrometer coupled to a multichannel analyzer, e.g., such as is normally used with an energy-dispersive x-ray detector;

(3) easily interchangeable photomultipliers having different spectral response characteristics;

(4) equipment for the study of CL decay time analysis;

(5) a means to cool specimens to liquid nitrogen or, preferably, to liquid helium temperatures;

(6) a means to definitely exclude light from the primary electron source from reaching either the specimen or the collector;

(7) signal mixing facilities allowing the CL signal to be combined with, e.g., the secondary electron signal; and, perhaps,

(8) a system of filters to combine the output of the CL detector with a color CRT.

Obviously, providing all of these items represents a major commitment to the CL mode. However, for merely examining light-emitting regions of the specimen surface, only items 1 and 6 are necessary, although item 7 is also desirable. The next step would be to provide item 3. If one already has a cold stage, the collection system could be mounted on it, thus providing item 5.

In organic materials, CL is usually produced by small amounts of impurities and/or crystal lattice distortions.[49] The color observed is most often determined by the nature of the impurity. Hence abrupt changes in color may indicate changes of crystal structure or lattice de-formation.[50] In the case of materials examined in the electron probe microanalyzer, the visible light emitted can be seen directly with the built-in light microscope and the observed colors correlated with x-ray data, e.g., banding observed by CL in cassiterite specimens was shown to be due to compositional differences.[51] Carrying this one step further, a photomultiplier can be introduced in place of the ocular in the optical microscope. The output of this PM can then be made to modulate the CRT brightness. Such a system was used to indicate the distribution of nickel-coated sapphire (Al_2O_3) whiskers in a sapphire whisker–Al base matrix composite. Combining these CL results with x-ray area scans of Ni led to the conclusion that poor sapphire–matrix bonding occurred. This poor bonding was in turn responsible for unacceptable inconsistencies in the mechanical properties of the composite material.[52] Furthermore, the distribution of sapphire whiskers was shown to be poor, with the aid of the CL results. Figure 38 shows the whiskers in the composite as given by the CL signal. Figure 39 shows the distribution of ZrO_2 in the grain boundaries of a high-temperature alloy obtained by CL in the same way from the electron probe microanalyzer.[53]

Muir, Holt, and co-workers have discussed the problems associated with obtaining quantitative results from CL spectra.[54,55] The major question is how low a CL efficiency can be evaluated. In the case of strong CL emitters, impurities may be detected at lower levels than those obtainable with electron probe microanalysis; Wittry detected Te concentration variations in GaAs by CL, but not by microprobe techniques.[56] Work on biological material is possible, but to date no quantitative results have been published; qualitative spectra have been shown.[46] All work with nonconducting samples must be carefully evaluated; a 200-Å carbon coating reduces CL intensities by 50% at all wavelengths.[46] Hence one must trade off intensity loss and possible artifacts due to charging by attempting to operate with uncoated specimens at lower primary electron beam voltages and currents as opposed to coating the specimen.

Examples of the uses of CL in solving a variety of problems include differentiation of anatase and rutile in paints,[48] observation of ferroelectric domains in sodium niobate,[48] location of subsurface crazing in glacial sands and other sediments,[57] and use in forensic evaluation of glasses.[45]

Metals also exhibit an effect analogous to CL known as Lilienfield radiation. This radiation is caused largely by bremsstrahlung and surface plasmon effects.[58] Hörl and Mügschl have shown, by using an ellipsoidal mirror–optical pipe arrangement very similar to that described for CL, that light-emission micrographs from metal specimens can be obtained.[47] These authors claim that such light-emission micrographs may be useful in cases where other contrast mechanisms may be weak, where conventional CL is weak, as is often the case when transition metal impurities are present, and, finally, for obtaining images of strongly radioactive specimens.

 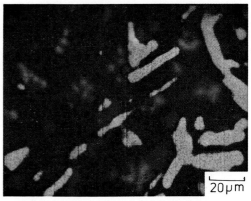

CHAPTER V, FIGURE 37. (left) Color composite of emitter base of a *p–n–p* transistor (2n2905A, ordinary amplifier, small switching device) taken in EBIC mode at three different SEM operating potentials. Blue represents the emitter base at 5 keV beam energy; green represents the emitter base at 10 keV; and red represents the emitter base at 20 keV beam energy. The effect of beam penetration on the active layers can be deduced from the colors: white region active at all three operating voltages; yellow active at 10 and 20 kV; cyan region active at 5 and 10 kV; red region active only at 20 kV. Absence of blue indicates that the region active at 5 kV is also active at 10 and 20 kV.

CHAPTER V,   FIGURE 38.
CHAPTER XI, FIGURE 9. (right) Composite map of scanning microprobe photographs of a composite material containing 15 vol % Ni-coated Al_2O_3 whiskers in a matrix of Al–10% Si alloy. The material is in the as-pressed condition. The yellow–green areas are the whiskers recorded with the cathodoluminescence signal. The red represents the Ni distribution. The pale blue represents the Si distribution. Note the nickel clumps (solid red blobs) at or very near the whisker edges. The remainder of the Ni is distributed throughout the matrix without regard for the presence of Si clumps. Color-addition methods were used to combine separate cathodoluminescence, Ni $K\alpha$ x-ray distribution, and Si $K\alpha$ x-ray distribution photographs of the same area. See Chapter XI for a more complete discussion of this specimen. (From Yakowitz, Jenkins, and Hahn.[13])

CHAPTER VI, FIGURE 7c. Color composite synthesized from Figs. 7a and 7b.

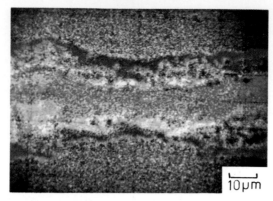

CHAPTER XI, FIGURE 8. Color composite synthesized from Figures 7 b–d: Ni—blue, Cr—green, Al—red.

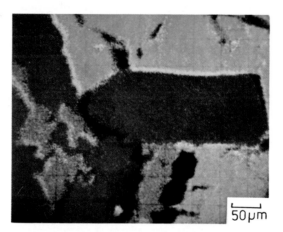

CHAPTER XI, FIGURE 10. Color composite showing elemental distribution for Pb (blue), S (green), and Si (red) in a galena ore. Compare with Figure 3 of Chapter XI.

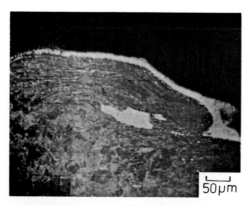

CHAPTER XI, FIGURE 11. Color composite showing undesired constitutent at interface of titanium–steel explosive weld—Fe (green) and Ti (red); the yellow is TiFe plus Ti$_2$Fe, brittle intermetallic compounds. The grain structure of the steel is also shown; note that the welding process has pulled the grains into long stringers. (From Weiss.[15])

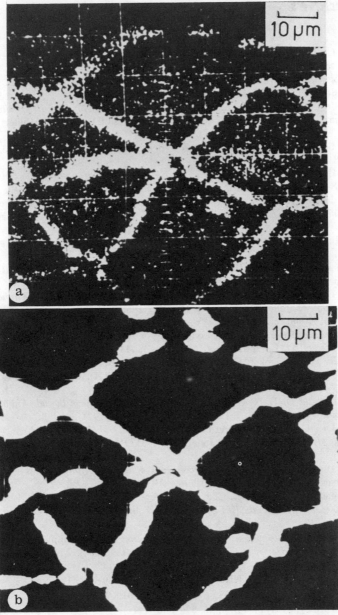

FIGURE 39.   X-ray area scan and cathodoluminescence images of a high-temperature alloy in which ZrO₂ has accumulated in the grain boundaries.
(a) X-ray area scan using Zr $L\alpha$ at 20 keV beam energy. (b)
Cathodoluminescence image of same region as (a).

REFERENCES

1. D. G. Coates, *Phil. Mag.*, 16, 1179 (1967).
2. C. G. van Essen and E. M. Schulson, *J. Mat. Sci.*, 4, 336 (1969).
3. C. G. van Essen, E. M. Schulson, and R. H. Donaghay, *J. Mat. Sci.*, 6, 213 (1971).
4. D. C. Joy and D. E. Newbury, *J. Mat. Sci.*, 7, 714 (1972).
5. D. C. Joy, D. E. Newbury, and P. M. Hazzledine, in *SEM/1972 Proceedings of the 5th Annual SEM Symposium* (O. Johari, ed.), IITRI, Chicago, Illinois (1972), p. 97.
6. G. R. Booker, A. M. B. Shaw, M. J. Whelan, and P. B. Hirsch, *Phil. Mag.*, 16, 1185 (1967).
7. P. B. Hirsch and C. J. Humphreys, in *SEM/1970 Proceedings of the 3rd Annual SEM Symposium* (O. Johari, ed.), IITRI, Chicago, Illinois (1970), p. 449.
8. D. C. Joy, G. R. Booker, E. O. Fearon, and M. Bevis, in *SEM/1971 Proceedings of the 4th Annual SEM Symposium* (O. Johari, ed.), IITRI, Chicago, Illinois (1971), p. 497.
9. C. S. Barrett and T. B. Massalski, *Structure of Metals*, McGraw-Hill, New York (1966).
10. G. Thomas, in *Modern Diffraction and Imaging Techniques in Material Science* (S. Amelinckx *et al.*, eds.), North-Holland, Amsterdam (1970), p. 159.
11. G. R. Booker and R. Stickler, in *SEM/1972 Proceedings of the 5th Annual SEM Symposium* (O. Johari, ed.), IITRI, Chicago, Illinois (1972), p. 225.
12. C. G. van Essen, in *Proceedings of the 25th Anniversary Meeting of EMAG, Institute of Physics*, Institute of Physics, London, (1971), p. 314.
13. D. C. Joy and D. E. Newbury, *J. Mat. Sci.*, 7, 714, (1972).
14. D. C. Joy, G. R. Booker, E. O. Fearon, and M. Bevis, in *SEM/1971 Proceedings of the 4th Annual SEM Symposium* (O. Johari, ed.), IITRI, Chicago, Illinois (1971), p. 497.
15. D. C. Joy and D. E. Newbury, in *SEM/1971 Proceedings of the 4th Annual SEM Symposium* (O. Johari, ed.), IITRI, Chicago, Illinois (1971), p. 113.
16. R. Stickler, C. W. Hughes, and G. R. Booker, in *SEM/1971 Proceedings of the 4th Annual SEM Symposium* (O. Johari, ed.), IITRI, Chicago, Illinois (1971), p. 473.
17. D. L. Davidson, in *SEM/1974 Proceedings of the 7th Annual SEM Symposium* (O. Johari, ed.), IITRI, Chicago, Illinois (1974), p. 927.
18. J. P. Spencer, G. R. Booker, C. J. Humphreys, and D. C. Joy, in *SEM/1974 Proceedings of the 7th Annual SEM Symposium* (O. Johari, ed.), IITRI, Chicago, Illinois (1974), p. 919.
19. J. R. Dorsey, in *Advances in Electrons and Electron Physics*, Supplement 6, *Electron Probe Microanalysis* (A. J. Tousimis and L. Marton, eds.), Academic Press, New York (1969), pp. 291–312.
20. J. R. Banbury and W. C. Nixon, *J. Sci. Instr.*, 44, 889–892 (1967).
21. D. C. Joy and J. P. Jakubovics, *Phil. Mag.*, 17, 61–69 (1968).
22. G. A. Wardley, *J. Appl. Phys.*, 42, 376–386 (1971).
23. G. W. Kamlott, *J. Appl. Phys.*, 42, 5156–5160)1971).
24. D. C. Joy and J. P. Jakubovics, *J. Phys. D*, 2, 1367–1372 (1969).
25. D. J. Fathers, D. C. Joy, and J. P. Jakubovics, in *SEM/1973 Proceedings of the 6th Annual SEM Symposium* (O. Johari, ed.), IITRI, Chicago, Illinois (1973), p. 259.

26. J. Philibert and R. Tixier, *Micron*, 1, 174 (1969).

27. D. J. Fathers, J. P. Jakubovics, and D. C. Joy, *Phil. Mag.*, 27, 765 (1973).

28. D. E. Newbury, H. Yakowitz, and R. L. Myklebust, *Appl. Phys. Lett.*, 23, 488 (1973).

29. D. J. Fathers, J. P. Jakubovics, D. C. Joy, D. E. Newbury, and H. Yakowitz, *phys. stat. sol. a*, 20, 535 (1973).

30. D. J. Fathers, J. P. Jakubovics, D. C.· Joy, D. E. Newbury, and H. Yakowitz, *phys. stat. sol. a*, 22, 609 (1974).

31. D. E. Newbury, H. Yakowitz, and N. Yew, *Appl. Phys. Lett.*. 24, 98 (1974).

32. T. E. Everhart, Ph.D. Dissertation, Cambridge University (1958), Chapter 7.

33. T. E. Everhart, in *SEM/1968 Proceedings of the 1st Annual SEM Symposium* (O. Johari, ed.), IITRI, Chicago, Illinois (1968), p. 1.

34. O. C. Wells, in *SEM/1969 Proceedings of the 2nd Annual SEM Symposium* (O. Johari, ed.), IITRI, Chicago, Illinois (1969), p. 397.

35. G. S. Plows, Ph.D. Dissertation, Cambridge University (1969).

36. M. C. Driver, in *SEM/1969 Proceedings of the 2nd Annual SEM Symposium*, (O. Johari, ed.), IITRI, Chicago, Illinois (1969), p. 403.

37. J. R. Banbury and W. C. Nixon, in *SEM/1970 Proceedings of the 3rd Annual SEM Symposium*, (O. Johari, ed.), IITRI, Chicago, Illinois (1970), p. 473.

38. G. S. Plows and W. C. Nixon, *J. Sci. Instr.*, 1, 595 (1968).

39. J. P. Fleming and E. W. Ward, in *SEM/1970 Proceedings of the 3rd Annual SEM Symposium* (O. Johari, ed.), IITRI, Chicago, Illinois (1970), p. 465.

40. H. Yakowitz, J. P. Ballantyne, E. Munro, and W. C. Nixon, in *SEM/1972 Proceedings of the 5th Annual SEM Symposium* (O. Johari, ed.), IITRI, Chicago, Illinois (1972), p. 33.

41. K. Kanaya and S. Okayama, *J. Phys. D*, 5, 43 (1972).

42. J. F. Bresse, in *SEM/1972 Proceedings of the 5th Annual SEM Symposium* (O. Johari, ed.), IITRI, Chicago, Illinois (1972), p. 106.

43. W. Czaja and J. R. Patel, *J. Appl. Phys.*, 36, 1476 (1965).

44. N. C. Macdonald and T. E. Everhart, *Appl. Phys. Lett.*, 7, 267 (1965).

45. L. Carlsson and C. G. van Essen, *J. Phys. E.* 7(2), 98 (1974).

46. M. D. Muir and D. B. Holt, in *SEM/1974 Proceedings of the 7th Annual SEM Symposium* (O. Johari, ed.), IITRI, Chicago, Illinois (1974), p. 135.

47. E. M. Hörl and E. Mügschl, in *Electron Microscope 1972, Proceedings of the 5th European Congress of Electron Microscopy*, Institute of Physics, London (1972), p. 502.

48. G. Pfefferkorn and R. Blaschke, in *SEM/1974 Proceedings of the 7th Annual SEM Symposium* (O. Johari, ed.), IITRI, Chicago, Illinois (1974), p. 143.

49. G. F. J. Garlick, in *Luminescence of Inorganic Solids* (A. Goldberg, ed.), Academic Press, New York (1966), p. 685.

50. D. B. Holt and M. Culpan, *J. Mat. Sci.*, 5, 546 (1970).

51. G. Reymond, S. Kimoto, and H. Okuzumi, in *SEM/1970 Proceedings of the 3rd Annual SEM Symposium* (O. Johari, ed.), IITRI, Chicago, Illinois (1970), p. 33.

52. H. Yakowitz, W. D. Jenkins, and H. Hahn. *J. Res. NBS*, 72A, 269 (1970).

53. K. F. J. Heinrich, D. E. Newbury, and H. Yakowitz, in *Proceedings of the 1973 Army Materials Research Conference at Sagamore, New York*, in press.

54. M. D. Muir and P. R. Grant, in *Quantitative Scanning Electron Microscopy* (D. B. Holt, M. D. Muir, P. R. Grant, and I. M. Boswarva, eds.), Accademic Press, New York (1974).

55. D. B. Holt, in *Quantitative Scanning Electron Microscopy* (D. B. Holt, M. D. Muir, P. R. Grant, and I. M. Boswarva, eds.), Academic Press, New York (1974).
56. D. B. Wittry, *Appl. Phys. Lett.*, **8**, 142 (1966).
57. D. H. Krinsley and P. W. Hyde, in *SEM/1971 Proceedings of the 4th Annual SEM Symposium* (O. Johari, ed.), IITRI, Chicago, Illinois (1971), p. 409.
58. H. Boersch, C. Radeloff, and G. Sauerbrey, *Z. Physik*, **165**, 464 (1961).

VI

SPECIMEN PREPARATION, SPECIAL TECHNIQUES, AND APPLICATIONS OF THE SCANNING ELECTRON MICROSCOPE

D. E. Newbury and H. Yakowitz

In the first ten years of commercial availability of the scanning electron microscope (SEM), the instrument has been applied in many fields in the physical and biological sciences. In this chapter, we shall present illustrative examples of the types of problems in materials science for which the SEM can provide useful information to the analyst. By no means are these examples a complete description of the scope of SEM applications. A more adequate appreciation of the wide range of SEM studies that have been carried out can be obtained by consulting the proceedings of the yearly conferences on scanning electron microscopy.[1,2]

Before describing particular examples of SEM applications, it is appropriate to consider certain aspects of generalized SEM techniques that are fundamental to many studies. Specimen preparation, stereomicroscopy, and techniques for dynamic experiments will be described. Following these general techniques, examples of specific techniques emphasizing various features of the SEM, including resolution, depth of focus, and contrast mechanisms, will be given.

D. E. NEWBURY and H. YAKOWITZ—Institute for Materials Research, National Bureau of Standards, Washington, D.C.

I. SPECIMEN PREPARATION FOR MATERIALS EXAMINATION IN THE SEM

One of the great strengths of scanning electron microscopy is the fact that many specimens can be examined with virtually no specimen preparation. The SEM is a surface examination tool; the information is carried by the backscattered and secondary electrons and other emitted signals. Specimen thickness is not a consideration as is the case in transmission electron microscopy, where the information is carried by transmitted electrons that penetrate 1 μm or less. Thus bulk specimens can be examined in the SEM with a size limited only by considerations of accommodation in the specimen stage. For the examination of images of topography contrast from conducting specimens, the only specimen preparation that is necessary is to ensure that the specimen is thoroughly degreased so as to avoid hydrocarbon contamination. Since we wish to examine the surface of the material, it is important to remove contaminants that may have an adverse effect on secondary electron emission. The electron beam can cause cracking of hydrocarbons, resulting in the deposition of carbon and other breakdown products on the specimen during examination. Contamination during operation frequently can be detected by making a magnification series from high magnification (small scanned area) to low magnification (large scanned area). The deposit forms quickly at high magnification because of the increased exposure rate. When the area is observed at low magnification, a "scan square" of contamination is observed (Figure 1). It is thus important to avoid introducing volatile compounds into the SEM. Residual hydrocarbons from the diffusion pump oil can also produce contamination under the influence of the beam. This problem can be avoided for the most part by using traps cooled with liquid nitrogen to condense hydrocarbon vapors. In the case of extreme sensitivity to contamination, such as in Auger electron analysis, a "dry (oil free) pumping" system using ion pumps may be necessary.

Specimen preparation becomes an important consideration under certain circumstances. As explained in Chapter IV, a weak contrast mechanism, such as electron channeling, is frequently impossible to detect in the presence of a strong contrast mechanism, such as topography contrast. It is thus necessary to eliminate specimen topography when we desire to work with electron channeling contrast, types I and II magnetic contrast, and other weak contrast mechanisms. Chemical polishing or electropolishing produces the mirror surface nearly free from topography that is required to obtain satisfactory images with weak contrast mechanisms. A large amount of literature describing such polishing techniques exists for most

FIGURE 1.   Formation of contamination under electron bombardment. The dark square is a result of hydrocarbon cracking built up during scanning at a higher magnification. The hydrocarbon layer changes the secondary electron emission characteristics. Material: $YFeO_3$; beam: 30 keV.

metals and alloys.[3] Metallographic mechanical polishing also removes topography and gives a high-quality mirror surface, but such mechanical polishing results in the formation of a shallow layer ($\sim$100 nm; 1000 Å) of intense damage in most metals.[3] Such a layer completely eliminates electron channeling contrast, and in magnetic materials the residual stresses in the layer result in the formation of surface magnetic domains characteristic of that particular stress state. If we are interested in domains characteristic of the bulk state of the material, such a residual stress layer must be avoided. Mechanical polishing to give a flat surface followed by brief electropolishing to remove the damaged layer often gives optimum results, since electropolishing alone can occasionally result in a polished but wavy surface. In general, specimen preparation remains an art, with each material presenting a different problem to the investigator. However, adequate specimen preparation is absolutely vital to certain kinds of SEM studies, particularly where competing contrast mechanisms are involved.

Insulating materials, for example, Al_2O_3, form a class of special difficulty for SEM examination. When the electron beam strikes an insulator, the absorbed electrons accumulate on the surface since no conducting path to ground exists. The accumulation of electrons builds up a space charge region. This charge region can actually deflect the incident beam in an irregular manner on subsequent scans, leading to severe distortions of the image.[4] Moreover, the existence of a surface charge changes the secondary electron emission greatly. The charging effect is also time dependent, since the surface charge will frequently build up to some value at which breakdown occurs, usually along some surface conduction path, after which charge buildup again occurs. Examples of SEM images with charging effects are shown in Figure 2.

Three techniques to avoid charging effects are available to the microscopist who must examine insulators in the SEM: (1) applying a conducting coating, (2) operating at low accelerating potential, and (3) using single-frame exposure techniques. Of these three techniques, the application of conducting coatings is by far the most widely used.[5–8] A thin layer of a conducting element such as carbon, aluminum, gold, or Au–Pd alloy is applied to the specimen by high-temperature evaporation or by plasma discharge techniques. The layer is made as thin as possible under the constraint that it be continuous in order to provide a conducting path to ground. Layers of the order of 10–100 nm (100–1000 Å) in thickness are usually suitable; it is desirable that the layer be as thin as possible so that it does not obscure fine details of the specimen that we wish to observe. Control of coating thickness can be obtained within the vacuum evaporator with piezoelectric crystal monitors, which measure the amount of material deposited by monitoring the change in oscillation frequency of a crystal due to increased mass from the coating. The choice of coating materials can be optimized for maximum secondary electron production. Based on considerations of physical aspects of the secondary electron emission process, Everhart concludes that a gold coating 8–10 nm (80–100 Å) thick should produce optimal secondary electron emission.[9] Thus, if high-quality images are desired, gold is the best choice for a coating. Often we wish to perform qualitative and/or quantitative x-ray analysis of the specimen. The scattering of electrons increases markedly with increasing atomic number. Thus a conductive coating of a low-atomic-number element is desirable so that the scattering effect is minimized; carbon or aluminum is usually chosen as the coating material in x-ray analysis situations (see Chapter XI).

If it is necessary to examine an insulating material in an uncoated condition, the second and third techniques listed above can be used. Reduction of charging at low accelerating voltages is due to the characteristics of electron emission from solids, as indicated in the following arguments due to Oatley.[10] If the electron emission coefficient, considering both backscattered primary electrons and secondary electrons, is plotted as a function of beam incident energy E_0, the relationship shown in Figure 3 results. For insulators, a region exists for which the number of emitted

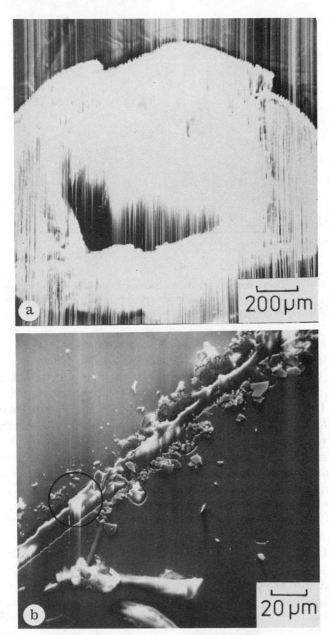

FIGURE 2.   Charging effects in the examination of insulators. (a) Gross charging effects during examination of bare Teflon; beam: 30 keV. (b) Minor charging effects. The specimen is carbon-coated glass scratched through the conducting layer. Note local charging effects (circled) due to the exposure of bare glass by the scratch.

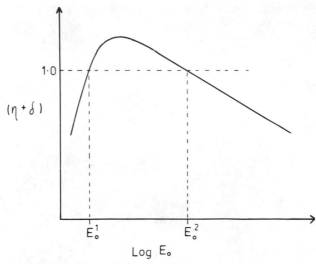

FIGURE 3.   Total electron emission (backscattered and secondary electrons) as a function of beam incidence energy E_0. E_I and E_{II} are the first and second crossover points.

electrons exceeds the number of incident electrons, i.e., $\delta + \eta > 1$. This region is defined by two values of the incident energy E_I and E_{II} for which $\delta + \eta = 1$; these values are referred to as the first and second crossover points. E_I is of the order of several hundred electron volts and E_{II} ranges from 1 to 10 keV, depending on the material. If the incident beam energy is less than E_I, then $\delta + \eta < 1$ and fewer electrons leave the specimen than enter it, resulting in a buildup of negative charge. This charge lowers the effective energy of the incident beam, producing a further decrease in $\delta + \eta$. This situation continues until the specimen is charged to a sufficient level to totally deflect the beam. If the beam energy is between E_I and E_{II}, then more electrons leave the specimen than enter it, i.e., $\delta + \eta > 1$. This charges the specimen positively, and the positive charge acts to decrease the effective value of δ, since the low-energy secondary electrons are attracted back to the specimen. The effective value of $\delta + \eta$ becomes unity due to these processes, and a dynamic equilibrium is set up with the emitted current equal to the incident current at a small, positive, constant surface charge. Slight variations in the curve of Figure 3 at different places in the field of view are not significant, since these will result in slight differences in the equilibrium surface potential. Operation with $E > E_{II}$ again results in negative surface charging, which decreases the effective value of E, raising $\delta + \eta$ until an equilibrium is established with the effective E equal to E_{II}. Oatley points out that this equilibrium is unsatisfactory, since E_{II}

can vary significantly about the surface, giving rise to large variations in the final surface potential from place to place. Slight leakage through surface conduction can greatly disturb such a situation, producing complicated image behavior as a function of time. It is thus desirable to operate with $E_I < E < E_{II}$, which is often achieved with E set to approximately 1 keV. Under such conditions, insulators can be examined uncoated. However, SEM performance is usually significantly poorer at such low accelerating potentials, since source brightness is greatly decreased under such conditions. Recently, quite satisfactory images have been obtained with field emission sources operating at accelerating potentials in the kilovolt range.[11]

Single-frame exposure is another technique which should be considered when insulating materials must be examined uncoated. This technique requires a high-quality storage oscilloscope which may not be available to many SEM users. The technique consists simply in scanning the field once and recording the image on a storage CRT. The beam is then blanked off the specimen. Normally, the first pass of the beam is satisfactory. During this scan, surface charges build up which deflect the beam in subsequent scans. By recording only one scan and then blanking the beam, the surface charge is allowed to dissipate while the microscopist examines the image. Such a storage CRT can offer possibilities of signal processing with the stored image so that the microscopist can optimize the image after recording it.[12] After several minutes, the charge will have dissipated from the specimen, and another frame can be exposed. Although somewhat tedious, such a procedure does allow the use of high-energy beams with uncoated insulators. This procedure would not allow sufficient time for recording an adequate number of x-ray counts in most analytical situations.

II. STEREOMICROSCOPY

The large depth of focus, typically equal to the field width or more, and the good resolution that are obtainable in the SEM make possible the effective use of stereo techniques in the examination of rough surfaces. Stereo-pairs of micrographs are prepared by recording the area of interest twice with the specimen tilted to different angles relative to the beam in the two images. The angular separation of the images of the stereo-pair is typically 5–10°, depending on the magnification.[13] This change in angle is usually accomplished by mechanically tilting the specimen in a goniometer stage, but the action can also be carried out by beam tilting operations with the specimen fixed. The stereo-pair is then examined with a suitable optical system ("stereo viewer") that helps the observer form an integrated image from the two photographs giving an appearance of depth through a parallax effect. It must be recognized that the depth effect is an illusion. A third dimension normal to the plane of the photographs is formed by the observer's visual process. A small amount of training with the optical equipment is usually necessary to observe the stereo effect; however, some people are unable to observe stereo effects because of the physiological makeup of their eyes. An example of a stereo-pair obtained from a rough surface is shown in Figure 4. When this figure is examined with a stereo

FIGURE 4. Stereo-pairs from fractured iron. (a) Tilt axis perpendicular to the ET detector–specimen axis. The apparent illumination is from the left side when arranged for stereo viewing. (b) Tilt axis parallel to the ET detector–specimen axis. The apparent illumination is from the top when arranged for stereo viewing. This is the preferred direction of apparent illumination.

viewer, the portions of the specimen that are elevated and those that are depressed can be easily discerned. Such information is invaluable for determining the elevation character of an object, i.e., whether it is above or below adjacent features. Frequently, such interpretation is difficult from a single image. For proper interpretation, the image recorded at the lower tilt angle must be placed at the left in the stereo viewer.

In an SEM equipped with the conventional arrangement of the specimen relative to the Everhart–Thornley emissive electron detector, the illumination appears to come from the top of the photograph. This illumination direction is the desired convention in presenting SEM images. However, when a stereo-pair is prepared by changing the tilt angle about a tilt axis that lies perpendicular to the Everhart–Thornley detector, the illumination in the stereo-pair appears to come from the side of the integrated image. Effective illumination from the top of the stereo-pair can be obtained provided that the tilt is produced about an axis parallel to the specimen–detector line. This tilt axis is normally not available in SEM stages. Lane has described an alternative procedure to achieve top illumination in stereo-pairs with the usual goniometer stage (tilt axis normal to the detector specimen line and rotation about the normal of the specimen stub).[13] Lane suggests the use of rotation rather than tilt. The stage rotation R, tilt θ, and desired parallax angle P are related by

$$\sin R = (\sin P)/\sin \theta \qquad (1)$$

As an example, for a tilt $\theta = 34°$ and a desired parallax of $P = 6°$, the rotation R must be $11°$. With a stage giving clockwise motion for an increasing rotation numerical value, the left member of the stereo-pair is the image taken at higher rotation reading. Also, the left photo must be rotated clockwise relative to the right photo by this rotation angle in order to match the fields. Rotation of the left member can be avoided if electronic image rotation is used to compensate for the mechanical stage rotation.

Quantitation of the topography of features in SEM micrographs can be carried out by measurements with stereo-pairs. The techniques for data taking and reduction are lengthy, and the reader is referred to details of the procedures given in the literature.[14–17] A means to observe stereo images in real time on the SEM has been outlined by Dinnis.[18–20] In this work, the two stereo views are displayed on two screens, the two viewing angles being obtained by double deflection of the beam by a set of coils placed between the final lens and the specimen. The stereo angle for viewing is 0–8°; for photography, angles up to 13° are possible with an operating voltage of 20 kV. Dinnis claims 500 Å resolution under these conditions.[19]

III. DYNAMIC EXPERIMENTS IN THE SEM

The SEM can be readily applied to *in situ* dynamic experiments, i.e., those experiments in which one or more variables are changed during observation. The resulting changes in structure, as revealed in the image, are then of interest. Special stages have been developed which change the specimen environment *in situ*, involving temperature,[21,22] applied stress,[23] magnetic field,[24] electric field,[25] and liquid surrounding media,[26] while allowing continual observation of the specimen. Such stage design is facilitated by having a large volume of space available in the specimen chamber. This volume is made possible by using a pinhole-type lens with a long focal length (common to all standard SEM's), so that the specimen is at least 10 mm clear of the lens. This design also provides a very low magnetic field at the specimen, which allows applied magnetic field experiments to be carried out (Chapter II).

When the experimental variable is changed as a step function to a new, constant value and the accompanying change in the specimen is completed quickly, the various states of the specimen image can simply be recorded by the normal photographic process in serial fashion. Cases of such experiments are illustrated in the next three examples. In Figure 5, the appearance of a group of grains in superplastic lead–tin eutectic alloy after successive strains is illustrated.[27] The grains in the unstrained condition were observed by atomic number contrast (bright regions are lead rich and darker regions are tin rich). Slight topographic contrast due to differential polishing rates in the two phases is also shown in Figure 5a. The surface roughness increases markedly as a function of tensile strain (Figures 5b, c). This roughness results from grain boundary sliding, which causes the grains to move different amounts in a direction perpendicular to the original surface. It is interesting to note that the atomic number contrast present in Figure 5a is lost in the image having extensive topography (Figure 5c). The topographic contrast is stronger than the atomic number contrast. In these micrographs, prepared with conventional differential amplification of the emissive mode signal and with no signal

---

FIGURE 5.   *In situ* deformation of lead-tin eutectic superplastic alloy. (a) 0% tensile strain; note atomic number contrast (bright areas are lead rich, dark areas are tin rich) and precipitates within lead-rich phase. (b) 12% tensile strain. Surface topography has developed due to grain boundary sliding. (c) 90% tensile strain surface topography further developed. Voids have formed between sliding grains. Note the weakening of atomic number contrast in (b) and (c) due to development of topography contrast.

2 µm

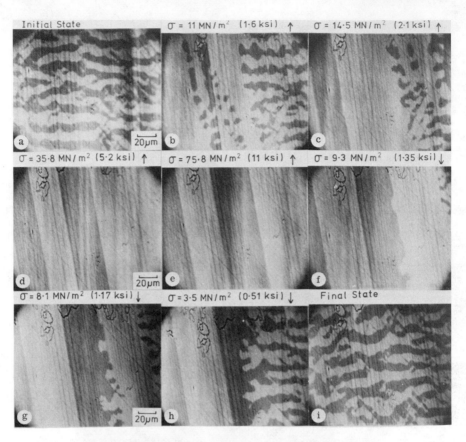

Initial State   σ = 11 MN/m² (1·6 ksi) ↑   σ = 14·5 MN/m² (2·1 ksi) ↑

a   20μm   b   c

σ = 35·8 MN/m² (5·2 ksi) ↑   σ = 75·8 MN/m² (11 ksi) ↑   σ = 9·3 MN/m² (1·35 ksi) ↓

d   20μm   e   f

σ = 8·1 MN/m² (1·17 ksi) ↓   σ = 3·5 MN/m² (0·51 ksi) ↓   Final State

g   20μm   h   i

10 mm   j

saturation allowed, the atomic number contrast is forced into one or two gray levels, thus making it effectively invisible.

The second example (Figure 6) is an *in situ* straining experiment in which the magnetic domain configuration in Fe–$3\frac{1}{4}\%$ Si, observed with type II magnetic contrast, changes as a function of applied elastic stress. The domain configuration is determined by a complicated energy balance, in which the applied stress exerts an effect through magnetostrictive energy.[22] In Figure 6a, the initial domain configuration is characteristic of residual stress in the material. When a stress is applied in the direction indicated, the domains reorganize into stripe domains with a long axis nearly parallel to the axis of applied stress. Stress values were obtained from load measurements determined with a load cell in series with the specimen. After formation of the stripe domains, further stress increases cause an increase in the width of the stripe domains. When the stress is released, the stripe domains persist to lower stress values, indicating a stress hysteresis effect. When the applied stress is totally removed (Figure 6i), domains of the initial form are found. Close examination of Figures 6a and 6i reveals that the original domain pattern is not exactly restored by stress removal, which again is a result of a stress hysteresis effect.

A third serial experiment is illustrated in Figure 7. In this experiment a magnetic field is applied to a specimen of Fe–$3\frac{1}{4}\%$ Si *in situ*. As explained above, the pinhole design of the final lens reduces the magnetic field at the specimen due to the lens field to low values, typically for a 10 mm working distance. *In situ* applied field experiments were carried out by surrounding an electromagnet with an iron circuit; the specimen formed the closure element of this circuit. A field could thus be applied to the specimen without significantly degrading the electron beam. Figure 7 shows a region of the specimen before (Figure 7a) and after (Figure 7b) applying a field. Certain domains are seen to have grown at the expense of their neighbors of opposite magnetization. The interaction of moving domains with inclusions and residual stress fields can thus be studied.

FIGURE 6. (a–i) Series showing response of magnetic domain structure to applied stress in Fe–$3\frac{1}{4}\%$ Si transformer alloy oriented (100) [001], i.e., (110) plane in the sheet surface and [001] rolling direction. Load values obtained from bridge-type load cell; arrows indicate increasing or decreasing applied stress. Transverse domains re-form into elongated domains whose walls are parallel to the tensile axis; on load release the transverse domains reappear but alter due to hysteresis effects. All micrographs taken using 30 kV operating potential, 55° specimen tilt, and 2×10^{-7} A specimen current. (j) Straining stage with specimen and load cell in place.

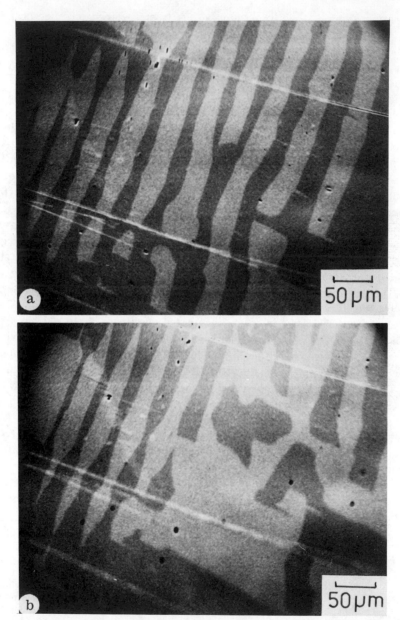

FIGURE 7.   *In situ* magnetization experiment: (a) No field applied; (b) 100 Oe
applied and removed. Field applied perpendicular to specimen sur-
face. Color composite synthesized from (a) and (b) is shown as (c)
(see color insert): cyan, no field; and red, 100 Oe applied and removed.

The results of serial experiments can be presented very effectively through the use of color photographic techniques to combine several black and white images. The technique is explained in detail in Chapter XI. Some examples of the color combination of serial images are given in Fig. 7c* and in Figs. 8–11 of Chapter XI. Differences between two states of the specimen are easily observed in the superimposed images.

Serial experiments actually involve a series of static states of the specimen; the specimen does not undergo any changes in time during photographic recording. When it is desired to study events that change significantly in the minimum time that is required to obtain photographic recordings (10 sec is a good estimate of the minimum practical frame time for photographs), then rapid scanning techniques must be employed. Recording is then carried out by direct signal storage on magnetic videotape. Virtually all commercial SEM's are now equipped with flicker-free (television rate) scanning systems, and commercial videotape recorders can be interfaced directly to the SEM–TV video output. Besides recording rapid dynamic changes, such TV rate scanning is a great aid in focusing and searching the specimen for areas of interest. Because the picture point time is greatly reduced at TV rates, it is necessary to increase the beam current significantly to achieve acceptable low-noise images. Increased beam current results in an increased probe size and a decrease in resolution (see Chapter II). Thus dynamic experiments at TV rates are limited to relatively low magnifications, usually $10,000\times$ or less. These difficulties notwithstanding, excellent TV rate images can be obtained with high contrast mechanisms operating, and even type II magnetic contrast (0.3% contrast) has been successfully studied in dynamic experiments with TV rate images.[29]

Events that occur in a time period less than the frame time of TV rate scanning, i.e., 0.04 sec, are for the most part inaccessible to study. Faster rate scanning is possible to a certain extent, but further decreases in the picture point time lead to extremely noisy images which are unacceptable. The one notable exception to this situation is the case in which the event is cyclical in nature with a high frequency. By employing the techniques of stroboscopy, events occurring with a frequency of 9 GHz have been successfully studied in the SEM.[30] The basic concept employed in stroboscopy is that the signal is sampled in phase with the event of interest. By sampling repeatedly at each picture point, an image with an acceptably high signal-to-noise ratio is built up. The principle is shown in Figure 8. If, in a cycle, the event of interest takes place between π and

---

*Figure 7c of this chapter and Fig. 8–11 of Chapter XI are color illustrations and will be found following p. 206.

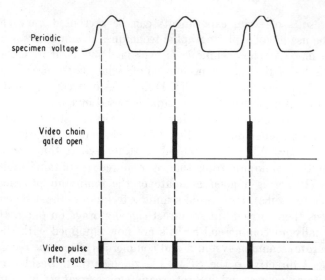

FIGURE 8.   Principle of stroboscopy experiment; video chain gate opens on application of strobed specimen voltage.

$3\pi/2$, then the image signal must be either created or sampled between π and $3\pi/2$ of the cycle. For example, with a picture point time of 1 μsec (10^6 picture points and a frame time of 1 sec) and an event frequency of 10^9 Hz, the signal must be sampled 1000 times per picture point at the correct phase. Two separate approaches to achieve useful stroboscopic operation are possible. In the gated video signal SEM stroboscope, the specimen is continuously illuminated by the scanned beam, and the signal from one of the detectors is sampled in synchronism with the event of interest. In the pulsed beam SEM stroboscope, the detector signal is monitored continuously while the electron beam is pulsed on and off the specimen in synchronism with the event. In general, the gated video signal SEM stroboscope is limited to lower frequency studies than the pulsed beam stroboscope because of distortion effects associated with the video signals and amplifiers, whereas beam pulsing can be successfully carried out at GHz rates. Gopinath and Hill[30,31] have published reviews of both techniques. Stroboscopy studies have been carried out on ladder circuits, Gunn devices, oscillators, and thin-film resistors.[31] Figure 9 shows a series of stroboscopic micrographs of a metal–oxide–semiconductor–transistor (MOST) ladder. In this case, a 7-MHz, 5-V peak-to-peak distorted sine wave was applied to gate 1 (the electrodes are, from the bottom, source 1, gate 1, drain 1, source 2, gate 2, etc. for the four devices on the chip), and a series of micrographs were recorded at various phases of a complete cycle. In

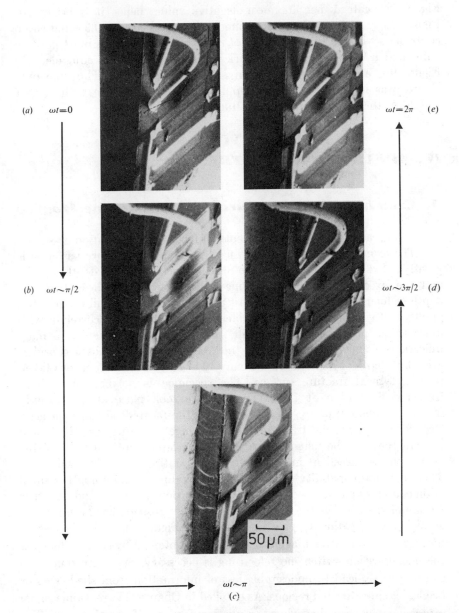

(*a*) $\omega t = 0$

(*b*) $\omega t \sim \pi/2$

$\omega t = 2\pi$ (*e*)

$\omega t \sim 3\pi/2$ (*d*)

50μm

$\omega t \sim \pi$
(*c*)

FIGURE 9.   Stroboscopy of metal–oxide–semiconductor–transistor (MOST) device. Phase is shown. See text for details; strobed at 7 MHz with 5.5-V distorted sine wave.

Figure 9a, gate 1 has its most negative value; hence it is bright. In Figure 9b, gate 1 is about at ground potential for $\omega t = 0.5\pi$ and so no contrast is visible. In Figure 9c, ωt is π and gate 1 has its most positive value and is dark compared to background. Figure 9d is equivalent to Figure 9b, and Figure 9c is equivalent to Figure 9a. Thus the only change was in the phase of the applied sine wave voltage with respect to that of the specimen periodic voltage.[32]

IV. APPLICATIONS OF THE SEM

A. Examination of Fractured Polycrystalline Iron

Special advantages of SEM: depth of focus, high resolution.

The examination of rough surfaces, particularly fracture surfaces, is greatly aided by two features of the SEM, namely the large depth of focus and the excellent resolution of fine-scale features. Of these two, the depth of focus is frequently the most useful. Fine-scale detail often varies greatly with position in the specimen and is difficult to correlate with macroscopic properties. As a result, most images are taken at a magnification of 5000× or less where the resolution limit is not a consideration but depth of focus is vital. Figure 10 is a series of micrographs taken from a typical fracture surface. The specimen is polycrystalline iron, fractured in tension at 78°K. The fracture took place almost entirely by intergranular fracture, as evidenced by the faceted, angular surfaces. Occasional regions of transgranular (cleavage) fracture can be noted (Figure 10a). In the magnification series, the large depth of focus of the SEM gives an excellent image; virtually the entire field is in good focus. The resolution capability of the SEM is of value in examining the small grain boundary particle seen at high magnification in Figure 10d. Another feature of the SEM is the fact that once the microscope has been focused at high magnification to give the image of Figure 10d, the images of Figures 10a–c at lower magnification can be obtained by simply changing the magnification switch; no refocusing is necessary. Specimen roughness can frequently lead to undesirable loss of information from shadowing or bright, signal-saturated regions. As detailed in Chapter IV, real-time signal processing can usually reduce or eliminate such information loss. The examination of specimens with extensive topographical detail provides a challenge to every function of the SEM and provides a good example of the flexibility of the instrument.

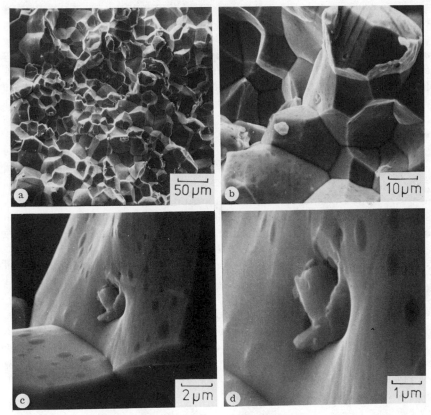

FIGURE 10.   Magnification series on a fracture surface. Specimen: polycrystalline iron fractured under elastic loading at 77°K. (a) Low-magnification view; most of the specimen fractured by an intergranular mechanism. (b) Well-developed facets, which result from intergranular fracture. (c) Detail on an intergranular surface showing a carbide particle partially reduced during a hydrogen anneal. (d) Enlarged view of carbide particle; note the domelike structures on the boundary, which may be precipitates.

B. Failure Analysis of a Composite Consisting of Borsic (Boron–Silicon Carbide) Rods in a Matrix of Plasma-Sprayed Aluminum

Special advantage of SEM: large depth of focus.

In order to enhance the mechanical properties of aluminum, but to retain low overall weight, boron–silicon carbide (borsic) rods arranged in a square matrix were plasma sprayed with aluminum. The major desired result was a large increase in strength-to-weight ratio for a material having a density of about 3 g/cm³. One reason for plasma spraying was to provide a strong bond between the aluminum and the borsic without introducing any other material to promote wetting of the borsic by the aluminum. Furthermore, since the aluminum did not need to be melted, the square array of borsic rods could be accurately maintained while the matrix was built up around the array. Borsic and aluminum have similar densities as well.

The preliminary mechanical tests of the composite were not entirely encouraging. Figure 11 shows one of the broken test specimens. Clearly, poor bonding of the 0.1-mm borsic rods and the Al matrix exists. The rods are very smooth, as is the aluminum surrounding the rods. The evidence shown by Figure 11 indicates that the Al did not wet the borsic when originally plasma-sprayed; on cooling, the Al seems to have shrunk slightly away from the rods. Thus the final composite apparently consisted of rods sitting in cylindrical holes slightly larger than the rod diameter. Lack of strong bonding would reduce mechanical strength of the material. Figure 11 also gives indications that the matrix is porous.

Evaluation of the failed composite with optical microscopy would have been nearly impossible. Some of the broken rods stand nearly 0.5 mm above the surface of the matrix. The rods are curved as well. Hence the small depth of focus of the optical microscope would have been a serious handicap in this investigation. The SEM was able to indicate several sources of difficulty in the composite in a short time.

C. Investigation of Returned Lunar Material— Glassy Spherules from Apollo 11

Special advantages of SEM: nondestructive examination, x-ray capabilities, large depth of focus.

Figure 12 shows a glassy spherule, 0.31 mm diameter, on which metal is spread on the surface, presumably as a result of the same impact event that produced the spherule. Other impacts may have formed the craters on the surface. Cracks can also be seen on the surface. These cracks are probably formed during the solidification process when the surface collapsed inward as the central areas began to solidify.

This particle was sectioned, and the metal areas (10×20 μm in size) on the outside of the particle were analyzed by electron probe microanalysis. The metal contains 3.5–4 wt % Ni, 95 ± 2 wt % Fe, 0.3 wt % Co, and 0.15 wt % P. The most prominent feature of this metallic area and most of the

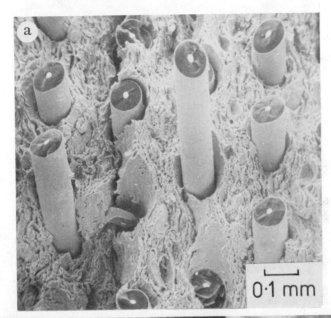

FIGURE 11.   Fracture surface of borsic rod–aluminum matrix composite. (a) Poor bonding of reinforcing rods to plasma-sprayed matrix is apparent. (b) Matrix shrank away from rod on cooling, leaving a space. The bright central core in (a) is tungsten onto which the basic rod was grown.

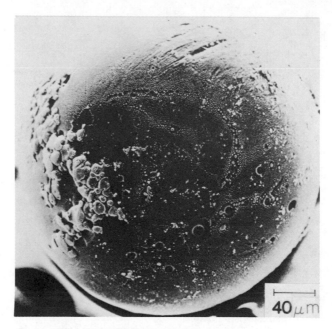

FIGURE 12.   Lunar glassy spherule collected by Apollo 11 crew. Particles at left are iron–nickel alloy of meteoritic origin. Other particles contain silicon, calcium, and titanium.

other ones analyzed is that it is surrounded by FeS (troilite). Presumably, the metal and troilite were originally part of a stony meteorite which impacted on the moon. After impact, the molten metal and troilite were probably splashed onto the glass sphere. The Fe–Ni froze first and pushed the FeS liquid to the outside of the metal area. The glass composition in this particle was not determined, so its origin is unknown.[33]

D. *Analysis of Corrosion Mechanism in a Steam Boiler Tube*

Special advantages of SEM: ability to examine the "as-corroded" piece, x-ray capability.

Corrosion products had built up on the condenser tubes of an SAE 1020 steel steam boiler, thus seriously impairing the boiler's efficiency. Examination with the SEM showed large groups of nodules deposited on the inner surface of the tube. X-ray area scans prepared with the energy-dispersive x-ray detector (see Chapter XI for details) attached to the SEM revealed that these nodules were rich in copper (Figure 13). This finding

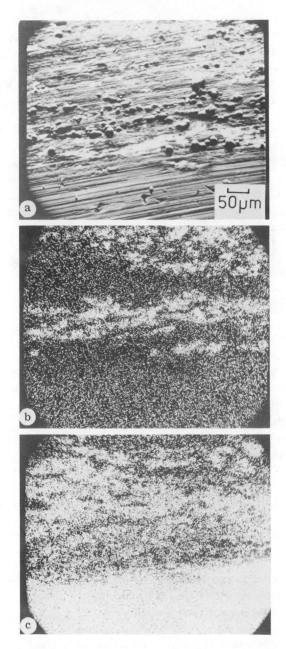

FIGURE 13. Corrosion product in condenser tube SAE 1020 steel from steam boiler. (a) Secondary electron micrograph. (b) EDS x-ray area scan showing copper distribution. (c) EDS x-ray area scan showing iron distribution.

was somewhat surprising but led to the deduction of the source of the problem. A circulating rust-inhibitor was used in the steel tubes; the steel tubes in turn were connected to copper pipes. The Fe–Cu combination set up a galvanic cell, and copper ions were plated onto the inside of the tube. The rust inhibitor carried the Cu ions into the steel tube. Insulating joints installed between the steel tubes and the copper pipes ended the difficulty.

E. *Rapid Phase Delineation in a Mineral Specimen*

Special advantages of SEM: ability to delineate phases in as-polished mineral thin sections by means of specimen current differences (atomic number contrast).

The number and identity of phases in a basalt sample were desired. The specimen current signal can be used to determine the number of phases present and arrange them in ascending atomic number order. The primary electron backscatter coefficient η is directly proportional to the atomic number Z and complementary to the specimen current i_s (see Chapter III). Hence, in a specimen current image, the region having the lowest atomic number will be the brightest and the region of the highest atomic number will be darkest. Figure 14 shows the basalt imaged in the specimen current mode.[34] Four distinct shades of gray are apparent, indicating that four phases of significantly different composition are present.

A mineralogist, knowing the basic makeup of basalts, was able to identify the lightest gray (region A) as ore minerals, the pearl gray region (region B) as pyroxene, and region C as plagioclase, and to deduce that the darkest region (region D) is a mineral alteration product.[35] More detailed examination with the x-ray units attached to the SEM would aid in the full characterization of this basalt. Nevertheless, the number of phases, their microstructural characteristics, and their preliminary identification were determined in a few minutes with the aid of the SEM and a trained mineralogical observer.

F. *Characterization of Wear Particles and Surface Degradation Produced by Wear in Bearing and Gear Tests*

Special advantages of SEM: high resolution, good depth of focus, x-ray capabilities.

The study of debris particles recovered from the lubricating fluid in wearing contact systems is carried out to identify specific wear modes

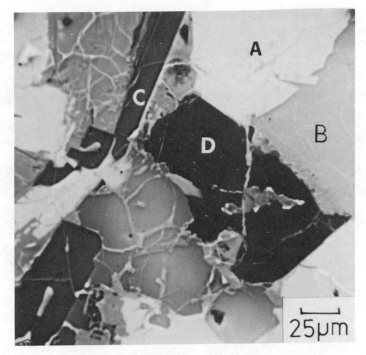

FIGURE 14.   Specimen current image of Disko Island Basalt. Four phases are distinguished by atomic number contrast. In ascending atomic number these are white, pearl gray, gray, and nearly black.

in industrial situations. One way of obtaining specimens directly suitable for SEM examination is to collect the particles from the lubricating oil on the basis of particle magnetic moment, thus saving analytical effort by rejection of nonmagnetic species. The resulting specimen has the wear particles collected into groups (strings) the largest particles (high magnetic moment) being deposited first on a glass substrate. Data on particle size, shape, degree of oxidation, contamination, and deformation can be obtained from SEM studies of such specimens.

Figure 15 shows a string of particles resulting from a ball-bearing bench test of wear. A spheroidal particle plus particles of random morphology are present; the sphere diameter is about 4.5 μm. X-ray spectra were prepared with the energy-dispersive system of the SEM in order to identify the particles; the results are shown in Figure 16.

Note in Figure 15 that the particle labeled c shows significantly lower image contrast; secondary electron imaging was used to prepare Figure 15. The spheroid and particle b are very rich in iron, but particle c is lower in iron and contains significant amounts of silicon, potassium,

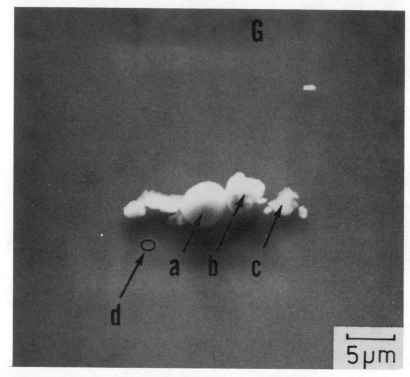

FIGURE 15.   String of wear particles recovered from a ball-bearing bench test. Spheroidal particle is 4.5 μm in diameter; note contrast differences between particles. EDS x-ray spectra were obtained from marked particles. Spheroidal particle is iron; particle c contains oil residue. (Courtesy of A. W. Ruff, Jr.)

and zinc. The latter three elements are found in oil residue. Thus particle c may be an iron oxide which has absorbed lubricating oil. The contrast difference in Figure 15 noted above may well depend on the iron/iron oxide ratio of the particles.[36]

There are difficulties inherent in attempts to unambiguously analyze small particles in close proximity by x-ray analysis methods. For instance, point-probe analysis may excite small surface contaminant particles lying atop larger particles. The beam may penetrate through the particle of interest and excite substrate radiation. For spheres of iron, an 0.5-μm diameter yields an x-ray intensity of about 75% that of bulk; for a 1.5-μm diameter, the figure is 95%.[36] Bayard has discussed particulate analysis

FIGURE 16.   EDS x-ray spectra from particles (a–c) and background (d) as shown in Figure 15.

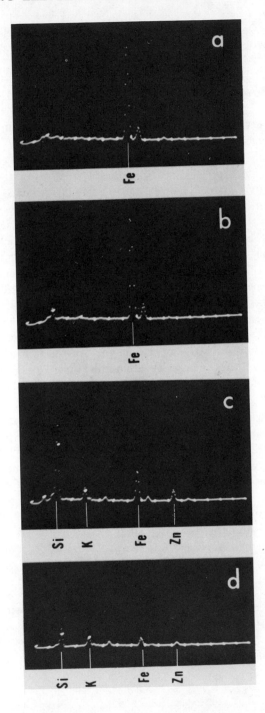

in detail.[37] Despite the difficulties, a great deal of useful information concerning wear particles can be obtained in a relatively short period.

G. Examination of Human Teeth

Special advantage of SEM: large depth of focus, contrast formation, ability to image at low magnification.

In a number of cases, it is often desirable to view the object of interest as a whole before proceeding to magnify small, and therefore perhaps unrepresentative, areas. The SEM can be made to image objects at very low magnification without undue modification. Ballard has described a low-magnification stage consisting of a vacuum flange fitted with a commercial, vacuum-tight ball feedthrough manipulation shaft which allows rotary motion of 360° about the shaft as well as motion in all planes about an arc up to 75°. This device replaces the standard specimen stage.[38] All of the signal processing capabilities of the SEM are unaffected, i.e., TV mode, secondary electron collection, etc. The way to obtain low magnification is to increase working distance from the objective lens and to reduce primary beam voltage. Ballard has used working distances of 13 cm; minor image distortion results from operating the SEM in this mode. By reducing the primary voltage to 1.5 kV, magnifications of 1.5 diameters were obtained.

Figure 17 shows a group of extracted human adult lower molar teeth. These teeth were etched with phosphoric acid solutions to remove contaminants. They were then carbon- and palladium-coated prior to introduction to the SEM. A great deal of dental calculus (calcified ingestants) is apparent on two of the teeth. Dental anatomical features such as deep crevices and cracks are apparent; teeth with such features are much more likely to retain food and so are cavity prone. Dental researchers are seeking means to impregnate teeth *in vivo* with plastics so as to reduce the tendency toward cavities in patients having such teeth. The molars shown in Figure 17 will be used in a pilot study of impregnation. (Sample courtesy of Dr. R. Bowen, NBS.)

H. Applications of Electron Channeling Contrast to Materials Problems

Special advantages of SEM: determination of orientation and crystal perfection.

FIGURE 17.   Low-magnification micrograph of adult human molar teeth. Fissures are typical of caries-prone teeth. Gray matter on one tooth is calcified detritus.

1. Orientation Determinations

The selected area electron channeling pattern (SACP) can be used to determine the orientation of particular features in a microsctructure. An example of the application of the SACP technique to an analysis of an annealing twin in Pb–1.5 wt % Sn is illustrated in Figure 18. Electron channeling contrast reveals the annealing twin in the micrograph (Figure 18a). SACP's taken on either side of the twin show identical orientations (Figures 18b, d). With the specimen placed normal to the beam, these SACP's give the orientation of the surface plane of the grain. The SACP from the annealing twin (Figure 18c) is observed to be a mirror reflection about the plane in this section. A complete solution of the character of the annealing twin requires a two-surface analysis, since the trace of the twin boundary, which lies out of the plane of the section through the grain in Figure 18a, must be determined. The details of how such a crystallographic analysis can be carried out are found in texts on x-ray diffraction analysis.[39] With a properly sectioned and polished specimen, such an

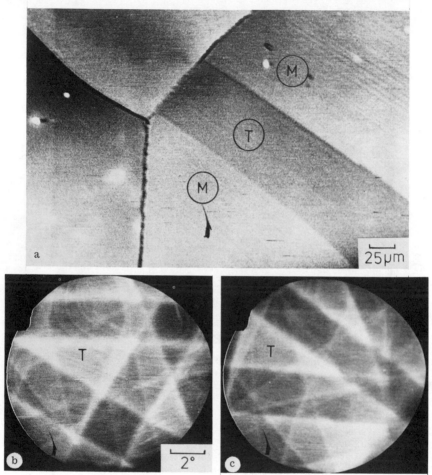

FIGURE 18.   Use of selected area channeling patterns for orientation studies. (a–d). Partial analysis of annealing twin in Pb–1.5% Sn. The SACP's are obtained from the regions indicated. The matrix patterns (b) and (d) are identical, whereas the twin pattern (c) is a mirror reflec-

analysis can be quickly carried out with SACP's and a tilting goniometer stage.

A second example of an application of orientation determination with SACP's is the determination of local texture, i.e., the orientations of a group of adjacent grains. The application makes use of the fact that SACP's can be recorded rapidly (typically one per minute for a 30-sec frame photographic recording) and analyzed quickly with a channeling map

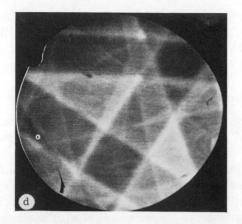

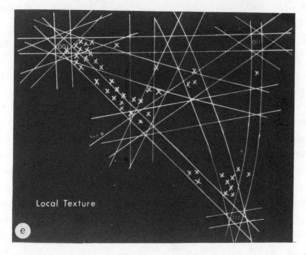

Local Texture

tion of (b) and (d). Note the similar triangles T marked in the SACP's. (e) Local texture pattern from annealed Fe-$3\frac{1}{4}$% Si. The crystallographic normals of a group of 50 grains are plotted from their SACP's into the unit crystallographic triangle. The normals tend to cluster near [001], [111], and [110].

(Chapter V). For cubic systems, the orientations can frequently be recognized directly on the viewing screen with the aid of the channeling map. Thus the orientations of each grain in a group of 25 or more can be quickly obtained. A simple texture determination can be made by plotting only the normal of each grain into the basic repeat unit of the stereogram, e.g., the unit triangle for cubic materials.[39] Such a plot of normals for a group of adjacent grains in Fe-$3\frac{1}{4}$ wt % Si is given in Figure 18e. The grains are not randomly oriented, but their normals are associated with the (110) bands, which indicates that [110] directions must lie in the plane of the specimen. This technique could be of use in studying local texture banding in certain systems.

2. CRYSTAL PERFECTION STUDIES

As discussed in Chapter V, the determination of absolute measurements of crystal perfection with electron channeling patterns is not yet practical due to the complex nature of changes in the pattern quality as a function of crystal–defects relationship. A second problem with such measurements is the quantification of pattern quality, itself a complicated factor to define.[40] Despite these difficulties involving absolute measurements, relative measurements are practical with channeling patterns, and very useful. The next two examples indicate the types of problems that have been studied to date.

Fracture surfaces can be successfully studied with channeling patterns providing the surface is not so badly distorted that the ECP is totally degraded and so undetectable. A successful study has been carried out in fractured polycrystalline iron with channeling patterns.[41] High-purity iron was fractured in tension in liquid nitrogen (77°K). The fracture occurred under nominally elastic loading; the fracture stress was 20% below the extrapolated plastic flow stress determined from higher temperature tests and the stress–strain plot showed no evidence of macroplasticity. The fracture surface (Figure 19a) shows that most of the fracture occurred by intergranular fracture processes with occasional cleavage fracture. Selected area channeling patterns (SACP) were obtained from areas of cleavage and of intergranular fracture. In general, the SACP's from cleavage surfaces were severely degraded (Figure 19b, c) and in some cases unobservable. The SACP's from intergranular surfaces were of high quality (Figure 19d, e) and were indistinguishable from patterns obtained from electropolished surfaces of well-annealed iron. In certain cases (e.g., Figure 19b) areas of local deformation were found on intergranular surfaces, apparently associated with grain boundary precipitates. SACP's from such areas on intergranular surfaces showed considerable degradation. These results indicate that the cleavage and intergranular fracture mechanisms result in surfaces having greatly differing perfection. Intergranular surfaces are essentially undisturbed by the process, at least to the sensitivity of SACP degradation, which has a minimum detectability of about 1% plastic strain.[40] The cleavage process clearly results in a much more degraded surface.

The plastic zone surrounding the roots of an arrested crack has been mapped with SACP's by Stickler *et al.*[42] and Davidson.[40] An example taken from Davidson's results is given in Figure 20. The micrograph with channeling contrast (Figure 20a) shows the position of the crack in the grain of interest. SACP's were taken at the positions indicated (Figure 20). Davidson has interpreted the pattern quality by visual comparison with a calibrated series of SACP's obtained for a range of strains produced in simple tension.[40] This comparison procedure was used to produce the graph of strain vs. distance from the crack tip. Useful engineering data can thus be obtained by this means.

In addition to the SACP technique, crystal perfection information can be obtained from channeling contrast in the micrograph. Bend contours reveal the presence of elastic and plastic bending strains.[43] Since large fields can be examined at low magnifications, and large areas can be evaluated by specimen traverses, the approximate proportion of distorted material in a microstructure can be quickly evaluated. An example of strain field mapping by direct imaging is given in Figure 21. Annealed polycrystalline nickel was electropolished to produce a mirror surface, and a hardness indentation mark was made with a diamond indenter. When

FIGURE 19.   Application of electron channeling patterns to deformation studies. Specimen: polycrystalline high-purity iron fractured under elastic loading at 77°K. (a) Low-magnification view of surface showing intergranular and cleavage fracture processes. (b, c [page 244]) Cleavage region and degraded SACP obtained from area indicated. Fine lines in the SACP are lost and band contrast is low. (d, e [page 245]) Intergranular region and SACP. The SACP is of a quality equal to that obtained from annealed iron.

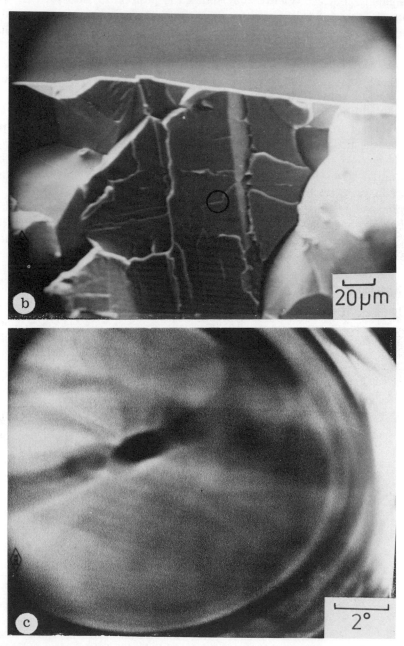

FIGURE 19 (cont'd).

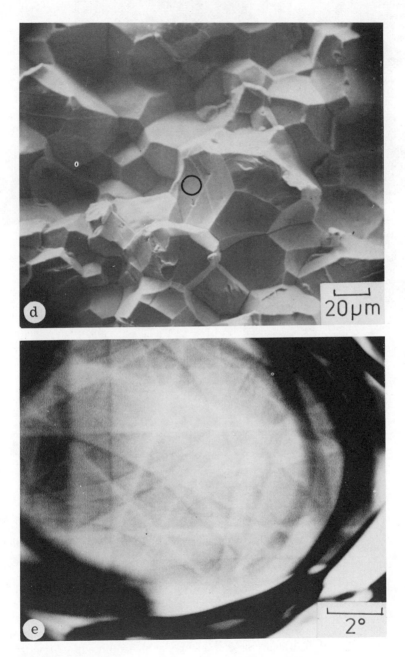

FIGURE 19 (cont'd).

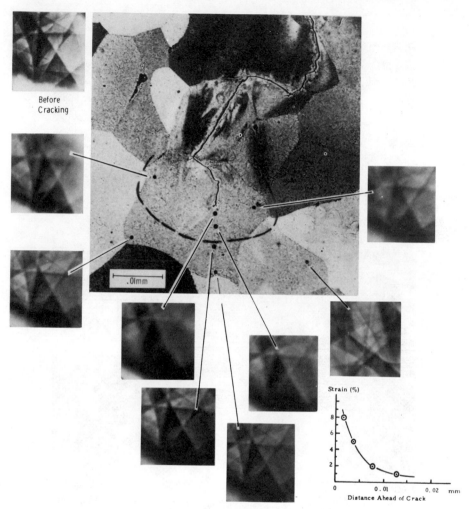

FIGURE 20. Application of electron channeling patterns to deformation studies. Arrested crack in Fe-3% Si transformer alloy. The crack can be seen in the micrograph with bend contours emanating from it. The SACP's are obtained from the indicated regions. By relating pattern quality to a calibration series, the strain as a function of position ahead of the crack can be estimated. This example was supplied by courtesy of Dr. D. Davidson, Southwest Research Institute.

imaged under conditions for channeling contrast (Figure 21), bend contours are observed to emanate for a considerable distance from the points of the indentation mark. In the arrested crack specimen examined by Davidson,[40] bend contours were observed along the crack (Figure 20a). This application of channeling contrast provides a unique technique for direct imaging of

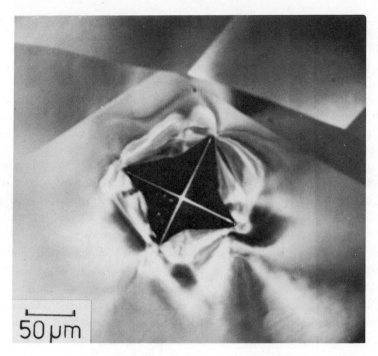

FIGURE 21. Diamond hardness indentation in electropolished polycrystalline pure nickel. Electron channeling contrast, recorded in specimen current mode, shows bend contours indicative of extent and location of strains introduced by indenting process.

strain fields in bulk specimens. It should be noted, however, that both SACP's and micrograph bend contours are only sensitive to bending-type strains. Simple tensile strains do not produce significant effects on channeling contrast, since the electron rocking curves (intensity vs. angle relative to the crystal) are virtually insensitive to small changes in lattice parameter. Bending strains, however, have the effect of increasing the beam divergence (Chapter V) and thus degrading channeling pattern quality. In the micrograph, bending of the crystal lattice leads directly to the formation of contrast contours (Chapter V).

I. Examination of Magnetically Written Information with Type I Magnetic Contrast

Special advantage of SEM: ability to observe magnetic information from tape or disk memories without disturbing the magnetic signatures.

The leakage fields associated with magnetically recorded information can disturb secondary electron trajectories, thus resulting in type I magnetic contrast (see Chapter V for full mechanistic details). The ultimate aim of such a procedure is to obtain and read the magnetic information without disturbing that information. Ferrier and Kyser, in a study of the minimum detectable magnetic object that could be conveniently studied with type I magnetic contrast, stated that slightly over 300 flux reversals per millimeter of tape could be detected.[44] Disk pack memories can presently be examined with success.

A low-frequency square wave signature is shown in Figure 22; the tape is commercially available sound recording tape. No specimen preparation was carried out; the tape was merely placed onto the SEM stub with a dab of silver paint. The operating conditions were 2 kV accelerating potential and 4×10^{-11} A specimen current. The unmodified Everhart–Thornley detector system was used to record the data; the specimen was normal to the electron beam. Very little signal processing was needed to obtain the image; γ was 1.0 and a small amount of differential amplification was used.

Figure 23 shows the same tape at higher magnification. Topographic contrast is competing strongly with the magnetic contrast at this level. Nevertheless, the magnetic details are clear and can be studied. Figure 24 is the same area as Figure 23 but photographed using the primary backscattered electron signal. The magnetic contrast vanishes since all the mag-

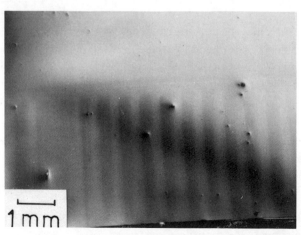

FIGURE 22.   Coarse square wave recorded on ordinary audio quality tape; contrast is type I magnetic contrast made visible with secondary electron signal. Specimen uncoated, operating voltage 2 kV.

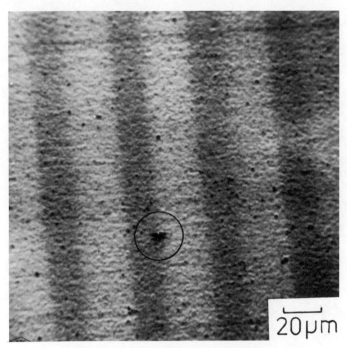

20μm

FIGURE 23.   Type I magnetic contrast in same specimen as Figure 22; photographed under same conditions. Topographic contrast is competing strongly with the magnetic contrast. Note particle at lower center.

netic information is derived from secondary electron trajectories being altered by the leakage field above the specimen. Note especially the comparison of the image of the tiny central particle. This particle may be a bit of magnetic oxide and thus directly responsible for the odd contrast in the secondary electron micrograph. Since magnetic fields play no role in Figure 24, the particle is revealed in its true topographic form on the surface of the tape.

In the study of real information-bearing systems, one wishes to improve the magnetic contrast if possible. Joy suggested the use of two Everhart–Thornley detectors and a subtractive circuit to increase the magnetic contrast at the expense of total signal collected.[45] Ferrier[48] has implemented such a system, combined with a cylindrical secondary electron detector,[46,47] in order to achieve enough sensitivity to successfully examine disk pack memory devices. The devices need only to be cleaned free of any surface film before being introduced into the SEM.[48]

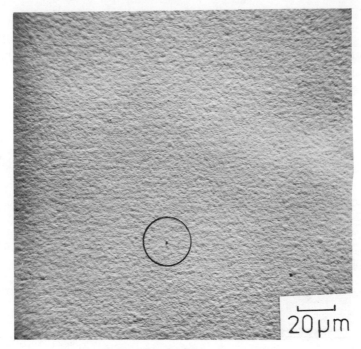

20μm

FIGURE 24.   Same area as Figure 23 but photographed using primary back-
                    scattered electron signal. Trajectory-dependent magnetic contrast
                    completely gone. Note appearance of particle at lower center;
                    particle is probably a magnetic oxide.

J. Applications of Type II Magnetic Contrast

Special advantage of SEM: ability to observe magnetic domains in
materials.

Type II magnetic contrast provides a new technique for the examination of magnetic domains in materials.[49] Images of domains in bulk
specimens of iron and iron alloys can be obtained with an edge resolution of
400 nm (4000 Å); instrument modifications offer the opportunity for considerable improvement of this value to 100 nm (1000 Å) or less. Differential
amplification (black level processing) can enhance the contrast in the final
image to levels which are easily observed. Figures 25–28 are examples
of the types of domain configurations that have been observed in commercial Fe–$3\frac{1}{4}$% Si transformer core steel. In Figure 25, the domain configuration near a grain boundary triple junction is shown. The regular
stripe domains of the interior of the major grain are disturbed near the

triple triple junction. This situation most likely results from residual stresses due to differential crystal contraction of the three grains involved in the junction during the annealing cycle. Residual stresses near a surface scratch (Figure 26) and the inclusion particle (Figure 27) again produce irregularities in an otherwise regular stripe domain pattern. The domain patterns in two of the three grains seen in Figure 28 vary widely with position, again a result of varying long-range stress fields.[50] Such residual stresses in transformer core steels can lead to increased transformer power losses through effects on the domains.[28] The SEM provides an excellent means of studying fine-scale details of the domain configuration while retaining the ability to examine large areas of bulk specimens.

Through analysis of the contrast as a function of rotation about the specimen normal, the local direction of the magnetization can be deduced (Chapter V). This fact, coupled with orientation information obtained from selected area channeling patterns, provides the information necessary to determine the type of domain walls present in a local area. Figures 29 and

FIGURE 25.   Type II magnetic contrast from Fe-$3\frac{1}{4}\%$ Si. Beam: 30 kV; $i =$ 2×10^{-7} A, tilt: 55°, signal: specimen current differentiated. The regular domain pattern in the interior of the grain is perturbed near the triple junction probably as a result of residual stresses.

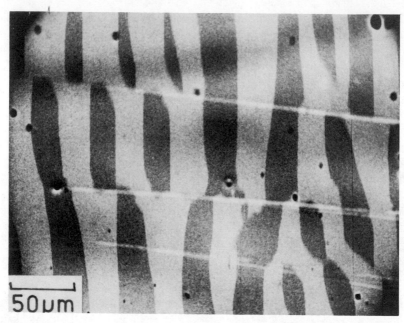

FIGURE 26.   Type II magnetic contrast: Conditions identical to Figure 25,
except backscattered electron signal. The shearing of domains due
to the stress fields of surface scatches is observed.

30 contain examples of such determinations. In each case the contrast
is maximized by rotation, so that the magnetization lies along the tilt
axis or at least a large component is resolved on that axis. The specimen
is then set normal to the beam, and an SACP is obtained from the area
to give the surface plane orientation and crystallographic directions in the
surface plane. The domains of Figure 29a are thus found to have 180°
domain walls in a (100) surface, and the domains of Figure 30a are
"dagger domains" characteristic of an orientation near the [110] pole
(Figure 30b).

K. Study of Reliability of Integrated Circuit Chips— Quality Control

Special advantage of SEM: ability to observe voltage contrast effects
and provide a high-resolution micrograph of the structure of circuits.

The area occupied by an individual integrated circuit is often less
than 1 mm². Hence more than 50 such circuits can be placed on 6 cm² of

surface. The SEM offers a practical way to statistically sample such an array for quality control purposes. One monitors the operation of the integrated circuit with the aid of the voltage contrast produced when various portions of the circuit are energized.

A quadrupole two-input positive NAND gate (SN74H00) is shown in Figure 31. The dark areas are at a positive potential relative to light areas. The operation of the NAND gate for all possible conditions is given in a truth table (Table I). An input "zero" means that the designated input is at ground potential, while a "one" indicates that the input is at the supply potential V_{cc} of +5.5 V. The predicted variations in potential are clearly observed. Thus this particular NAND gate was operating properly at the time of testing. Note that the area occupied by the circuit is about 0.75 mm².

L. Observation of Ferroelectric Domains

Special advantage of SEM: ability to observe voltage contrast effects, capability of carrying out dynamic experiments.

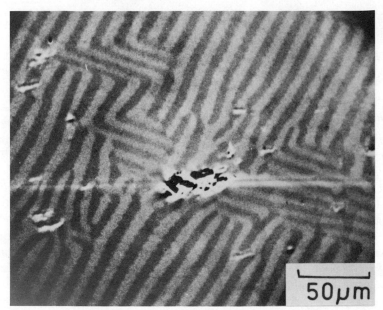

FIGURE 27.   Type II magnetic contrast: Conditions identical to Figure 25. Irregularities in a regular stripe domain structure are observed near an inclusion. "Islands" of domains at right angles to the main pattern are formed.

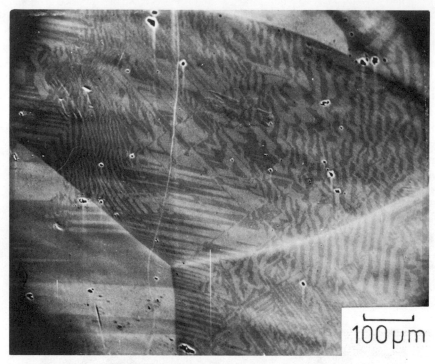

FIGURE 28.   Type II magnetic contrast: conditions identical to Figure 25. An extremely varied domain pattern is observed in a single grain, presumably due to residual stresses.

The voltage contrast mechanism can be used to directly observe ferroelectric domain structures associated with surface charge perturbation induced by the electric polarization. The application to triglycerine sulfate of this method has been outlined.[51]

Figure 32 illustrates ferroelectric domains in lithium niobate. The sample was uncoated; the accelerating potential was 30 kV. The contrast is made visible by accumulation of charge carriers due to impinging electrons in various regions. Dynamic experiments in which the beam voltage or current is varied enable one to observe domain growth, wall switching, and domain migration directly. Such information is useful in characterizing the polarization state of the specimen.

M.   Electron-Beam Writing

Special advantage of SEM: ability to produce circuit components by electron-beam exposure; the components can be less than 1 μm wide.

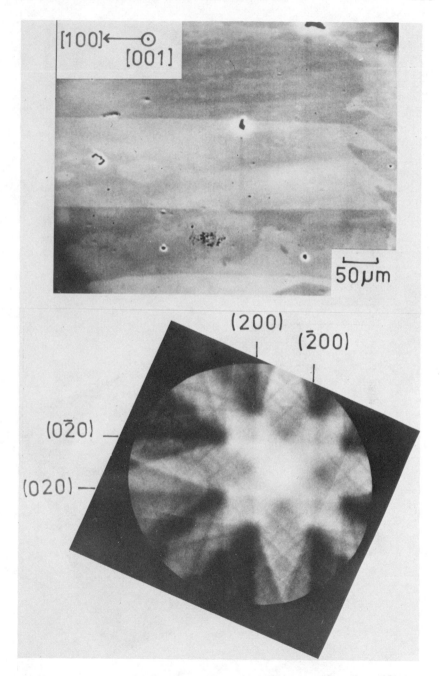

FIGURE 29. Use of SACP's to identify domain walls. The orientation of the region identifies the stripe domain walls as 180° walls. The surface plane is (100).

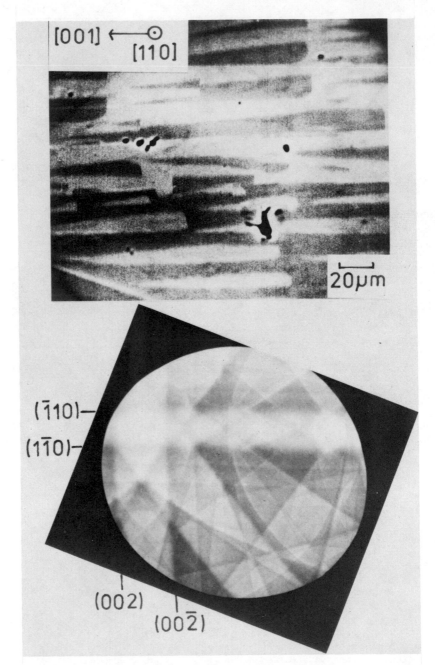

FIGURE 30.   Use of SACP's to identify a domain structure. The orientation of the surface plane is about 5° off (110), which results in the formation of "dagger" domains.

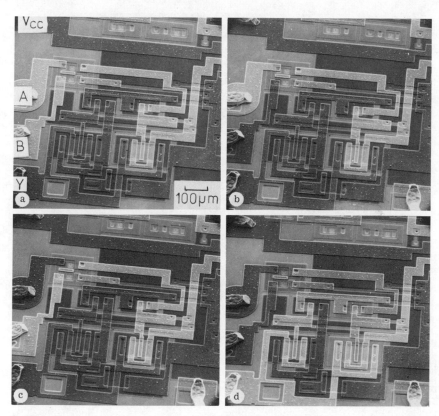

FIGURE 31. Operation of integrated circuit observed using voltage contrast. Circuit is a quadruple two-input positive NAND gate, SN 74H00. The dark areas are at a positive potential relative to light areas. The operation of the NAND gate for all possible conditions is given in a truth table (see Table I). An input "zero" means that the designated input is at ground potential, while a "one" indicates that the input is at the supply potential V of $+5.5$ V. The predicted variations in potential are clearly observed. (Figure courtesy of W. J. Keery and K. O. Leedy, Institute of Applied Technology, NBS.)

TABLE I. Truth Table for NAND Gate

Condition	Inputs		Output
	A	B	Y
a	0	0	1
b	0	1	1
c	1	0	1
d	1	1	0

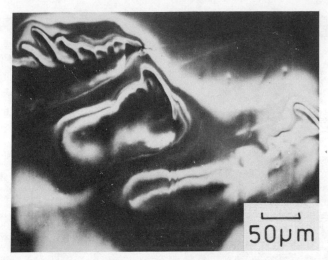

FIGURE 32.   SEM micrograph showing ferroelectric domains in lithium niobate. Sample was uncoated and subjected to an accelerating potential of 30 kV. Contrast made visible by buildup of charge carriers (due to impinging electrons) in various regions.

There are two major ways in which the SEM can be used for preparing electronic devices: beam exposure of photoresist materials and direct metallic deposition (DMD). The former has been used in production, while DMD is still in the experimental stage. Both techniques use the SEM to provide a fine, controllable writing device to produce the desired circuit pattern.

The DMD process relies on the ability of an electron beam to cause molecular dissociation of a metal compound. In practice, the metal compound is usually a thin film deposited on a substrate. When a beam of electrons strikes the thin film, molecules in the compound may be excited into a dissociative state by electron impact. The molecule dissociates into a metallic component, which remains on the substrate, and into a volatile component, which is driven off. After sufficient dissociation has occurred, the remaining film may become an electrical conductor having a finite resistance. Resistance values of such a film depend on the electron beam exposure. The size and shape of the pattern of resistors created will be determined by the pattern scanned by the beam. Later, if unexposed material is removed by solvent action, a pattern of controlled dimensions and electrical properties remains on the substrate.

If the electron beam is the only source of energy for the process, the exposures required are large, typically 0.1–1 coul/cm², compared to present

FIGURE 33.   (a) Pattern of silver on silicon substrate deposited from AgCl by a punch-tape-controlled electron beam. Conducting silver layer approximately 1 μm thick; edge definition about $\frac{1}{2}$ μm. (b) Pattern of silver resistors on silicon substrate prepared in same fashion as (a). Width of finest silver line about 3 μm.

electron-resist materials (10^{-5} coul/cm^2) and high-resolution photographic emulsions (10^{-9} coul/cm^2). The increased exposure necessary for DMD results from the larger number of bonds that must be broken by the electron beam to cause conduction.

The quality of the deposited metal films can be degraded if reactions occur with residual gases in the specimen chamber. In particular, when large exposures (1 coul/cm^2) are necessary, the polymerization of residual hydrocarbons must be avoided if the films are to be of consistently high quality. Hence the partial pressures of the residual gases should be kept as low as possible.

If DMD is to become practical on a large scale in the manufacture of microcircuits, the process must be rapid, reproducible, suitable for automatic control, and capable of high resolution. To date, resistors less than 1 μm wide have been obtained by DMD in high vacuum (10^{-8} Torr).[52]

To achieve reproducibility in the DMD process, a functional relationship between resistance and exposure must be outlined. This relationship has been investigated by monitoring resistance values during DMD. The relationship between composition and resistance has also been studied for deposition of silver from AgCl, by simultaneously monitoring x-ray intensities and resistance values. A clear correlation between the shapes of the resistance vs. exposure curve and the chlorine content vs. exposure curve was obtained.[52,53]

Figure 33 shows two examples of silver deposition from AgCl films. The patterns were produced with a punch-tape-controlled electron beam. The silver conductive layer is about 1 μm thick. Fine details in the representation of King's College Chapel are apparent in Figure 33(a); the entire silver-on-silicon Christmas card is only 325 μm from top to bottom.[54,55] The finest lines are about 2 μm wide. These results indicate that extremely small shapes can be produced by electron-beam writing techniques.

REFERENCES

1. O. Johari (ed.), *Proceedings of the Annual SEM Symposium*, 1968–1974, IITRI, Chicago, Illinois (1968–1974).
2. *Scanning Electron Microscopy: Systems and Applications 1973*, Conference Series No. 18, Institute of Physics, London (1973).
3. G. L. Kehl, *The Principles of Metallographic Laboratory Practice*, McGraw-Hill, New York (1949).
4. J. B. Pawley, in *SEM/1972 Proceedings of the 5th Annual SEM Symposium*, IITRI, Chicago, Illinois (1972), p. 153.
5. G. E. Pfefferkorn, in *SEM/1973 Proceedings of the 6th Annual SEM Symposium*, IITRI, Chicago, Illinois (1973), p. 751.

6. P. Echlin and P. J. W. Hyde, in *SEM/1972 Proceedings of the 5th Annual SEM Symposium*, IITRI, Chicago, Illinois (1972), p. 137.
7. G. E. Pfefferkorn, H. Gruter, and M. Pfautich, in *SEM/1972 Proceedings of the 5th Annual SEM Symposium*, IITRI, Chicago, Illinois (1972), p. 147.
8. P. Echlin, in *SEM/1974 Proceedings of the 7th Annual SEM Symposium*, IITRI, Chicago, Illinois (1974), p. 1019.
9. T. E. Everhart, in *Proceedings of the 3rd Annual Stereoscan Colloquium Morton Grove, Illinois*, Kent Cambridge Scientific (1970), p. 1.
10. C. W. Oatley, *The Scanning Electron Microscope*, Cambridge, The University Press (1972), p. 165.
11. L. M. Welter and V. J. Coates, in *SEM/1974 Proceedings of the 7th Annual SEM Symposium*, IITRI, Chicago, Illinois (1974), p. 59.
12. N. C. Yew and D. E. Pease, in *SEM/1974 Proceedings of the 7th Annual SEM Symposium*, IITRI, Chicago, Illinois (1974), p. 191.
13. W. C. Lane, in *Proceedings of the 3rd Annual Stereoscan Colloquium Morton Grove, Illinois*, Kent Cambridge Scientific, (1970), p. 83.
14. O. C. Wells, *Brit. J. Appl. Phys.*, **11**, 199 (1960).
15. G. S. Lane, *J. Phys. E*, **2**, 565 (1969).
16. A. Boyde, *J. Microscopy*, **98**, 452 (1973).
17. A. Boyde, in *SEM/1974 Proceedings of the 7th Annual SEM Symposium*, IITRI, Chicago, Illinois (1974), p. 93.
18. A. R. Dinnis, in *SEM/1971 Proceedings of the 4th Annual SEM Symposium*, IITRI, Chicago, Illinois (1971), p. 41.
19. A. R. Dinnis, in *Electron Microscopy 1972, Proceedings of the 5th European Congress of Electron Microscopy*, Institute of Physics, London (1972), p. 178.
20. A. R. Dinnis, in *Scanning Electron Microscopy: Systems and Applications 1973*, Conference Series No. 18, Institute of Physics, London (1973), p. 76.
21. B. W. Griffiths and J. A. Venables, in *SEM/1972 Proceedings of the 5th Annual SEM Symposium*, IITRI, Chicago, Illinois (1972), p. 9.
22. R. M. Fulrath, in *SEM/1972 Proceedings of the 5th Annual SEM Symposium*, IITRI, Chicago, Illinois (1972), p. 17.
23. D. J. Dingley, *Micron*, **1**, 206 (1969).
24. D. E. Newbury and H. Yakowitz, in *Proceedings of the 19th Annual Conference on Magnetism and Magnetic Materials, New York*, Vol. II, American Institute of Physics, New York (1974) p. 1372.
25. M. C. Driver, in *SEM/1969 Proceedings of the 2nd SEM Symposium*, IITRI, Chicago, Illinois (1969), p. 405.
26. R. S. Morgan, J. Lebiedzik, and E. W. White, in *SEM/1973 Proceedings of the 6th Annual SEM Symposium*, IITRI, Chicago, Illinois (1973), p. 205.
27. D. E. Newbury, "An Investigation into Mechanisms of Superplasticity," Ph.D. Thesis, the University of Oxford (1972).
28. D. J. Craik and R. S. Tebble, in *Ferromagnetism and Ferromagnetic Domains*, Wiley, New York (1965), p. 1.
29. D. E. Newbury and H. Yakowitz, unpublished results.
30. A. Gopinath and M. S. Hill, in *SEM/1973 Proceedings of the 6th Annual SEM Symposium*, IITRI, Chicago, Illinois (1973), p. 197.
31. A. Gopinath and M. S. Hill, in *SEM/1974 Proceedings of the 7th Annual SEM Symposium*, IITRI, Chicago, Illinois (1974), p. 235.
32. G. S. Plows and W. C. Nixon, *J. Sci.. Inst. (J. Phys. E)*, Ser. 2, **1**, 595 (1968).
33. J. I. Goldstein, E. P. Henderson, and H. Yakowitz, in *Proceedings of the Apollo 11 Lunar Science Conference*, (1970), Vol. 1, p. 499.

34. H. Yakowitz, C. E. Fiori, and D. E. Newbury, in *SEM/1973 Proceedings of the 6th Annual SEM Symposium*, IITRI, Chicago, Illinois (1973), p. 173.

35. W. G. Melson and G. Switzer, *Am. Mineralogist*, **51**, 644 (1968).

36. A. W. Ruff, Jr., National Bureau of Standards Report, NBSIR 74-474 (1974), p. 15.

37. M. Bayard, in *Microprobe Analysis* (C. A. Andersen, ed.), Wiley, New York (1973), p. 323.

38. D. Ballard, *Proc. Electron Microscope Soc. Am.* (1974), p. 446.

39. C. S. Barrett, *Structure of Metals*, McGraw-Hill, New York (1952).

40. D. L. Davidson, in *SEM/1974 Proceedings of the 7th Annual SEM Symposium*, IITRI, Chicago, Illinois (1974), p. 927.

41. D. E. Newbury, B. W. Christ, and D. C. Joy, *Met. Trans.*, (1974), **5**, 1505.

42. R. Stickler, C. W. Hughes, and G. R. Booker, in *SEM/1971 Proceedings of the 4th Annual SEM Symposium*, IITRI, Chicago, Illinois (1971), p. 473.

43. D. C. Joy, D. E. Newbury, and P. M. Hazzledine, in *SEM/1972 Proceedings of the 5th Annual SEM Symposium*, IITRI, Chicago, Illinois (1972), p. 97.

44. R. P. Ferrier and D. F. Kyser, in *Proc. EPASA*, Vol. 8 (1973), Paper 4).

45. D. C. Joy, Ph.D. Thesis, The University of Oxford (1969), Chapter 4.

46. J. R. Banbury and W. C. Nixon, *J. Phys. E*, **2**, 1055 (1969).

47. J. P. Ballantyne, H. Yakowitz, E. Munro, and W. C. Nixon, in *Proceedings of the 25th Anniversary Meeting of E.M.A.G. 1971*, Conference Series No. 10, Institute of Physics, London (1971), p. 194.

48. R. P. Ferrier, The University of Glasgow, personal communication (1973).

49. D. J. Fathers, J. P. Jakubovics, D. C. Joy, D. E. Newbury, and H. Yakowitz, *phys. stat. sol. a*, **20**, 535 (1973).

50. H. Yakowitz, Strain Contours in Iron-3 wt % Silicon Transformer Sheet, Ph.D. Thesis, Univ. of Maryland (1970).

51. C. Michel and A. Sicignano, in *Proc. EPASA*, Vol. 8 (1973), Paper 48.

52. J. P. Ballantyne, Ph.D. Thesis, Cambridge University (1972).

53. J. P. Ballantyne, H. Yakowitz, and W. C. Nixon, in *Proceedings of the Sixth International Conference, X-ray Optics and Microanalysis*, University of Tokyo Press, Tokyo, (1972), p. 219.

54. J. P. Ballantyne, C. Dix, and W. C. Nixon, in *29th Annual Proceedings of the Electron Microscope Society of America*, Boston, Massachusetts (1971).

55. J. P. Ballantyne and W. C. Nixon, *J. Vac. Sci. Technol.*, **10**, 1094 (1973).

VII

X-RAY SPECTRAL MEASUREMENT AND INTERPRETATION

E. Lifshin, M. F. Ciccarelli, and R. B. Bolon

I. INTRODUCTION

Chemical analysis in the scanning electron microscope and electron microprobe is performed by measuring the energy and intensity distribution of the x-ray signal generated by a focused electron beam. The subject of x-ray production has already been introduced in the chapter on electron beam–specimen interactions (Chapter III), which describes the mechanisms for both characteristic and continuum x-ray production. This chapter is concerned with the methods for detecting and measuring these signals as well as converting them into a useful form for qualitative and quantitative analysis.

II. CRYSTAL SPECTROMETERS

A. Basic Design

Until 1968, when solid state detectors were first interfaced to microanalyzers, the crystal diffraction spectrometer (CDS) or wavelength-

E. LIFSHIN, M. F. CICCARELLI, and R. B. BOLON—General Electric Corporate Research and Development, Schenectady, New York.

dispersive spectrometer (WDS) was used almost exclusively for x-ray spectral characterization. The basic components of the WDS are illustrated in Figure 1. A small portion of the x-ray signal generated from the specimen passes out of the electron optical chamber and impinges an analyzing crystal. If Bragg's law is obeyed,

$$n\lambda = 2d \sin \theta \tag{1}$$

(where n is an integer, 1, 2, 3, . . . ; λ is the x-ray wavelength; d is the interplanar spacing of the crystal; and θ is the angle of incidence of the x-ray on the crystal), the x-rays will be diffracted and detected by a proportional counter. The signal from the detector is amplified, converted to a standard pulse size by a single-channel analyzer (SCA), and then either counted with a scaler or displayed as a ratemeter (RM) output on a strip chart recorder. A typical qualitative analysis therefore involves obtaining a strip chart recording of x-ray intensity as a function of crystal angle, converting peak positions to wavelengths through Bragg's law, and then using the Moseley relationship [equation (9), Chapter III] to relate the detected wavelengths to the presence of specific elements. In practice, crystal spectrometer readings are either proportional to wavelength or can be read directly in wavelength and then standard tables can be used for elemental identification.

Bragg's law can be easily derived with the aid of Figure 2. A beam of coherent (in phase) x-rays is assumed to be specularly reflected from parallel crystal planes spaced d units apart. Of the two x-ray beams shown in Figure 2, the lower one will travel an additional path length $ABC = 2d \sin \theta$ prior to leaving the sample. If this distance equals an integral number of wavelengths $n\lambda$, then the reflected beams will combine in phase and an intensity maximum will be detected by the proportional counter.

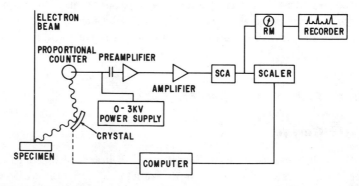

FIGURE 1. Basic components of a crystal diffraction spectrometer system.

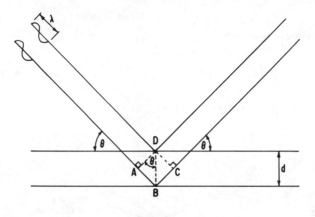

FIGURE 2. Diffraction according to Bragg's law.

The x-ray signal in focused electron beam instruments is fairly weak and can be thought of as originating from a point source; therefore curved-crystal, fully focusing x-ray spectrometers are used in preference to flat-crystal spectrometers of the type normally associated with tube-excited x-ray emission analysis. In a fully focusing spectrometer of the Johansson type, illustrated in Figure 3, the x-ray point source, the specimen, the analyzing crystal, and the detector are all constrained to move on the same circle with radius R, called the focusing circle. Furthermore, the crystal planes are bent to a radius of curvature of $2R$ and the surface of the crystal itself is ground to a radius of curvature of R. As a result of this geometry, all x-rays originating from the point source will have the same incident angle θ on the crystal and will be brought to a focus at the same point on the detector, thereby maximizing the overall collection efficiency of the spectrometer without sacrificing good wavelength resolution. Clearly, if a flat crystal were used, the angle of incidence of the x-ray beam would vary across the length of the crystal, giving rise to broad and possibly overlapped peaks with reduced maximum intensity and peak-to-background ratio. Although Soller slits could be used to obtain a more parallel beam striking the crystal, they would have the adverse effect of reducing the peak intensity.

Figure 3 also illustrates the additional geometric requirement of a constant x-ray takeoff angle Ψ imposed by having a small, fixed x-ray port for the x-ray signal to leave the electron optical chamber. The fully focusing requirement is maintained by moving the analyzing crystal along a straight line away from the sample while rotating the crystal, and moving the detector through a fairly complex path so as to make the focusing circle

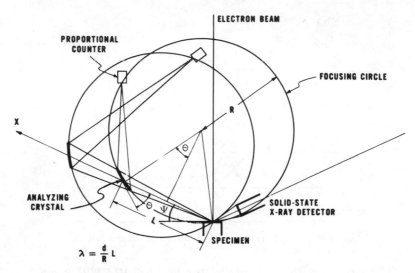

FIGURE 3. Fully focusing crystal spectrometer geometry.

rotate about the point source. An interesting consequence of this arrangement is that the crystal-to-source distance L is directly proportional to the wavelength. This can be shown with the aid of Figure 3: First we write

$$L/2 = R \sin \theta \quad \text{or } L/2R = \sin \theta$$

Combining this with Bragg's law, equation (1), gives

$$n\lambda = 2d \sin \theta = 2dL/2R$$

or for first-order reflections

$$\lambda = (d/R)L \tag{2}$$

with higher order reflections occurring simply at multiplies of the L value for the first-order reflections. Most spectrometers read out directly in L. Actually, the majority of fully focusing spectrometers use crystals that are only bent to a radius of curvature of $2R$ but not ground to be completely tangent to the focusing circle, since grinding a crystal will tend to degrade its resolution by causing increased defects and a broader mosaic spread. This compromise, known as Johann optics, results in some defocusing of the image at the detector, but does not seriously impair resolution. Another type of Johann optics spectrometer involves maintaining a constant crystal-to-source distance and bending the crystal so that λ varies with R as given by equation (2). Although the mechanics of this type of spectrometer are somewhat simpler than the linear spectrometer, only a few crystals, such as mica and LiF, can tolerate repeated flexing without being seriously damaged, and for this reason bending-crystal spectrometers are no longer commonly used in microanalysis. A further variation of the use of Johann optics is incorporated in the "semifocusing spectrometer," which also uses a fixed source-to-crystal distance, but in this arrangement several crystals bent to various fixed radii of curvature are mounted on a carousel so that each can be switched into position rather than using a single bending crystal. However, since the focusing condition is strictly obeyed for only one wavelength per crystal, a certain degree of defocusing and consequent loss of resolution and peak intensity will be experienced at wavelengths other than the optimum value. The advantage of this approach is that placement of the x-ray source on the focusing circle is less critical, so that when x-ray images are obtained by scanning the electron beam over the surface of the sample, they are less susceptible to defocusing effects since the entire image is, in fact, defocused.

In fully focusing instruments in which the electron beam is in the plane of the focusing circle as shown in Figure 3, exact placement of the source on the focusing circle is critical to obtaining both optimum wavelength resolution and accurate wavelength values. In instruments with coaxial light optics, proper placement of the source is achieved by bringing the sample into optical focus in a prealigned position by adjusting the Z axis control of the specimen stage. In instruments without light optics it is necessary to preset the crystal spectrometer to the wavelength of a known element in the sample and adjust the Z axis of the specimen stage to obtain a maximum x-ray signal. Vertical placement of the sample is less critical in instruments in which the plane of the focusing circle is rotated 90° about the axis X, shown in Figure 3. However, displacement of the electron beam in the $X-Y$ plane will lead more rapidly to defocusing effects in scanning images.

Most electron microprobes and scanning electron microscopes can be equipped with more than one crystal spectrometer. Multiple spectrometers, each containing several crystals, are necessary not only for analyzing more than one element at a time, but also to include the variety of crystals required for optimizing performance in different wavelength ranges. Table I lists some of the most commonly used analyzing crystals, showing their comparative resolution, reflectivity, and $2d$ spacings. Since $\sin \theta$ cannot exceed one, Bragg's law establishes an upper limit of $2d$ for the maximum wavelength diffracted by any given crystal. More practical limits are imposed by the spectrometer design itself, since it is obvious from Figure 3 that for $\sin \theta = 1$ and therefore $\theta = 90°$, the detector would have to be at the x-ray source point inside the electron optical column. A lower wavelength limit is imposed by equation (2) since it becomes impractical to physically move the analyzing crystal too close to the specimen.

TABLE I. Crystals Used in Diffraction

Name	$2d$, Å	Lowest atomic number diffracted	Resolution	Reflectivity
α-Quartz($10\bar{1}1$)	6.687	$K\alpha_1$ 15-P $L\alpha_1$ 40-Zr	High	High
KAP($10\bar{1}0$)	26.632	$K\alpha_1$ 8-O $L\alpha_1$ 23-V	Medium	Medium
LiF(200)	4.028	$K\alpha_1$ 19-K $L\alpha_1$ 49-In	High	High
PbSt	100.4	$K\alpha_1$ 5-B	Medium	Medium
PET	8.742	$K\alpha_1$ 13-Al $L\alpha_1$ 36-Kr	Low	High
RAP	26.121	$K\alpha_1$ 8-O $L\alpha_1$ 33-As	Medium	Medium

B. The X-Ray Detector

The most commonly used detector with microanalyzer crystal spectrometer systems is the gas proportional counter shown in Figure 4. It consists of a gas-filled tube with a thin wire, usually tungsten, held at a 1–3 kV potential, running down the center. When an x-ray photon enters the tube through a thin window on the side it causes a photoelectron to be ejected from the gas, which then loses its energy by ionizing other gas atoms. The electrons thus released are then attracted to the central wire, giving rise to a charge pulse. If the gas fill used is P10 (90% argon–10% methane), approximately 27 eV is absorbed per electron–ion pair created. For Cu $K\alpha$, which has an energy of slightly over 8 keV, about 300 electrons will be directly created by the absorption of a single photon. This would be an extremely small amount of charge to detect without a special cyrogenic preamplifier system (low temperature, low noise). But if the positive potential of the anode wire is high enough, secondary ionizations occur which can increase the total charge collected by several orders of magnitude. Figure 5 schematically shows the effect of increasing the bias voltage applied to a tube on the gas amplification factor. The initial increase corresponds to increasing primary charge collection until it is all collected (a gas amplification factor of one) and the curve levels off in the "ionization" region. Increasing the potential beyond this point initiates secondary ionization, the total charge collected increases drastically, and the counter tube enters what is termed the "proportional" region, because the collected charge remains proportional to the energy of the incident photon. Further increasing the voltage causes the tube to enter the Geiger region, where each photon causes a discharge giving rise to a pulse of a fixed size independent

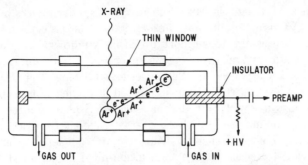

FIGURE 4. A gas flow proportional counter.

of its initial energy, thereby losing the information necessary for pulse height analysis. A further disadvantage is that the counter "dead time," which is the time required for the tube to recover sufficiently to accept the next pulse, increases from a few to several hundred microseconds. Any increase in applied voltage beyond the Geiger region will result in permanent damage to the tube. In practice, operation in the lower part of the proportional region is preferred to minimize the effect of gain shifts with counting rate.

The proportional counter shown in Figure 4 is of the gas flow type, normally used for detecting soft x-rays (>3 Å). A flowing gas, usually P10, is chosen because it is difficult to permanently seal the thin entrance windows necessary to reduce absorption losses. Hendee *et al.*[1] have shown that window transmission for Al $K\alpha$ using 4.5-mil Be foil is 1.2%, for 1-mil Be foil is 55%, for 0.2-mil Mylar is 30%, and for Formvar thin enough to

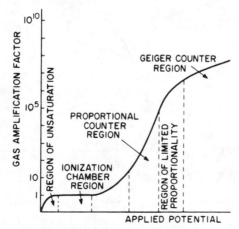

FIGURE 5.   The effect of applied counter tube bias on the gas amplification factor. (Courtesy of E. Bertin.)

give interference fringes is 84%. Since crystal spectrometers are normally kept under vacuum to eliminate absorption of the x-ray beam in the air, it is usually necessary to support ultrathin windows like Formvar and cellulose nitrate on fine wire screens in order to withstand the pressure differential of 1 atm; however, this causes an additional decrease in detector collection efficiency. Recently, unsupported stretched polypropylene films have come into use with considerable success. Sealed counters containing krypton or xenon are used for shorter x-ray wavelengths since, as shown in Figure 6, they have a higher quantum counter efficiency (the fraction of input pulses detected $\times$ 100%) than argon-filled detectors at 1 atm. The efficiency of argon-filled detectors for shorter x-ray wavelengths can, however, be increased by increasing the gas pressure to 2 or 3 atm.

C. Detector Electronics

Since the high voltage is applied directly to the counter tube anode, the preamplifier input is isolated by ac coupling as shown in Figure 1.

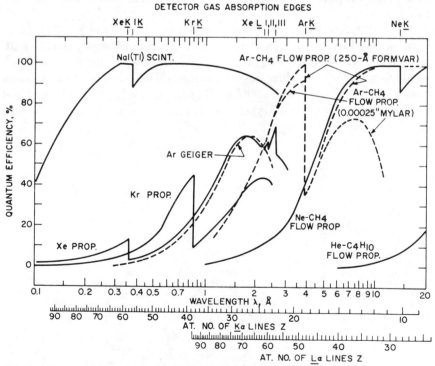

FIGURE 6.   Representative collection efficiencies for proportional counters filled with different gases. (Courtesy of E. Bertin.)

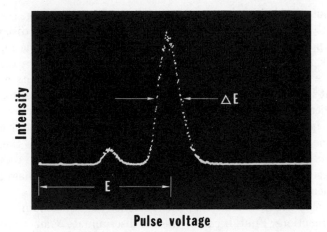

Pulse voltage

FIGURE 7. Multichannel analyzer display of Cr $K\alpha$ pulse amplitude distribution from a flow proportional counter on a crystal spectrometer.

Charge-sensitive preamplifiers, physically located as close as possible to the detector to reduce noise, are normally used to convert the collected charge into a negative voltage pulse with a fast rise time and relatively long decay time (typically about 50 μsec). The size of the preamplifier output pulse is controlled by the applied counter tube voltage so as to be of a sufficient amplitude above the background noise that it can be transmitted several feet via a coaxial cable to the main amplifier. There it is further shaped and amplified to give rise to an almost Gaussian output pulse with a magnitude typically ranging from a few tenths of a volt to 10 V and a total duration of a few microseconds or less. It is important to note at this point that because of the overall linear gain of the detection system, the main amplifier output pulses will still be proportional to the energy of the incident photons striking the detector. This point is further emphasized with the aid of Figure 7, which shows a multichannel analyzer (to be discussed later) display of the main amplifier output collected for several thousand photons of Cr $K\alpha$ radiation. Each successive pulse entering the detector is processed individually and assigned to a storage location in the multichannel analyzer according to the voltage of the main amplifier output pulse. The resulting histogram of pulse energy distribution shows a large main peak and a smaller, lower voltage "escape peak." The natural spread of the main peak arises from the fact that each monoenergetic photon entering the detector does not give rise to the same number of electron–ion pairs since there are several competing processes available by which the initial photo-electron may dissipate its energy. A similar point will be made again in connection with solid state x-ray detectors. The percentage resolution of a detector is defined as 100 times the width of the pulse distribution curve at half-maximum divided by the mean peak voltage. The percentage resolution of a properly functioning tube is about 15–20%. The pulse distribution should be approximately Gaussian and free of large, asymmetric tails. The escape peak is caused when the initial photoelectron ionizes a core shell electron in a counter tube gas atom, which subsequently emits an x-ray photon that leaves the tube without being reabsorbed. The position of the peak will therefore occur at a voltage proportional to that of an x-ray with an energy equal to that of the initial x-ray photon minus the excitation energy of the counter tube gas, which in the case of Figure 7 is the Cr $K\alpha$ energy minus the argon K excitation potential. It is difficult to reduce the size of the escape peak because elements normally have a low mass absorption coefficient for their own radiation.

The single-channel analyzer (SCA) following the main amplifier serves two functions. First, it can be used to discriminate against unwanted pulses by rejecting those that are not within a preselected voltage window above a preselected baseline. Second, any pulse that is not rejected is converted to a fixed amplitude and time duration output pulse, independent of its original value and compatible with the scaler or ratemeter input requirements. There has always been some confusion about the precise role of the energy discrimination capability of the SCA. It cannot improve the energy selection of the spectrometer system for wavelengths close to that of the characteristic line being measured, for this has already been done by the analyzing crystal itself, which has much higher energy resolution than the flow proportional counter. It can, however, eliminate both low- and high-energy noise in the detector electronics as well as higher order reflections ($n > 1$ in Bragg's law) arising from the diffraction of higher energy characteristic lines or continuum that can occur at the same spectrometer setting as the line being measured. Considerable discretion should be used in setting up an SCA because even though it may improve the peak-to-background ratio in some cases, shifts in the pulse distribution out of the energy window with count rate can result in the loss of pulses and serious errors in line intensity measurement for quantitative analysis.

Typical crystal spectrometer scans of ratemeter output vs. wavelength for a nickel-base superalloy are illustrated in Figures 8a for LiF and 8b for RAP. Separation of $K\alpha_1$ and $K\alpha_2$ peaks (6 eV for vanadium) for the major elements in Figure 8a demonstrates the high-energy resolution that can be expected with a crystal spectrometer. Two other capabilities, namely light element detection and peak shift measurement, are illustrated in Figure 9, which shows superimposed computer-controlled scans for boron $K\alpha$ in pure boron, cubic boron nitride, and hexagonal boron nitride. The peak shifts and satellite lines are due to shifts in the outer electron energy states associated with differences in the chemical bonding. Measurements of this type can also be used to fingerprint various cation oxidation states

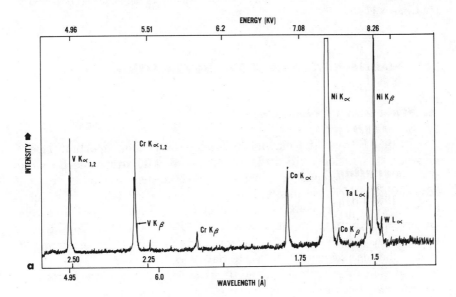

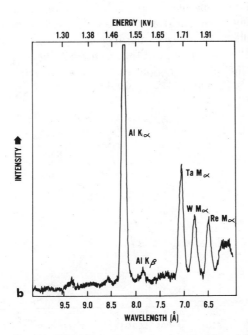

FIGURE 8. Crystal diffraction spectrometer scans of a nickel base superalloy. (a) Scan using an LiF crystal, (b) scan using an RAP crystal.

in metal oxides.[2] A more complete discussion of this effect is given in Chapter XII.

III. SOLID STATE X-RAY DETECTORS

A. *Operating Principles*

In 1968 Fitzgerald, Keil, and Heinrich published an important paper[3] in which they first described the use of a solid state x-ray detector on an electron beam microanalyzer. Although their system was barely capable of resolving adjacent elements, it did demonstrate the feasibility of inter-facing the two instruments, and the next few years saw a period of rapid improvement in detector resolution for Fe $K\alpha$ (6.4 keV) of from 500 eV to less than 150 eV to better accommodate microanalysis requirements. Today the idea of using solid state detectors for low-energy x-ray spec-troscopy is no longer a novelty, and they can be found on a large percentage of scanning electron microscopes and electron microprobes.

The operating principles of the solid state detector are illustrated in Figure 10. The x-ray signal from the sample passes through a thin beryl-ilum window into an evacuated chamber containing a cooled, reverse-bias p–i–n (p-type, intrinsic, n-type) lithium-drifted silicon crystal. Absorption of the x-rays in the intrinsic region results in the formation of electron–hole pairs, which are collected by the applied bias to form a charge pulse, which

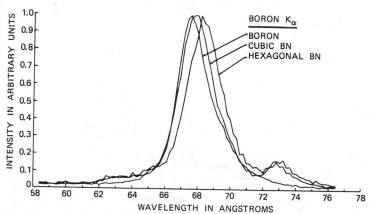

FIGURE 9.   Boron $K\alpha$ scans obtained from pure boron, cubic boron nitride, and hexagonal boron nitride.

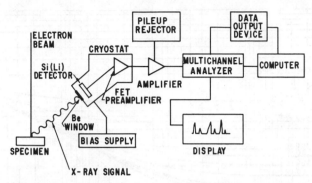

FIGURE 10. Operating schematic of a Si(Li) detector system.

is then converted to a voltage pulse by a charge sensitive preamplifier. The signal is further amplified and shaped by a main amplifier and passed to a multichannel analyzer (MCA) where the pulses are sorted by voltage (which remains proportional to their incident energy). The energy distribution is then displayed on a cathode ray tube or an $X–Y$ recorder or is transmitted to a computer for further processing.

As in the case of the proportional counter, the energy of the incident photon is distributed between useful charge production and other dissipative processes. A stream of essentially monoenergetic photons will therefore, give rise to an inherent spread in the detector output pulse distribution. The observed natural line width can then be obtained by combining the electronic noise contribution and the inherent line width according to the following expression:

$$\text{FWHM(eV)} = \{(\text{FWHM})^2_{\text{noise}} + [2.35(F\epsilon E)^{1/2}]^2\}^{1/2} \tag{3}$$

where FWHM is the observed full-width at half-maximum, $(\text{FWHM})_{\text{noise}}$ is the electronic noise contribution, F is the Fano factor, ϵ is the number of eV per electron–hole pair (about 3.6 eV for Si), and E is the initial x-ray energy. Figure 11 shows values calculated by Woldseth[4] of the observed FWHM as a function of energy for different contributions of electronic noise. The two principal methods of noise reduction presently used are to cool the detector and first-stage preamplifier FET to near liquid nitrogen temperatures and to also use pulsed optical feedback (POF) systems of the type illustrated in Figure 12. With this method the noise normally associated with resistive feedback in preamplifiers is eliminated by simply not using any feedback to drain the charge accumulated from the detector until the preamplifier output reaches a predetermined voltage level. A light-emitting diode is then turned on, which causes a leakage current to

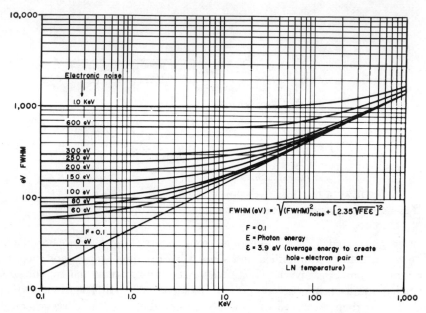

FIGURE 11.   Si(Li) energy resolution, including intrinsic and electronic noise effects as a function of energy (from Woldseth,[4] p. 2.6).

flow in the FET, restoring it to its original operating point. Since considerable noise is generated when the POF is turned on, it is necessary to gate the MCA off during that period. POF preamplifiers are now in common use by most manufacturers, with the exception of ORTEC, who accomplishes

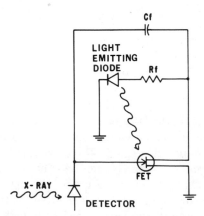

FIGURE 12.   Schematic of pulsed optical feedback in a Si(Li) detector preamplifier system.

the same effect with a proprietary technique called "dynamic charge restoration" which does not require special MCA gating.

The operating characteristics of the main amplifier in a Si(Li) detector system are more critical than with the WDS because all of the spectral dispersion is done electronically. Special circuitry must be used to ensure maximum linearity, low noise, rapid overload recovery, and stable high-count-rate performance. Most commercial amplifiers use pole zero cancellation networks to compensate for pulse overshoot when internal ac coupling is used, and dc restoration circuits to clamp the pulse baseline to a stable reference voltage.

B. Pulse Pileup

High-count-rate effects are also more serious in Si(Li) detector systems where loss of resolution and peak shifts due to pulse pileup can occur, causing increased peak overlap and inaccurate count rate and energy determinations. Furthermore, unlike the WDS, the limiting count rate is the total system count rate for all energies, or more correctly, the energy product, equal to the integrated product of energy times count rate. This can be a particularly serious problem when attempting to optimize sensitivity for a weak peak, since the maximum acceptable energy product of the system will be determined by the accompanying stronger peaks from other elements in the specimen.

The problem of obtaining high count rate capability without seriously impairing resolution and count rate accuracy has become a major concern of detector manufacturers now that the resolution race has slowed down. Two important developments are the use of pulse pileup rejectors and live-time correctors. Pulse pileup occurs when the time duration between pulses approaches the processing time of the main amplifier. It shows up as energy distortion, either of the second of two closely spaced pulses or of both pulses, if the peak amplitude of the first is not reached before the arrival of the second. The resulting effect is to subtract pulses from their proper channels and to add them to channels of higher energy. This point is illustrated in Figure 13, taken from work by Lifshin and Ciccarelli,[5] which shows a spectrum of iron obtained with and without pulse pileup rejection. The difference is most noticeable between the Fe $K\alpha$ and 2 Fe $K\alpha$ peaks. Pulse pileup rejection is accomplished by placing a circuit in parallel with the main amplifier, which can gate the MCA input off if a second pulse arrives within a predetermined time after the main amplifier begins to process a previous pulse. The 2 Fe $K\alpha$ and Fe $K\alpha$ + Fe $K\beta$ are also artifacts which result from pulse on pulse pileup, for two pulses arriving

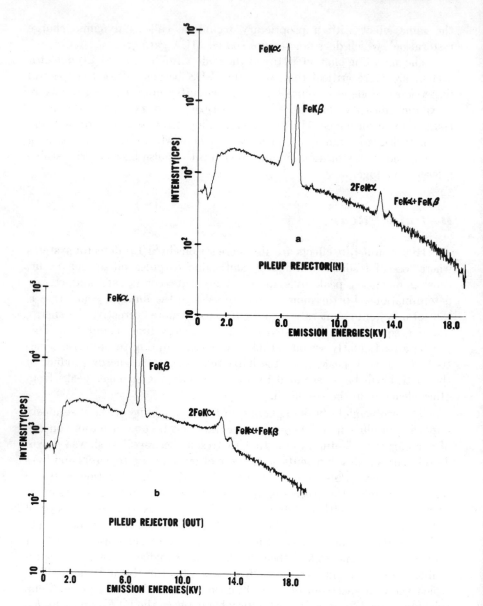

FIGURE 13.   The use of pileup rejection to improve an iron spectrum. Figure from
Lifshin and Ciccarelli.[5]

at the detector simultaneously cannot be distinguished from a single pulse with their combined energy. In performing a quantitative analysis the $K\alpha + K\beta$ and twice the $2K\alpha$ intensities should be added to the $K\alpha$ peak.

Live time refers to that portion of time during an analysis in which the system can accept and process x-ray pulses. Most MCA's provide the option of preselecting either live time or clock time as the period for data collection. If live time is chosen, then the counting periods will be extended over clock time to compensate for any process that temporarily gates the MCA off. These processes include pulsed optical feedback, pulse pileup rejection, and pulse measurement and storage in the MCA. Clearly, any accurate comparative measurements such as those necessary for quantitative analysis must be performed in the live-time mode.

Figure 14 shows several curves of MCA input count rate as a function of main amplifier input count rate for selected amplifier time constants in a Si(Li) detector system using pulse pileup rejection and live-time correction. The end points of the different curves correspond to the maximum count rates possible before serious degradation in resolution occurs when the amplifier can no longer maintain its proper dc reference level. In practice there is little reason to go to this point because the maximum output count rate for each curve is achieved at a lower input counting rate. Figure 14 also illustrates the inevitable tradeoff between good resolution,

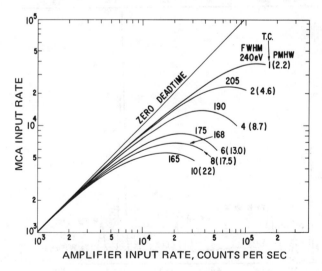

FIGURE 14.   MCA input count rates as a function of main amplifier input count rate for several main amplifier time constants (from Lifshin and Ciccarelli,[5] courtesy of Princeton Gamma-Tech).

associated with a large amplifier time constant, and the high count rate capability associated with a small time constant.

C. *The Multichannel Analyzer*

The function of the multichannel analyzer is to collect, sort, store, and display the pulses received from the main amplifier. In practice, this is accomplished by using an analog-to-digital converter (ADC) interfaced to a hard-wired computer memory. Pulses are first given an acceptance test using an upper and lower level discriminator to see if they are within a preset level of acceptability and also to determine whether any internal events cause the system to be busy. If they are accepted, the standard method of conversion is from pulse amplitude to time. This is accomplished by charging a special stretcher capacitor to a voltage proportional to the peak value of the input signal and counting the number of clock pulses that occur while the capacitor discharges to its original level. In this manner an input signal of from 0 to 10 V can, for example, be linearly converted to some number between 0 and 512. This number then serves

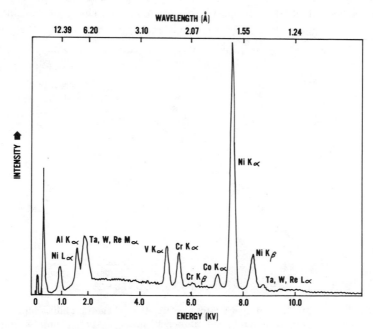

FIGURE 15.   A Si(Li) spectrum of the same nickel-base superalloy sample used to give Figures 8(a) and 8(b).

as an address of storage location in memory where an add-one operation occurs. Thus, in the preceding example, a 5-V pulse would produce an add one in channel 256. Most analyzers offer considerable flexibility in data collection by making various combinations of conversion gains (the resolution of the ADC, one part in 256, 512, 1024, etc.), digital offsets, and memory groups available. In other words, it is possible to take an analyzer with 1024 channel memory and divide it into halves or quarters for storing or displaying multiple spectra.

Although earlier MCA's normally displayed data as a dot histogram of x-ray intensity vs. atomic number, most newer instruments use combined graphical and alphanumeric displays to read out energy and in some cases element identification directly. Figure 15 is an example of a solid state detector spectrum obtained from the same sample used to obtain the crystal spectrometer scans given in Figure 8.

IV. A COMPARISON OF CRYSTAL SPECTROMETERS WITH SOLID STATE X-RAY DETECTORS

An ideal x-ray detector would be small, inexpensive, easy to operate, collect most of the x-rays emitted from a sample, have a resolution better than the natural x-ray line width being measured (a few electron volts), and be capable of collecting spectral data rapidly without losing information. As shown in Table II, neither crystal diffraction spectrometers nor Si(Li) detectors individually have all of these characteristics, but when used together the two techniques do, in fact, complement each other. The superior energy resolution of the crystal spectrometer results in significantly higher peak-to-background ratios and better spectral dispersion,

TABLE II. Comparison of Spectrometers

	Spectrometer	
	Crystal diffraction	Energy dispersion Si(Li)
Energy resolution at 5.9 keV	<10 eV	150 eV
Solid angle	~0.001 ster, variable	~0.01 ster, fixed
Detector efficiency	<30%	100% (for ~3–15 keV)
Focusing requirements	Focusing circles	Minimal
Mechanical design	Complex	No moving parts
Scanning	Required	Not required
Speed of analysis	Minutes to hours	Minutes

thereby minimizing the possibility of peak overlap. This can be readily seen by comparing spectra from the same superalloy standard obtained using a crystal spectrometer (Figures 8a, b) with that obtained from a Si(Li) detector (Figure 15). Figure 8a clearly distinguishes among the Ta $L\alpha$ line, the Ni $K\beta$, and the $W L\alpha$, while the same region of the spectrum is quite unclear in Figure 15. Similarly, the Ta, W, and Re $M\alpha$ lines are easily separated using an RAP crystal but remain unresolved with the Si(Li) detector. The inferior resolution of the solid state detector often makes it necessary to establish the presence of a series of spectral lines for a given element when the identification of a particular peak is ambiguous or the peak of a suspected line is obscured by another element. In such cases it is common practice to collect one or more pure elemental spectra and display them simultaneously with that of the unknown in order to make a direct comparison.

In regions of a Si(Li) detector spectra where lines are clearly separated, a low peak-to-background ratio does not necessarily imply a low sensitivity for a given element. Ziebold[6] has shown that the minimum detectable concentration of a particular element in a given system can be estimated from

$$C_{\rm DL} = 3.29a/[(P/B)Pt]^{1/2} \tag{4}$$

where P/B is the peak-to-background ratio, P is the peak intensity in counts per second per nA (cps/nA), t is the total counting time in seconds, and the parameter a [see Chapter IX, equation (34)] depends on the overall composition of the system being analyzed. For equivalent counting times the factor $1/[(P/B)P]^{1/2}$ can be used as an indication of the relative sensitivity of the two x-ray detection methods. Table III shows a comparison of data obtained at 25 kV for a particular commercial instrument. It can be seen that the sensitivity of the two techniques can be quite similar because the lower peak-to-background ratios of the solid state detector are compensated for by higher cps/nA values made possible by having the detector located closer to the sample and thereby collecting x-rays over a larger solid angle. Two different types of peak-to-background ratio are indicated in Table III, namely measurement at the maximum peak channel divided by the value of an adjacent background channel, and an integrated value obtained by dividing the integrated peak counts by the integrated background. It is the latter, smaller number, which must be used in calculating the sensitivity factor if integrated counts are also used. Sutfin and Ogilvie[7] have shown that an optimum sensitivity will be obtained by using all of the counts contained within 1.2 times the full-width at half-maximum of the peak. It should also be mentioned that although normalized to cps/nA, the data given in Table III were collected

TABLE III. Comparison of EDS Si(Li) and WDS Sensitivity at 25 kV

Element	EDS P/B [a]	EDS P/B [b]	cps/nA [c]	$1/(P^2/B)^{1/2}$
Mg	104	23.2	9367	0.0016
Al	148	14.0	18669	0.0014
Si	112	25.1	18109	0.0011
S	76	21.3	19997	0.0014
Ti	52	13.4	18006	0.0016
Fe	44	17.4	7805	0.0022
Cu	41	12.5	8421	0.0026

Element	Crystal	WDS P/B	cps/nA [d]	$1/(P^2/B)^{1/2}$
Mg	RAP	1970	1680	0,0005
Al	RAP	1280	2180	0.0006
Si	RAP	570	1670	0.0010
Ti	LIF	1260	330	0.0016
Fe	LIF	986	1010	0.0010
Cu	LIF	352	1230	0.0015
Si	PET	570	1670	0.0013
Ti	PET	795	2130	0.0008

[a] P/B measured at peak channel.
[b] Integrated P/B.
[c] Measured at 10^{-10} A.
[d] Measured at 10^{-8} A.

at a different probe current for each detector. This was done because a serious loss of resolution and spectral distortion can occur in Si(Li) detector spectra at high count rates (see Figure 14). Therefore, although the comparison given in Table III may be valid at a probe current of 10^{-10} A, corresponding to a small electron beam size, the Si(Li) detector values for both P/B and cps/nA will be poorer than those given if higher beam currents and consequently a larger probe size could be used for a particular application.

One of the most advantageous characteristics of Si(Li) detector systems is their ability to collect and process x-rays simultaneously from all elements with atomic number above fluorine. Crystal spectrometers, on the other hand, must be scanned to perform a qualitative analysis, thereby limiting the data collection time for a given element to a fraction of the total scan time. Spectra of the type shown in Figure 15 can be collected in several minutes, while those of Figures 8a and 8b took more than $\frac{1}{2}$ hr to collect.

The higher cps/nA capability of Si(Li) detectors is partially due to a higher geometric collection efficiency and partially due to higher inherent detector quantum efficiency. Figure 16, calculated by Woldseth,[4] shows

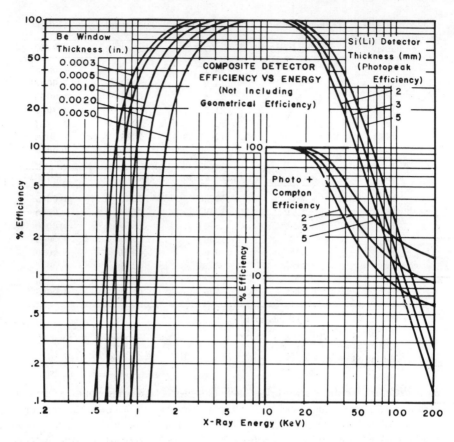

FIGURE 16.   Calculated silicon detector efficiency as a function of x-ray energy
(from Woldseth,[4] p. 2.8).

that for a 3-mm-thick detector sealed with a 0.3-mil Be window, close to
100% of the x-rays in the 2.5–15-keV energy range striking the detector
will be collected. At higher energies a certain percentage of x-ray photons
will be transmitted through the silicon crystal, while at lower energies a
certain percentage of photons will be absorbed in the Be window. Significant absorption of soft x-rays can also occur in the surface "dead layer"
or gold contact layer on the detector crystal. Absorption and detector
noise generally limit light element analysis to fluorine, although windowless
detectors capable of detecting carbon have been reported.[8] By comparison, the quantum counting efficiency of crystal spectrometers is generally
less than 30%, partially due to transmission losses in the proportional
counter tube (see Figure 6) and partially due to losses in the diffraction

crystal. Large-*d*-spacing crystals and thin-window proportional counters make it possible to detect elements down to beryllium.

V. THE ANALYSIS OF X-RAY SPECTRAL DATA

A. General Considerations

It has already been shown in this chapter that qualitative analysis is based on the ability of a spectrometer system to measure characteristic line energies and relate those energies to the presence of specific elements. This process is relatively straightforward if (1) the spectrometer system is properly calibrated, (2) the operating conditions are adequate to give sufficient x-ray counts so that a given peak can be easily distinguished from the corresponding background level, and (3) no serious peak overlaps are present.

Quantitative analysis, on the other hand, involves accurately measuring the intensity of spectral lines corresponding to preselected elements for both samples and standards under identical operating conditions, calculating intensity ratios, and then converting them into chemical concentration by the methods described in Chapter IX. Since quantitative analysis can now be performed with 2–5% relative accuracy, it is obvious that great care must be taken to ensure that the measured response of the x-ray detector system is linear over a wide range of counting rates, and also that the useful signal can be easily extracted from the background. In practice, a knowledge of the absolute spectrometer counting efficiency is generally not required since its effect cancels in establishing intensity ratios. It is necessary to make adjustments for changes in the spectrometer efficiency with count rate, hence the need for dead-time corrections for proportional counters [see Chapter IX, equation (42)] and live-time corrections for Si(Li) detectors. As would be expected, background measurements become increasingly important as peak-to-background ratios get smaller. For example, a 100% error in a background measurement of a peak 100 times larger than the background introduces a 1% error in the measured intensity, whereas the same error in the case of a peak twice background introduces a 50% error. Large peak-to-background ratios are not always the case, even with crystal spectrometers, and accurate background measurement becomes important particularly at low concentrations. The most commonly used method of background measurement with a crystal spectrometer is to detune to wavelengths slightly above and below the tails of the characteristic peak and then establish the value of the background at the peak setting by linear interpolation. It should be emphasized that the spectrometer should not be detuned between sample and standard measurements, because of problems associated with reproducing the spectrometer position to achieve identical operating conditions. The reason is that mechanical backlash normally prevents accurate repositioning to the previous peak value. It is also difficult to obtain 2% reproducibility in the intensity by observing a maximum ratemeter signal while manually adjusting the spectrometer control.

Background measurements with Si(Li) detectors are usually much more critical than with crystal spectrometers because of lower peak-to-background ratios and difficulties related to finding suitable background areas adjacent to the peak being measured. This point is, in fact, illustrated in Figure 15, where numerous peaks appear in close proximity and the background level between them is somewhat uncertain. An alternative approach using basic x-ray theory has been applied by Ware and Reed[9] and Lifshin *et al.*[10] to characterize the shape of background over a larger energy range than that just adjacent to the peak. The remainder of this chapter will be concerned with exploring this approach and its consequences in microanalysis.

B. The Background Shape

The principal source of the background is the x-ray continuum produced by the deceleration of the electron beam in the sample. Early theoretical work by Kramers[11] and experimental studies by Kuhlenkampff[12] concluded that the spectral distribution of the continuum can be given by

$$I_{E^g} = KZ(E_0 - E) \tag{5}$$

where I_{E^g} is the generated x-ray intensity per unit energy interval per incident electron, E_0 is the electron beam voltage, E is the photon energy, Z is the specimen atomic number, and K is a constant, often called Kramers' constant, which is supposedly independent of Z, E_0, and $E_0 - E$. Actually the term intensity in equation (5) refers to the x-ray energy rather than the number of photons per unit energy interval per incident electron N_{E^g}, which is given by

$$N_{E^g} = KZ(E_0 - E)/E \tag{6}$$

This point has been a source of considerable confusion in the literature since equation (5) is often used incorrectly to describe the shape of the spectral distribution of the background observed with a solid state detector. Equation (6) for the intensity of the continuum I_λ was also discussed in Chapter III. This misunderstanding of the definition of intensity was undoubtedly complicated by the fact that the first ionization detectors integrated total radiant energy collected rather than counting individual x-ray photons. The difference in the shape of the two curves is illustrated in Figure 17.

In order to determine the value of Kramers' constant, it is necessary to make absolute measurements of the continuum spectral distribution. It is extremely difficult to make this type of measurement with a crystal diffraction spectrometer on a microanalyzer because the spectrometer efficiency varies with energy, and, furthermore, is generally not known. In his dissertation several years before the availability of solid state x-ray detectors, Green[13] described a series of measurements of both continuum and characteristic x-ray production efficiencies in which he used a proportional counter of known collection efficiency to directly measure x-ray spectra. Figure 18, taken from this work, shows a spectrum obtained from copper. Because the broad characteristic and escape peaks mask much of the continuum, Green did not attempt to measure the distribution, but rather assumed Kramers' equation to be correct and calculated K by equating the area A under the dashed curve with an integrated form of equation (5), namely

$$A \cong \tfrac{1}{2}KZE_0^2 \tag{7}$$

As shown in Figure 23, the values he obtained for K were found to be atomic number dependent. Characteristic production efficiencies were then determined by subtracting the calculated background from the total

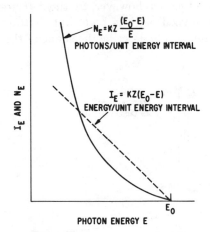

FIGURE 17. The theoretical shape of the x-ray continuum according to Kramers' equation (from Green[13]).

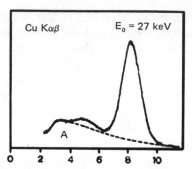

FIGURE 18.   Copper pulse distribution curve obtained by Green[13] using a proportional counter to measure x-ray production efficiencies (from Green[13]).

spectrum. More recently Lifshin *et al.*[10] have performed the same type of experiment, taking advantage of the better energy resolution and higher collection efficiency of a Si(Li) detector system. Figure 19 shows a similar spectrum obtained from copper using a 165-eV (at Fe $K\alpha$) Si(Li) detector 3 mm thick by 6 mm in diameter. The maximum values of the Cu $K\alpha$ and Cu $K\beta$ peaks have been truncated to emphasize the detailed structure of the background. The decrease in intensity at low energy is due to absorption in the detector window. The sharp discontinuity in the background intensity at slightly over 9 keV is caused by self-absorption effects due to the presence of a discontinuity in the copper absorption coefficient at the energy of the copper K excitation potential. The background at the exact energy, 8.979 keV, is obscured, however, by the detector broadening of the Cu $K\beta$ peak. This is a good place to point out that any attempt to apply computer fitting techniques to describe the shape of the observed spectrum

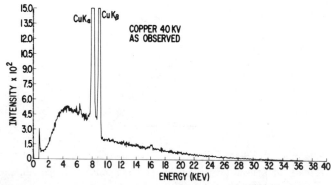

FIGURE 19.   A Si(Li) detector spectrum obtained from a pure copper standard.

for purposes of background subtraction must confront these details, which in the case of multielement samples can be quite complicated. As mentioned earlier, Kramers' equation does not apply to the observed background distribution, but rather to the continuum as it is generated in the specimen. The generated spectrum can be calculated directly from the observed spectrum by the following equations:

$$I_{E^g} = 4\pi n_E E/\Omega f(\chi) R_E iC \tag{8}$$

where I_{E^g} is the intensity of the generated continuum photons per incident electron per keV interval and n_E is the number of observed continuum x-ray photons per keV interval; for characteristic x-rays

$$N_A/4\pi = n_A/\Omega f(\chi_A) iC \tag{9}$$

where $N_A/4\pi$ is the number of generated photons per incident electron per steradian and n_A is the observed count rate for the characteristic line of element A. In these equations Ω is the solid angle for x-ray collection, $f(\chi)$ is the absorption factor (see Chapter IX), R is the backscatter factor (see Chapter IX), i is the probe current, and C is the conversion factor for probe current to electrons. The subscripts E and A refer to particular continuum energies or characteristic lines, respectively. Although these measurements are relatively straightforward, they will only be accurate if the following conditions are met.

1. The probe current is measured in a Faraday cage with a calibrated specimen current meter.

2. Spectral distortion is minimized by keeping the count rate low (see Figure 14).

3. Live-time correction is used.

4. The solid angle of the detector is defined by an aperture of known diameter [smaller than the Si(Li) crystal diameter] at known distance from the specimen.

5. n_E and n_A are corrected for window and detector absorption as well as transmission losses (see Figure 16).

6. Care is taken to prevent unwanted x-rays or electrons from entering the detector. Figure 20 illustrates some of these problems. Electrons scattered from the sample can generate x-rays from the objective lens, specimen chamber, or sample stage. These signals can be reduced or eliminated by a combination of collimation and carbon-coating the possible sources. Reflected electrons can also return to the specimen and excite x-rays which cannot be removed by collimation. In this case carbon coating of the various possible sources greatly reduces the fraction of these electrons because of the low backscatter coefficient of carbon. Electrons can also enter the detector directly through a 0.3-mil Be window, particularly

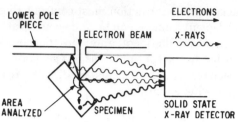

FIGURE 20.   Possible sources of signals detected by Si(Li) detectors.

when accelerating voltages greater than 30 kV are used. They can be eliminated by electrostatic filtering or using a thicker window. The x-rays generated by direct electron excitation are also capable of fluorescing other points in the specimen and specimen environment. This point will be discussed later in this chapter and in Chapter VIII.

Figure 21 shows the results obtained by calculating the generated spectrum from the data given in Figure 19 according to equation (8) and then plotting $I_E{}^g/Z$ $(=N_E{}^g E/Z)$ vs. $E_0 - E$. Again the scale has been truncated to accentuate the details of the background. It can be seen that, as expected, the jump at the K absorption edge is eliminated but that the curve, which should follow equation (5), is not linear. A forced least squares fit of data to Kramers' equation gives a value of $K = 1.89 \times 10^{-6}$ for Cu, which, as shown in Figure 23, is slightly less than that measured by Green. It should be emphasized again that the superior resolution of the Si(Li) detector compared to the proportional counter used in Greens' experiments makes it possible to observe the actual shape of the continuum distribution directly. Figure 23 also shows values of K obtained at 40 kV for other elements, which are compared to data of Green,[13] Dyson,[14] and Compton and Allison.[15] It is evident that Kramers' "constant" is atomic number dependent. Another way of looking at the validity of equation (8) is to plot $I_E{}^g$ vs. Z for fixed values of $E_0 - E$ as shown in Figure 24. Kramers'

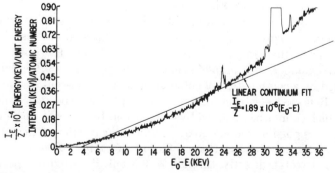

FIGURE 21.   Linear fit to generated continuum spectrum of copper at 40 kV.

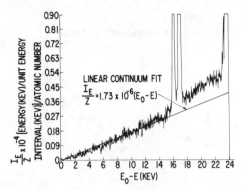

FIGURE 22. Linear fit to generated continuum spectrum of copper at 25 kV.

equation predicts that the slope should be one; however, it appears closer to 1.3. Rao-Sahib and Wittry[16] found a similar result using a crystal spectrometer tuned to a fixed wavelength and varying the atomic number of the sample.

Figure 22 shows results obtained from copper at 25 kV. A least squares fit between 25 kV and the absorption edge at ~9 kV ($E_0 - E$ varying between 0 and 16 kV) in this case shows a reasonably good fit to a straight line. However, it is clear in looking at Figures 21 and 25, the latter showing results obtained from Ag at 40 kV, that a second-degree term will greatly improve the fit over a broader energy range. For Ag at 40 kV the equation is

$$I_{E^g}/Z = 4.682 \times 10^{-8}(E_0 - E)^2 + 7.657 \times 10^{-7}(E_0 - E) - 7.838 \times 10^{-8} \tag{10}$$

Furthermore, as shown in Figure 26, different coefficients apply for different

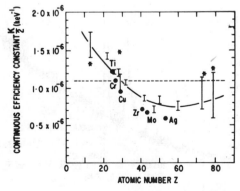

FIGURE 23. Kramers' constant, given as $K/2$, as a function of atomic number: *, Ref. 14; - - -, Ref. 15; vertical bars, Ref. 17; ●, Ref. 10.

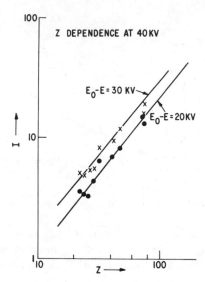

FIGURE 24.   Atomic number dependence of the continuum intensity at fixed
values of $E_0 - E$.

operating voltages. In summary, it appears, on the basis of the data presented in Figures 21–26, that a parabolic fit to the background is a better
way to mathematically describe the continuum than by the use of Kramers'
equation. Furthermore, if Kramers' equation is to be used at all, it should
be recognized that K is a function of Z, E_0, and $E_0 - E$.

Figure 27 shows the success achieved by fitting the 40-kV copper data
of Figure 21 to a parabola, which was then subtracted from the spectrum
and transformed back into the more familiar intensity vs. energy plot
given in Figure 28. It is evident from this figure that the method of back-

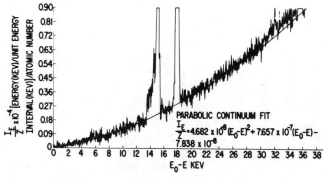

FIGURE 25.   Parabolic fit to generated continuum spectrum of silver at 40 kV.

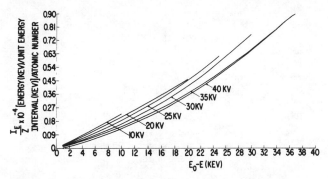

FIGURE 26. Voltage dependence of the silver continuum distribution.

ground subtraction used is extremely effective. The small residual background appearing below the Cu $K\alpha$ peak may be due to errors introduced by applying the same absorption correction for the continuum as for characteristic lines. This assumption is equivalent to postulating that the distribution of continuum production is the same as for characteristic x-rays. This is only approximately correct, however, because the ionization cross sections for the two processes differ slightly. These differences will be more significant at lower energies where absorption effects are greater.

C. Characteristic X-Ray Production Efficiencies

An added benefit of the method of analysis described in the previous section is that the characteristic line intensities that remain following peak subtraction are in an absolute form and can easily be converted into x-ray production efficiencies. Figure 29 shows results obtained for both $K\alpha$ and

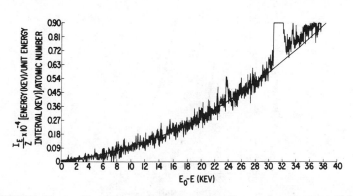

FIGURE 27. Parabolic fit to generated continuum for copper at 40 kV.

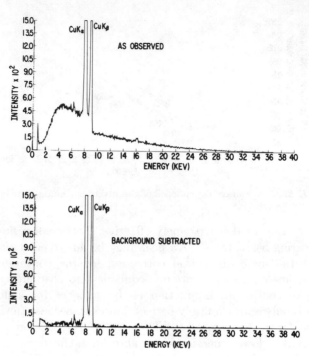

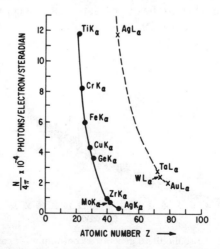

FIGURE 28.   Observed copper spectrum at 40 kV with and without background subtraction.

FIGURE 29.   X-ray production efficiency at 40 kV.

$L\alpha$ lines for a number of elements at 40 kV. Voltage dependence measurements are shown in Figure 30, where the x-ray yield of Cu $K\alpha$, Ti $K\alpha$, and Ag $L\alpha$ is plotted as a function of $U - 1$, where U is the overvoltage ratio, equal to the beam voltage E_0 divided by the critical excitation potential E_c. It can be seen that these results compare quite favorably with those reported by Green and Cosslett.[17] If the operating conditions for this type of measurement prove to be easily reproducible, it is not unreasonable to anticipate the possibility sometime in the future of quantitative measurements using tabulated standard intensities.

D. Indirect-to-Direct X-Ray Excitation Ratios

The characteristic line intensities given in Figure 30 are partially due to direct x-ray production by the primary electron beam, termed direct excitation, and partially due to secondary fluorescence by the continuum, called indirect excitation. Since indirect excitation can occur over a larger volume than direct excitation, it is important to know its relative contribution to observed x-ray intensities, particularly in the examination of fine structure. This point is further emphasized in Chapter VIII, where it is shown that the x-ray signal from small particles of unknown composition may be almost solely due to direct x-ray production. Indirect-to-direct excitation ratios are also required to perform the continuum fluorescence correction that will be described in Chapter IX. This correction is normally small but can be significant in certain cases. The continuum fluorescence models proposed by Henoc[18] and Springer[19] use theoretical expressions for both indirect and direct excitation. In each case, Kramers' equation is used to describe the continuum (see Figure 31) and an integration of

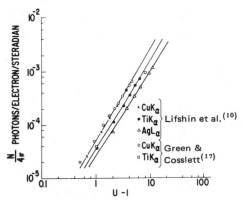

FIGURE 30.   Voltage dependence of x-ray production efficiency for Cu $K\alpha$, Ti $K\alpha$, and Ag $L\alpha$.

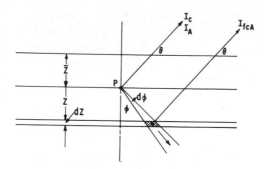

$$dI_{fcA} = \frac{1}{2} C_A \frac{r_k - 1}{r_k} W_k \frac{M_E^A}{M_E} exp(-\bar{Z} M_A CSC\theta) \frac{\ln(1+X)}{X} N(E) dE$$

$$X = \frac{M_A}{M_E} CSC\theta \qquad N(E) = KZ(E_0 - E)/E$$

FIGURE 31.   Model used for the derivation of indirect x-ray production equations (see Chapter IX for definition of terms).

indirect x-ray production is performed for all energies greater than the characteristic absorption edge.

Indirect-to-direct excitation ratios can also be determined experimentally using the methods described in this chapter. Figure 32 shows indirect-to-direct fluorescence ratios calculated by substituting the experimentally determined continuum distribution function into the equation given in Figure 31, using a computer to integrate the resulting expression over energy, and finally dividing the result by the measured characteristic line less the calculated indirect fluorescence contribution. The results are in reasonable agreement with theory but do not give quite as drastic an increase with atomic number as that predicted by Springer.[19]

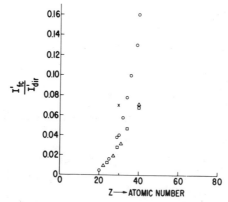

FIGURE 32.   The ratio of indirect to direct x-ray production as a function of atomic number. ☐, Present work; ◯, Ref. 19; △, Ref. 18; ✕, Ref. 17.

REFERENCES

1. C. F. Hendee, S. Fine, and W. B. Brown, *Rev. Sci. Instr.*, **27**, 531 (1956).
2. J. E. Holliday, in *The Electron Microprobe* (T. D. McKinley, K. F. J. Heinrich, and D. B. Wittry, eds.), Wiley New York (1963), p. 3.
3. R. Fitzgerald, K. Keil, and K. Heinrich, *Science*, **159**, 528 (1968).
4. R. Woldseth, in *X-Ray Energy Spectrometry*, Kevex Corp., Burlingame (1973).
5. E. Lifshin and M. F. Ciccarelli, in *SEM/1973 Proceedings of the 6th Annual SEM Symposium*, (O. Johari, ed.), IITRI, Chicago, Illinois (1973), p. 89.
6. T. O. Ziebold, *Anal. Chem.*, **39**, 858 (1967).
7. L. V. Sutfin and R. E. Ogilvie, in *Energy Dispersion X-Ray Analysis: X-Ray and Electron Probe Analysis* (J. C. Russ, ed.), ASTM, Philadelphia, Pennsylvania (1971), p. 197.
8. N. C. Barbi, A. O. Sandborg, J. C. Russ, and C. E. Soderquist in *SEM/1974 Proceedings of the 7th Annual SEM Symposium*, (O. Johari, ed.), IITRI, Chicago, Illinois (1974), p. 151.
9. N. G. Ware and S. J. B. Reed, *J. Phys. E*, **6**, 286 (1973).
10. E. Lifshin, M. F. Ciccarelli and R. B. Bolon, in *Proceedings Eighth National Conference on Electron Probe Analysis*, New Orleans, (1973), p. 29.
11. H. A. Kramers, *Phil. Mag.*, **46**, 836 (1923).
12. H. Kulenkampff, *Ann. Phys.*, **69**, 548 (1922).
13. M. Green, Ph.D. Thesis, University of Cambridge (1962).
14. N. A. Dyson, Ph.D. Thesis, University of Cambridge (1956).
15. A. H. Compton and S. K. Allison, in *X-Rays in Theory and Experiment*, D. Van Nostrand, New York (1943), p. 106.
16. T. S. Rao-Sahib and D. B. Wittry, in *Proceedings of the Sixth International Conference on X-Ray Optics and Microanalysis*, (G. Shinoda, K. Kohra and T. Ichinokawa, eds.), University of Tokyo Press, Tokyo (1972), p. 131.
17. M. Green and V. E. Cosslett, *J. Phys. E*, **1**, 425 (1968).
18. J. Henoc, in *Quantitative Electron Probe Microanalysis*, NBS Special Publication 298, Washington, D.C. (1968), p. 197.
19. G. Springer, *N. Jb. Miner. Abh.*, **106**, 241 (1967).

VIII

MICROANALYSIS OF THIN FILMS AND FINE STRUCTURE

R. B. Bolon, E. Lifshin, and M. F. Ciccarelli

I. INTRODUCTION

Traditionally, quantitative electron microanalysis was almost exclusively performed with normal electron beam incidence on flat samples containing phases of sufficient size (5–10 μm) to ensure complete containment of the volume of x-ray excitation. These restrictions were chosen to give a symmetric distribution of the detected x-rays about the electron optical axis, thereby simplifying the form of the quantitative correction expressions given in Chapter IX. Today, however, electron microprobes have evolved into dual-purpose, high-resolution SEM–EPMA's equipped with specimen stages capable of tilt and rotation in addition to X, Y, and Z translations. Questions therefore arise of how small a region can be analyzed chemically, what method of analysis should be used, and how data collection can be optimized. It will be shown that the principal factors affecting x-ray spatial resolution are (1) probe position and stability, (2) probe current as a function of diameter, (3) electron penetration and scattering, and (4) indirect x-ray excitation. These factors will then be related to the study of thin films and fine structure.

R. B. BOLON, E. LIFSHIN, and M. F. CICCARELLI—General Electric Corporate Research and Development , Schenectady, New York.

II. FACTORS AFFECTING X-RAY SPATIAL RESOLUTION

A. Probe Position and Stability

Although the analysis of micrometer and submicrometer structure has always been of major interest to metallurgists and biologists, earlier microanalyzers were not equipped with adequate methods for viewing specimens, since the resolution of their coaxial light optics and scanning systems was generally limited to about 1 μm. Problems in specimen viewing are less important now, however, due to the new generation of instruments, which combine high-resolution ($<$200 Å) secondary electron imaging capabilities with crystal spectrometers and solid state x-ray detection systems. As will be described in the following section, simply being able to observe a specific detail does not necessarily imply that it can be chemically analyzed. The basic prerequisites are that it be possible to physically locate the electron beam on the feature of interest and that the electronics and mechanical stage be sufficiently stable so as to ensure that the beam will remain in the preselected spot for the time necessary to collect an adequate x-ray signal.

B. Probe Current as a Function of Diameter

The generation of x-rays in a focused electron beam instrument is a relatively inefficient process since typically less than one characteristic x-ray photon is emitted per thousand electrons incident on a pure element sample. This means that a solid state detector subtending an angle of 0.01 ster will have an overall efficiency of about 10^{-5} photon/electron (5×10^4 photons/nA) as compared to a crystal spectrometer, which is typically two orders of magnitude less. A comparison of representative counting rates per nA for pure elements is presented in Table III, Chapter VII. At 25 kV a probe current of 1 nA (10^{-9} A) corresponds to a probe size of about 2000 Å; therefore, if the element of interest is present in amounts of less than 1%, the total x-ray signal collected in a reasonable time may be statistically insufficient to establish its presence above background. In practice, time limitations are imposed by the rates of specimen contamination and instrumental drift. However, because x-ray emission varies directly with beam current, the option always exists to increase the counting rate by increasing the beam current. As discussed in Chapter II, Smith[1] has shown, considering only diffraction and spherical aberration effects, that the relationship between probe diameter d and the maximum probe current i_{max} is given by

$$i_{max} = 1.26 \frac{J_c}{T} \left(\frac{0.51 d^{8/3}}{C_s^{2/3} \lambda^2} - 1 \right) \times 10^{-10} \qquad (1)$$

where J_c (A/cm^2) is the cathode emission current density, C_s (cm) is the spherical aberration coefficient, T (°K) is the cathode temperature, and λ is the electron wavelength $= 12.4/\sqrt{E_0}$, where E_0 is the incident beam energy in eV. This equation predicts a drastic decrease in maximum obtainable probe current, and therefore x-ray emission, with decreasing probe size. The only way of increasing i_{max} for a fixed probe size is to increase the source brightness or the incident beam voltage. It has been shown by Broers[2] that LaB$_6$ cathodes can give a brightness increase of as much as a factor of ten over conventional W sources, and in the case of very small probe sizes, field emission sources can give an improvement of as much as several orders of magnitude. The practical significance of equation (1) is that in order to halve the probe size to examine finer structure, the x-ray intensity is decreased by almost a factor of ten.

Equation (1) defines the probe–current relationship under ideal conditions free from astigmatism due to column contamination, stray fields, and mechanical vibration. A more realistic feeling for this relationship can be obtained from direct measurement of the probe diameter as shown in Figure 1. The electron beam is scanned over a knife edge and the transmitted electron signal displayed in a profile mode. If the probe is assumed to have a Gaussian distribution, then the current density is given by

$$j = (i/0.36\pi d^2) \exp(-r^2/0.36 d^2) \qquad (2)$$

where d is the beam diameter defined at FWHM, i is the total beam current measured in a Faraday cage, and r is the distance from the center of the beam; then d can be determined from

$$i_x/i = \tfrac{1}{2}[1 + \operatorname{erf}(x/0.36 d)] \qquad (3)$$

where i_x is the current detected when the knife edge is at a distance x from the center of the beam. From this distribution it can be shown that d can be approximated by 1.1 times the horizontal distance from the point of 10% intensity to that of 90% maximum intensity. Representative measured values are given in Table I for beam voltages of 10, 20, and 30 kV and objective lens working distances of 13 and 37 mm. Three points should be evident. First, the higher current and lower beam voltages necessary to examine light elements and L or M lines from certain higher atomic numbers are not compatible with obtaining the smallest possible beam size for a

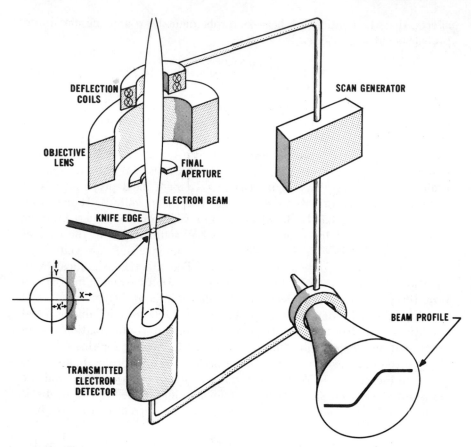

FIGURE 1. Schematic of experimental apparatus for beam diameter measurements.

given current. Second, increasing the working distance to accommodate retractable light optics and electron traps also degrades resolution. Third, under actual operating conditions resolution is often worse than predicted by equation (1), because the probe distribution is generally not Gaussian, due to instrumental design factors and column contamination. Therefore, in practice, the real distribution can be several times larger than the anticipated d value.

C. Electron Beam Penetration and Scattering

The most critical factors determining submicrometer x-ray spatial resolution are electron penetration and scattering. Figure 3 in Chapter III

TABLE I. Beam Diameter for Different Beam Currents, Voltages, and Working Distances [a]

Beam current, A	Beam diameter,[b] Å		
	at 10 kV	at 20 kV	at 30 kV
10^{-11}	720/830	333/736	405/738
10^{-10}	1124/3030	405/630	343/540
10^{-9}	1730/3420	1062/2250	594/1260
10^{-8}	14400/23850	3960/5004	1530/4780

[a] 200-μm aperture.
[b] The first value is for a working distance of 13 mm, the second for 37 mm.

shows that even if an infinitely small electron probe strikes a sample, electrons may travel up to several micrometers in all directions before losing all of their initial energy, or in this case, having energy below the critical excitation potential for x-ray production. The size of the excited x-ray volume increases with increasing beam voltage, as shown in Table II, where the electron range for x-ray production (last column) has been cal-

TABLE II. The Effect of Incident Beam Energy on EDS Sensitivity and Electron Range for X-Ray Production

	EDS P/B [a]	EDS P/B [b]	cps/nA	$1/(P^2/B)^{1/2}$	Range,[c] μm
10 keV					
Al	47	16.1	5,697	0.0024	0.850
Ti	31	11.1	2,197	0.0050	0.376
Cu	5	2.9	85	0.0460	0.076
20 keV					
Al	146	28.0	17,144	0.0011	2.499
Ti	53	12.9	13,583	0.0019	1.364
Cu	28	10.8	5,154	0.0036	0.572
30 keV					
Al	205	31.2	22,789	0.0009	4.634
Ti	62	12.1	24,728	0.0020	2.642
Cu	51	14.4	13,084	0.0020	1.216
40 keV					
Al	68	13.0	29,001	0.0012	7.163
Ti	36	5.8	43,925	0.0016	4.156
Cu	18	4.2	29,822	0.0025	1.978

[a] P/B measured at peak channel.
[b] Integrated P/B.
Range $= 0.077(E_o^{1.5} - E_c^{1.5})/\rho$. According to Reed.[3]

culated from the Reed[3] range equation. This table also shows that increasing the beam voltage can improve x-ray peak-to-background ratios and the x-ray intensity per unit beam current. The sensitivity, as described in Chapter VII, is indicated by $1/(P^2/B)^{1/2}$. It is difficult to make an exact prediction of the changes in peak intensity and peak-to-background ratio with accelerating voltage. For some elements peak intensity varies as $(E_0 - E_c)^{1.6}$,[4] as discussed in Chapter III, over a limited range of E_0, where E_c is the critical excitation potential, but the corresponding background varies almost directly with E_0. Measurements of this type have been almost exclusively made on pure elements and may not be valid for alloys, due to absorption and shifts in the distribution of the continuum that can result in decreasing sensitivity with increasing voltage. These effects are particularly important for the analysis of a light element in a heavy element matrix.

Understanding the relationships given in Tables I and II is an essential prerequisite to the quantitative analysis of micrometer and submicrometer structures, since the most commonly used method of analysis, the ZAF method, requires not only that the x-ray excited volume be contained within the phase being analyzed, but also that the generated x-ray signal exits through that phase. The present practical limit of x-ray spatial resolution by conventional analysis methods is therefore rarely less than 1 or 2 μm, and the analyst is often uncertain as to whether or not the beam is adequately contained within the region being analyzed.

D. Indirect X-Ray Production

As described in Chapter VII, the measured characteristic x-ray signal of a pure element is due to both direct excitation by the electron beam and indirect excitation by the continuum. In multielement systems it is also possible to have additional contributions from secondary fluorescence by the characteristic lines of other elements in the specimen. The important point to mention with regard to fine structure analysis is that the volume of direct x-ray production, which is defined by the extent of electron penetration and scattering, can be significantly smaller than that for x-rays produced by either the continuum or other characteristic lines. This difference is schematically illustrated in Figures 20 and 31 in Chapter VII, where it is shown that the volume of direct excitation serves as a source of x-rays which radiate out in all directions until they are absorbed, up to several micrometers away, causing secondary fluorescence.

There are two undesirable results from this effect. The first is the erroneous identification of elements in a phase that are actually present in adjacent phases. Duke and Brett,[5] for example, used an undiffused sample of pure copper bonded to pure iron to show that Fe $K\alpha$ radiation could be detected with the electron beam located up to 30 μm away from the interface on the pure copper. The second effect is that quantitative analysis is complicated by the fact that the line intensities measured from a small particle will be less than those obtained from a larger phase of the same composition measured under identical operating conditions, because the characteristic line intensities will lack the contribution due to indirect fluorescence. This point will be further illustrated in a later section. Generally, the relative contributions of direct and indirect excitation have not been measured experimentally but must be calculated theoretically. In Figure 32, Chapter VII, it can be seen that for Cu $K\alpha$ in pure copper, for example, about 4% of the observed characteristic line intensity is due to excitation by the continuum. Estimates of the magnitude of secondary fluorescence effects by characteristic lines can be made using the methods described in Chapter IX.

III. CHARACTERIZING THE X-RAY-EXCITED VOLUME

A. Depth Distribution Profile

Quantitative electron microanalysis involves converting measured intensity ratios (sample/standard) into chemical composition using either empirical or theoretical methods. Implicit in any theoretical scheme is the determination of the generated intensity ratios, which requires that the observed intensities be corrected for absorption effects. However, this can only be accomplished if the spatial distribution of x-ray generation is known. In the case of a bulk sample it is sufficient to describe only the distribution of x-ray production with depth by the function $\phi(\rho z)$. The intensity generated from a thin layer $d(\rho z)$ thick at a mass depth of ρz is given by

$$dI = \phi(\rho z)\, d(\rho z) \tag{4}$$

which can be integrated to give the total generated intensity:

$$I = \int_0^\infty \phi(\rho z)\, d(\rho z) \tag{5}$$

If an absorption correction is applied, then the observed intensity is given by

$$I' = \int_0^\infty \phi(\rho z)\, \exp[(-\mu/\rho)\rho z \csc \Psi]\, d(\rho z) \tag{6}$$

The geometric considerations are illustrated in Figure 2. The first measured values of $\phi(\rho z)$ curves were obtained by Castaing and Deschamps[6] for Al, Cu, and Au. The basis of the experiment was to measure the intensity of radiation emitted from layers of known thickness at different known depths relative to a similar layer isolated in space. To do so required the use of a tracer material for the layer with a different characteristic wavelength but with similar x-ray generation properties to the element being measured. An example of $\phi(\rho z)$ curve for Au $L\alpha$ is given in Figure 3. In practice, relatively few $\phi(\rho z)$ curves have been measured for pure elements and none have been measured for alloys. Instead, the methods developed in Chapter IX are routinely used for bulk materials.

B. Monte Carlo Calculations

An alternative method for determining the distribution of x-ray excitation is the use of Monte Carlo (MC) calculations. This technique, already mentioned in Chapter III, has the advantage that it describes x-ray production in three dimensions and is furthermore not limited to single-phase materials, thus making it useful for thin-film and fine-particle analysis. Although a number of different types of Monte Carlo calculations are presently in use,[7-9] they all have certain common characteristics. Initially a mean electron range is found for a given energy and target material, and then divided into a number of fixed- or variable-length segments. Finally a large number of simulated trajectories is generated segment by segment,

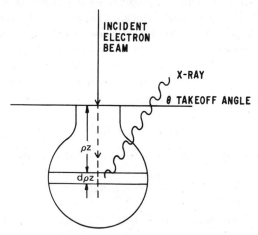

FIGURE 2.   Schematic representation of x-ray excited volume and x-ray emission from a layer $d(\rho z)$ at a depth ρz.

using random numbers in conjunction with scattering probability expressions.

A common but mistaken notion is that these trajectories are possible electron paths. This is of course wrong because, in the first place, the Bethe energy loss expression[10] (Chapter III), which is most commonly used to calculate electron ranges, describes only the average energy loss of a large number of electrons, not the energy loss of individual electrons. Second, the choice of either fixed- or variable-length segments is also arbitrary, since mean free path expressions are also derived from collective, not individual, electron behavior. A restriction is placed, however, by the dimensions of the structure to be analyzed, which requires that the step size be significantly smaller than the structure.

In other words, what is done is to first use equations characterizing the collective behavior of a large number of electrons to describe individual electron energy loss and scattering behavior in a particular sample geometry. The results of many trajectories are then used to define the collective behavior of a large number of electrons. During this double transformation a certain amount of information is lost, and this fact, combined with the inherent inaccuracies of the theoretical equations used, is the reason why most models must rely on some empirical parameters to adjust results to fit experimental data. Eventually, the most useful model will be one that best describes the greatest number of target materials, geometries, and operating conditions. Such a model will probably use several empirical factors so as to fit calculated results to a wide variety of experimental data.

The use of MC calculations for the study of small inclusions and single-element films can be illustrated by the work of Bolon and Lifshin.[11] The model used is a modified version of that proposed by Curgenven and Duncumb.[12] In single-phase materials it can be summarized in the following manner. Electrons striking a specimen are assumed to give up their energy in accordance with the Bethe retardation law:

$$-\frac{dE_m}{dX} = 7.85 \times 10^4 \frac{Z}{A} \frac{\rho}{E_m} \ln \left(\frac{1.166E_m}{J}\right) \qquad (7)$$

where E_m is the mean electron energy, X is the path length traversed, Z is the atomic number of the specimen, A is the atomic weight, and J is its mean ionization potential. The mass thickness range of an electron can then be calculated from

$$R = - \int_{E_R}^{E_0} \frac{1}{dE/d(\rho X)} \, dE \qquad (8)$$

where E_R is the cutoff energy below which the electrons are of no further interest. However, the average depth of penetration will be considerably less than R since the actual paths of electrons in the sample will not follow straight lines, due primarily to elastic scattering from the atomic nuclei. Simulated electron trajectories are computed by dividing the maximum range into 100 parts and calculating the scattering at each step from the Rutherford scattering equation:

$$\cot \frac{\beta}{2} = \frac{2P_0\sqrt{Y}}{b} \tag{9}$$

where β is the scattering angle, Y a random number from zero to one, P_0 the maximum impact parameter, in angstroms, chosen to give the proper back-scattered fraction, and $b = 1.44 \times 10^{-2}Z/E$. The azimuthal scattering direction is in turn selected from 2π radians by a random number. X-ray production along the trajectory is estimated by comparing the ionization cross section Q, calculated at the midpoint of each step by Bethe's equation as modified by Worthington and Tomlin,[13] with a random number for all $E > E_c$ where

$$Q = \text{const} \times (\ln U)/UE_c^2 \tag{10}$$

and $U = E/E_c$.

In order to describe electron trajectories and x-ray production in two-phase materials, the following modifications were made to the Curgenven and Duncumb program:

1. Maximum electron ranges were calculated for both phases, divided into 100 parts, and tables of energy as a function of path length stored for each phase.

2. An equation describing the phase boundaries was used to determine which phase the electron under consideration was in at the end of each step. The appropriate step size, scattering equation, and ionization equation for that phase were then used to calculated the next step.

3. An option was provided to include the effect of electron probe size. This was accomplished by assuming that the current density j in the probe could be described by the Gaussian distribution given in equation (2).

4. A different scattering expression, proposed by Roth,[14] was incorporated when it was found that the Rutherford equation [equation (9)] did not adequately describe the experimental data and is given by

$$\sin \frac{\beta}{2} = \frac{-\Delta s}{\lambda} \ln[e^{-\lambda/\Delta s} + (1 - e^{-\lambda/\Delta s})Y]^{1/2} \tag{11}$$

where β is the scattering angle, Δs the step size, and λ the mean free path of an electron, with

$$1/\lambda = 1.964 \times 10^4 (Z^2/AE^2) \ln(17.3E^{1/2}/Z^{1/3}) \tag{12}$$

as given by Bethe, Rose, and Smith.[10] In order to yield the correct backscattered electron fraction, however, it was found necessary to multiply the mean free path by the empirically determined factor

$$F = 1 + [Z/2(N_s E_0)^{1/2}] \tag{13}$$

where N_s is the number of steps.

5. A histogram giving, as a function of depth ρz, the ionization cross section normalized to its value at step 1 and by the total number of trajectories was introduced to give $\phi(\rho z)$. This replaced the process of converting from an ionization probability to an event via a random number and then back to a probability upon completion of the calculations. An example of a $\phi(\rho z)$ histogram calculated by Monte Carlo techniques is given in Figure 3, where it is compared to the experimental data of Castaing and Deschamps[6] obtained by the tracer technique. The lack of smoothness in the calculated curve is due largely to using only 500 electron trajectories. Note also that the curve starts at a value greater than one, then peaks at a value of about 2.3 before dropping off. This is a result of lateral electron scattering as well as backscattering from lower levels.

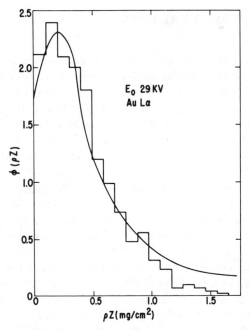

FIGURE 3. The function $\phi(\rho z)$ vs. ρz for Au $L\alpha$ comparing a Monte Carlo histogram with the experimental data of Castaing and Deschamps.

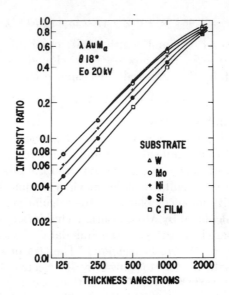

FIGURE 4.   Gold film calibration curves measured on various substrates. Take-off angle $\Psi = 18°$. (Taken from Bolon and Lifshin.[11])

IV. THIN-FILM ANALYSIS

From the standpoint of electron beam microanalysis, a thin film can be defined as a layer of material that is not sufficiently thick to totally contain all of the forward scattered electrons entering it capable of exciting the characteristic lines being measured. In other words, the x-ray intensities measured from such a layer would always be less than those emitted from a bulk sample of the same composition under identical operating conditions. This definition can even be expanded to include layers that do not include the volume of indirect excitation, since, as pointed out in a previous section, they too, should give intensities less than those obtained on bulk standards. Although considerable work remains to be done in this field, techniques have been developed for determining the thickness and composition of supported and unsupported films using both empirical and theoretical methods.

A calibration curve of x-ray intensity, usually normalized to a bulk standard, is made as a function of film thickness in the empirical method. Figure 4 shows a series of calibration curves of x-ray intensity ratios measured for gold films on various substrates for a takeoff angle Ψ of 18°.[11] It is clearly evident that the relative intensity increases both with film thickness and also with substrate atomic number for films of any given thickness. The latter effect is due to increased excitation from backscattered electrons as the substrate atomic number increases. Figure 5 shows the effect of beam voltage on the calibration curves obtained for gold films on a silicon substrate. As would be expected, the intensity ratio for a given film thickness decreases with increasing voltage due to increased penetration of the electron beam into the sample. One of the most difficult problems of getting good calibration curves is knowing the thickness of the deposited film. In the case of the data given in Figures 4 and 5, a special rotating shadow mask was designed so that seven films varying in thickness by factors of two could be vapor-deposited simultaneously. It was then only necessary to measure the thickest film by optical interference microscopy where the accuracy of that technique is greatest. In a similar study, Butz and Wagner[15] used a water-cooled quartz oscillator balance to measure W, Nb, Al, and Mo films on various refractory metal substrates so as to obtain calibration curves in the 2–300 Å range. Generally calibration curves are not available for most practical problems and it is necessary to resort to theoretical methods.

One of the first theoretical approaches, as reported by Sweeney *et al.*,[16] was to compute the ratio k of the intensity measured on a thin film $I_A{}^f$ for element A to the intensity measured on bulk standard of element A, $I_A{}^b$, according to the following equation:

$$k = \frac{I_A{}^f}{I_A{}^b} = \frac{C_A \displaystyle\int_0^{\rho t} \phi'(\rho z)\,\exp[(-\mu/\rho)_A \rho z\,\csc\Psi]\,d(\rho z)}{\displaystyle\int_0^{\infty} \phi(\rho z)\,\exp[(-\mu/\rho)_A{}^\circ \rho z\,\csc\Psi]\,d(\rho z)} \tag{14}$$

where C_A is the composition of element A, $(\mu/\rho)_A$ and $(\mu/\rho)_A{}^\circ$ are the mass

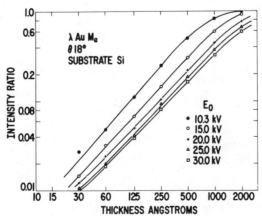

FIGURE 5. Gold film calibration curves measured on a silicon substrate for various voltages. Takeoff angle $\Psi = 18°$. (Taken from Bolon and Lifshin.[11])

absorption coefficients for A radiation in the sample and standard, respectively, and ρt is the mass thickness of the film. The accuracy of this method is fairly limited, however, since the $\phi'(\rho z)$ curve for the thin film will not be the same as the $\phi(\rho z)$ curve for the standard due to differences in electron scattering and stopping powers. As pointed out earlier, relatively few $\phi(\rho z)$ curves have been measured experimentally. Sweeney *et al.*[16] did not describe how such calculations could be done theoretically and futhermore assumed that $\phi(\rho z)$ and $\phi'(\rho z)$ are the same. More recently, however, Reuter[17] has developed a rigorous method for calculating $\phi(\rho z)$ curves for pure elemental films on substrates and bulk standards. He was able, with his technique, to predict intensity ratios with better than 10% accuracy for a number of different film–substrate combinations.

Figures 6 and 7 show some typical computer-drawn projections of MC calculations for bulk and thin (500 Å) gold films on various substrates corresponding to some of the data in Figures 4 and 5. Theoretical intensity ratios for some of the same data were calculated using the MC program to generate $\phi'(\rho z)$ and $\phi(\rho z)$ profiles for the thin films and bulk, respectively.

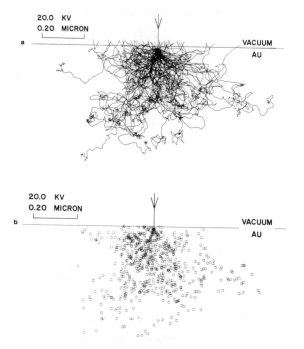

FIGURE 6.   Monte Carlo computer-simulated electron and x-ray distributions in bulk gold for 20-keV electrons.

TABLE III. Comparison of Monte Carlo Calculations and Experimental Data for 500-Å Au Films on Various Substrates [a]

Substrate	k Calc.	k Meas.
Unsupported	0.202	0.183
Si	0.220	0.221
Ni	0.290	0.257
Mo	0.311	0.290
W	0.361	0.310

[a] Operating conditions, $E_0 = 20.0$ kV, $\Psi = 18.0°$, Au $M\alpha$.

These profiles were then converted into intensity, k, ratios by integrating equation (6) to the film thickness (rather than to infinity) and dividing by the integrated bulk value. A comparison of experimental data with Monte Carlo calculations is presented in Table III, which shows the results from 500-Å gold films on different substrates, and in Table IV, which shows the results from different thicknesses of gold films on silicon at 20 kV. Although the agreement between experimental and theoretical values in Table III is good, the fairly large discrepancy in Table IV for the 0.2-μm film led the authors to try the scattering expression proposed by Roth.[14]

Additional comparisons are shown in Table V, in which MC values are compared with experimental data obtained by Reuter.[17] The agreement is good for Au on Pt at 30 kV and for Cu on Au at 10 kV; however, in the case of Cu on Au at 30 kV the MC value is significantly less. This is expected since there is a strong fluorescence effect of Cu by Au which is not taken into consideration by the MC method. The discrepancy in the case of B is not so easily explained but could be caused either by a failure of the model or due to experimental difficulties associated with low-voltage (4 kV) analysis.

TABLE IV. Comparison of Monte Carlo Calculations and Experimental Data for Au Films on Si Substrates [a]

Thickness, μm	k Calc. (Rutherford)	k Meas.	k Calc. (Roth)
0.025	0.074	0.104	0.096
0.05	0.22	0.22	0.23
0.10	0.50	0.44	0.50
0.20	0.90	0.78	0.77
0.40	1.0	1.0	1.0

[a] Operating conditions, Au $M\alpha$, $E_0 = 20.0$ kV, $\Psi = 18.0°$.

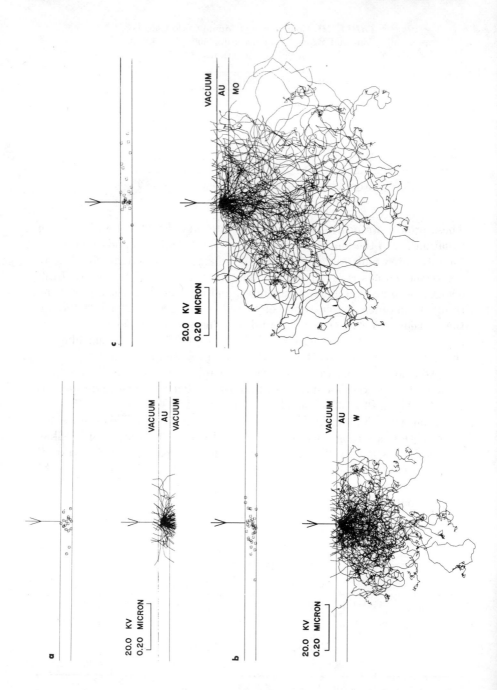

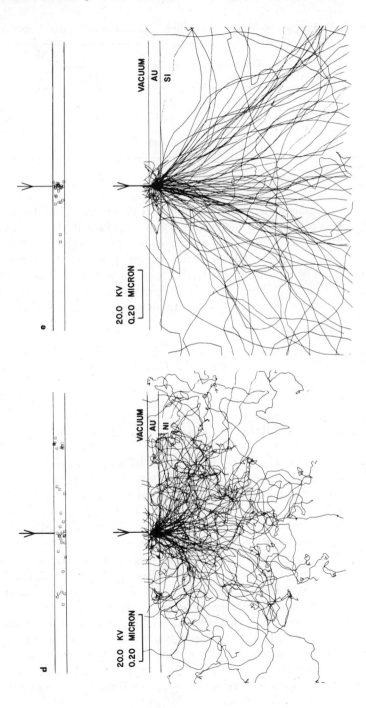

FIGURE 7.   Monte Carlo computer-simulated electron and x-ray distributions in 500-Å gold films on various substrates.

TABLE V. Comparison of Experimental vs. MC-Calculated X-Ray k Ratios [a]

E_0, kV	Target	Thickness, μm	Step size, μm	k Ratio Meas.	k Ratio Calc.
30.	Au on Pt	0.100	0.026	0.276	0.260
10.	Cu on Au	0.093	0.0056	0.590	0.613
30.	Cu on Au	0.093	0.036	0.104	0.077
4.	Au on B	0.0537	0.0027	0.407	0.509

[a] $\Psi = 52.5°$.

The thin-film analysis techniques described thus far have been principally concerned with determining film thickness if the composition is known. Other methods have been developed for determining the composition even if the film thickness is not known. For very thin films Philibert and Tixier[18] have shown that if fluorescence and absorption effects are assumed to be negligible, then the ratio of corrected relative intensities $(k = I'/I^b)$ for two elements A and B in an unsupported thin film sample is given by

$$\frac{k_A}{k_B} = \frac{C_A}{C_B} \frac{P_A}{P_B} \tag{15}$$

where

$$P_A = \frac{\log U_A}{E_c{}^A R_A / S_A}, \qquad \frac{1}{S_A} = \int_1^{U_A} \frac{\log U \, dU}{M_A \log U W_A}$$

with $M_A = Z_A/A_A$, $U_A = E_0/E_c{}^A$, $W_A = 1.166 E_c{}^A / J_A$, E_0 is the beam voltage, $E_c{}^A$ is the critical excitation potential of A, R_A is the backscatter factor, and S_A is the mean ionization potential of element A. (See Chapter IX for specific discussions of these factors.) They applied this technique to the study of a thin Ni–Al precipitate extracted from a stainless steel by carbon replication. They show that for an operating potential of 30 kV, the ratio P_{Al}/P_{Ni} equals 0.63. Therefore a k_{Ni}/k_{Al} value of 3.27 in equation (15) gives a C_{Ni}/C_{Al} ratio of 2.1, thus favoring NiAl as the composition of the precipitate. However, if the uncorrected k ratio had been used, the phase would have been identified as Ni_2Al or Ni_3Al.

Another method which has been successfully applied to the quantitative analysis of thin films is that proposed by Hall,[19] which will be described in detail in Chapter XIII on biological analysis. He assumes that the observed intensity from a thin film is proportional to the elemental mass per unit area S_x and that the x-ray continuum can be used as a measure of the total mass per area S in a sample. Therefore the concentration is given by

$$C_x = S_x/S \qquad (16)$$

The principal assumption of the model is that a measurement of the characteristic line intensity can be related to S_x while a simultaneous measurement of the continuum can be used to determine S. In his original work, characteristic line intensities were measured with a crystal spectrometer while the continuum was measured directly with a proportional counter using pulse height analysis to eliminate the characteristic peaks. It is now possible, however, to make both measurements simultaneously using a single Si(Li) detector system.

Quantitative analysis of supported films has been successfully performed by Kyser and Murata[20] using MC techniques. Their method consists in generating a series of calibration curves of relative intensity vs. composition for various thicknesses for each element and then using graphical analysis to give a unique fit for thickness and composition consistent with experimental data (see Chapter XII).

V. PARTICLES, INCLUSIONS, AND FINE STRUCTURES

Small particles, inclusions, and second-phase structures with dimensions less than the electron-beam-excited volume constitute a class of problems that practically every microprobe analyst has encountered at one time or another. Typical examples include the identification of airborne particulates, casting inclusions, precipitate particles, second-phase structure, and fine detail in thin biological sections. As described earlier, qualitative analyses on this scale are usually complicated because it is often difficult to determine whether or not the characteristic lines detected originate from elements in the region of interest or nearby phases. One method of circumventing this problem is to use extraction replication techniques to separate the particles from the matrix. If the particles are originally free of any matrix, such as atmospheric pollutants, they can be trapped on a filter and transferred to a low-background substrate such as a carbon supporting film. Maggiore and Rubin[21] have, for example, described a technique for collecting particles on a Nuclepore™ filter coated with a 200-Å film of carbon. After the particles are collected they are covered with another carbon film and the filter paper is cut into small squares and placed carbon side down on blank carbon-coated nylon transmission electron microscope grids resting on several layers of filter paper in a petri dish. A drop of chloroform is placed on each square, the filter paper saturated, and the petri dish covered and left for several hours until the polycarbonate film dissolves leaving the particles trapped in a carbon sandwich on the electron microscope grids. The samples can then be examined in an SEM equipped with a Si(Li) detector. They found that by using this method they were able to qualitatively analyze particles as small as 500 Å without having to contend with problems from extraneous lines.

Frequently particle analysis entails the examination and identification of many hundreds of particles. A computer-controlled SEM–EPMA described by White and co-workers[22] has been used to obtain particle sizes, shapes, and masses in addition to tentative identifications. The particles are dispersed and mounted on a Ge substrate and the beam scanned over a large number in raster form. When a monitor on the secondary electron signal indicates that the beam has encountered a particle, the beam is stopped, the coordinates recorded, and a step scan initiated across the particle, counting the characteristic x-rays. When the beam leaves the particle, the coordinates are again recorded and the fast raster scan continued. In addition to the characteristic spectra from the particles, the intensity of the substrate x-rays is also monitored. The decrease in intensity of the substrate is a measure of the energy lost by the beam penetrating the particle and can therefore be related to the mass thickness of the particle. Computer processing of the large amount of resulting data gives particle size, shape, and mass distributions in addition to tentative compositions.

If particles cannot be extracted from their matrices, thin sections prepared by electrothinning or ion milling can be used to improve x-ray spatial resolution over that obtainable from bulk samples. Figure 8 shows secondary and dark-field transmission scanning images obtained from a thinned section of a directionally solidified CoCr–TaC eutectic alloy, as well as the Si(Li) detector spectra measured at 45 kV from the TaC rod phase and the surrounding matrix. Note that even though the electron beam used was less than one-tenth of the rod diameter and the volume of direct x-ray excitation was believed to be well within the rod, Co and Cr peaks are still visible in the spectra. This could be interpreted as Cr and Co in solution; however, such an explanation would be inconsistent with known solubility data, and it is believed that the signals are due to indirect x-ray fluorescence and electron backscattering effects.

If fine particles are examined in bulk samples, the only approach available to the operator to improve x-ray spatial resolution is to lower the operating voltage and reduce the beam current while sacrificing the x-ray signal in the hope of containing the volume of direct excitation in the phase of interest. A more systematic approach, however, to define the volume of direct excitation is to use MC calculations.

Figures 9a–d illustrate the effects of changing the beam voltage from 15 to 30 kV for a beam with a Gaussian half-maximum diameter of 0.2 μm impinging on a hypothetical 1-μm-diameter hemispherical particle of TaC in a Ni–10 Cr matrix. Note that at 15 kV the beam is essentially contained, while at 30 kV extensive penetration occurs. The decreased number of circles, depicting possible generated x-rays, reflects the fact that only a fraction of the x-rays that would have been generated in a bulk sample are actually generated. Of course, in reality, one may detect a greater number due to the overvoltage effect, but the ratio to a bulk standard would be less.

Although fine structure geometry is usually unknown, Bolon and Lifshin[11] have had encouraging success in applying their method to the quantitative analysis of TaC rods in nickel-base superalloys. Experimental data were collected with a JEOL JXA50A SEM equipped with both crystal spectrometers and a Princeton Gamma-Tech energy-dispersive detector. Selective etching of the alloy to reveal the rods shows them to have an easily measurable, fairly uniform square cross section, as can be seen in Figure 10.

Measurements of the beam diameter were made by scanning the electron beam across a knife edge and recording the response of a transmission electron detector as described earlier. X-ray data were collected on 1.5-μm rods with both 0.25- and 1.5-μm-diameter beams at a current of 5×10^{-10} A and beam voltage of 20 kV. The energy-dispersive detector was used because of its higher sensitivity at the current necessary to obtain the desired probe size. According to Monte Carlo plots similar to those in Figure 11, the direct x-ray excited volume should be contained within the TaC rods for a 0.25-μm beam but not for a 1.5-μm beam. This point was tested experimentally by observing the x-ray profile while the electron beam was scanned across the rods in a polished cross section as shown in Figure 12. Profile A corresponds to the Ta $M\alpha$ intensity along line L for the 0.25-μm beam. The curve appears to rise about 0.5 μm from the carbide boundary and level off for about 1.0 μm within the rod. This flattening was taken as evidence that the direct x-ray excited volume was contained within the rod even though, as shown in Table VI, the average relative intensity ratio 0.84 (measured with respect to pure Ta) was lower than the 0.92 measured on the bulk TaC standard. This discrepancy is the result of three effects. First there is a missing contribution due to indirect fluorescence which would have occurred outside the confines of the rod. Second and possibly more important, there is a significant tail to the electron beam due to the effects described earlier. Lastly, fluorescence by the final molybdenum foil aperture was found to be an unwanted source of secondary x-ray excitation.

Profile B shows the Ta $M\alpha$ intensity for the 1.5-μm beam. In this

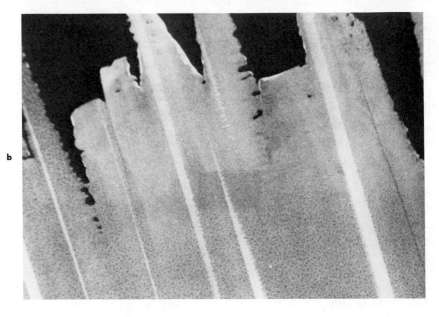

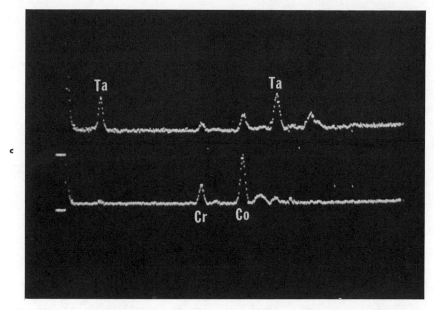

FIGURE 8.   Thin section of directionally solidified CoCr–TaC eutectic. (a) Scanning transmission (dark field), (b) secondary electron image, (c) x-ray spectra from (upper) TaC fiber and (lower) matrix 1 μm from fiber.

case, as predicted by the MC calculations, the x-ray excited volume is not contained within the rod since the peak intensity neither levels off nor reaches the maximum value of profile A. The slight displacement of profile B is due to a shift in the image associated with defocusing the beam to 1.5 μm. Table VI also contains a comparison of measured and calculated values of intensity ratios for Ta $L\alpha$ and $M\alpha$ lines determined at the centers of the rods. The measured values for the 1.5-μm beam are also lower than those predicted by the Monte Carlo calculation. The greater discrepancy in the case of the $L\alpha$ lines indicates again the probable influence of indirect fluorescence, which has been estimated to be as high as 20% of the direct intensity.[4] Although a certain degree of success has been attained by the use of MC calculations to characterize fine structures of known geometry, the basic problem of performing accurate *in situ* quantitative analysis on irregularly shaped particles remains unsolved.

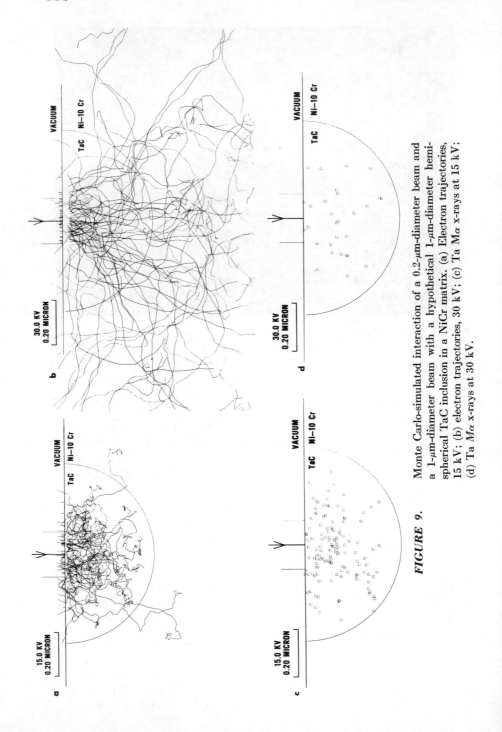

FIGURE 9. Monte Carlo-simulated interaction of a 0.2-μm-diameter beam and a 1-μm-diameter beam with a hypothetical 1-μm-diameter hemispherical TaC inclusion in a NiCr matrix. (a) Electron trajectories, 15 kV; (b) electron trajectories, 30 kV; (c) Ta Mα x-rays at 15 kV; (d) Ta Mα x-rays at 30 kV.

FIGURE 10.   Directionally solidified NiCr–TaC eutectic alloy etched to reveal the TaC rods.

TABLE VI. Tantalum Carbide: Relative Intensity Data from 1.5-μm TaC Rods

Rod #	0.25-μm Beam		1.50-μm Beam	
	Ta $M\alpha$	Ta $L\alpha$	Ta $M\alpha$	Ta $L\alpha$
Experimental				
1	0.85	0.81	0.44	0.44
2	0.83	0.79	0.47	0.46
3	0.84	0.77	0.44	0.40
4	0.86	0.82	0.45	0.43
Average	0.84	0.80	0.45	0.43
Monte Carlo	0.97	0.93	0.47	0.55

Data from bulk TaC	Ta $M\alpha$	Ta $L\alpha$
Average measured value	0.92	0.89
ZAF calculated value	0.92	0.90

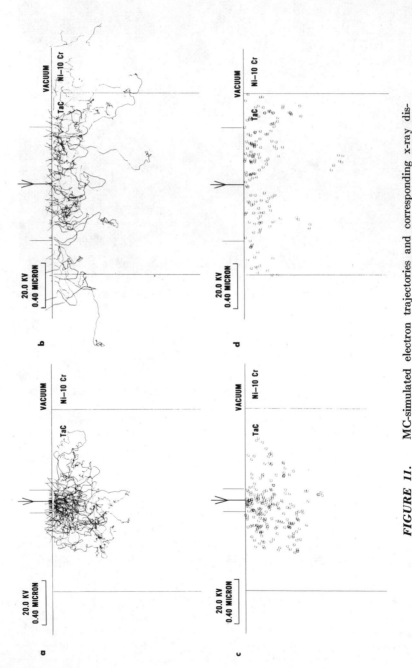

FIGURE 11.   MC-simulated electron trajectories and corresponding x-ray distributions for: (a, c) focused electron beam (0.2 μm), (b, d) defocused electron beam (1.0 μm), using Ta *M*α x-rays.

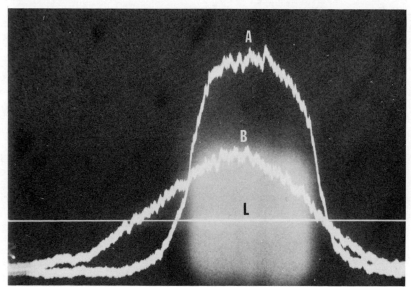

FIGURE 12. TaC rod (1.5 μm) with line of scan (L) and corresponding Ta $M\alpha$ x-ray line profiles for a focused (0.2 μm) beam (A) and a defocused (1.5 μm) beam (B).

REFERENCES

1. K. C. A. Smith, Ph.D. Dissertation, Univ. of Cambridge (1956).
2. A. N. Broers, in *SEM/1970 Proceedings of the 3rd Annual SEM Symposium* (O. Johari, ed.), IITRI, Chicago, Illinois (1970), p. 3.
3. S. J. B. Reed, in *X-Ray Optics and Microanalysis*, IVth International Congress on X-Ray Optics and Microanalysis, Orsay, 1965 (R. Castaing, P. Deschamps, and J. Philibert, eds.), Hermann, Paris (1966), p. 339.
4. M. Green and V. E. Cosslett, *Proc. Phys. Soc. (London)*, **78**, 1206 (1961).
5. M. B. Duke and R. Brett, *Geol. Surv. Res, B*, 101 (1965).
6. R. Castaing and J. Deschamps, *J. Phys. Radium*, **16**, 304 (1955).
7. H. E. Bishop, Ph.D. Dissertation, Univ. of Cambridge (1966).
8. G. Shinoda, K. Murata, and R. Shimizu, in *Quantitative Electron Probe Microanalysis* (K. F. J. Heinrich, ed.), NBS Special Publication No. 298 (1968), p. 155.
9. R. Shimizu, T. Ikuta, and K. Murata, *J. Appl. Phys.*, **43**, 4233 (1972).
10. H. A. Bethe, M. E. Rose, and L. P. Smith, *Proc. Am. Phil. Soc.*, **78**, 573 (1938).
11. R. B. Bolon and E. Lifshin in *SEM/1973 Proceedings of the 6th Annual SEM Symposium* (O. Johari, ed.), IITRI, Chicago, Illinois (1973), p. 285.
12. L. Curgenven and P. Duncumb, TI Research Laboratory Report No. 303, July 1971.
13. C. R. Worthington and S. G. Tomlin, *Proc. Phys. Soc.*, **A69**, 401 (1956).
14. Laura Roth, Dept. of Physics, MIT, private communication (1972).
15. R. Butz and H. Wagner, *Surface Science*, **34**, 693 (1973).

16. W. E. Sweeney, R. E. Seebold, and L. S. Birks, *J. Appl. Phys.*, **31**, 1061 (1960).
17. W. Reuter, in *Proceedings of the Sixth International Conference on X-Ray Optics and Microanalysis*, Osaka, 1971 (G. Shinoda, K. Kohra, and T. Ichinokawa, eds.), University of Tokyo Press, Tokyo (1972), p. 121.
18. J. Philibert and R. Tixier, in *Quantitative Electron Probe Microanalysis* (K. F. J. Heinrich, ed.), NBS Special Publication No. 298 (1968), p. 13.
19. T. A. Hall, in *Physical Techniques in Biological Research* (G. Oster, ed.), Academic Press, New York (1971), Vol. 1A, p. 157.
20. D. F. Kyser and K. Murata, *IBM J. Res. Dev.*, **18**, 352 (1974).
21. C. J. Maggiore and I. B. Rubin, in *SEM/1973 Proceedings of the 6th Annual SEM Symposium* (O. Johari, ed.), IITRI, Chicago, Illinois (1973). p. 129.
22. J. Lebiedzik, K. G. Burke, S. Troutman, G. G. Johnson, and E. W. White, in *SEM/1973 Proceedings of the 6th Annual SEM Symposium* (O. Johari, ed.), IITRI, Chicago, Illinois (1973), p. 121.

METHODS OF QUANTITATIVE X-RAY ANALYSIS USED IN ELECTRON PROBE MICROANALYSIS AND SCANNING ELECTRON MICROSCOPY

H. Yakowitz

I. INTRODUCTION

In his thesis,[1] Castaing proposed that quantitative electron probe analysis could be carried out using pure elements as standards. This proposal is accepted, of choice or necessity, by most analysts; the accuracy of the procedure is, however, the subject of much debate. Castaing further stated that the ratio of characteristic x-rays generated from element A in the specimen to pure A was equivalent to the concentration of element A in the specimen. This statement is the basis for quantitative electron probe microanalysis. Castaing's treatment can be represented by the following considerations. The average number of ionizations n per primary beam electron incident with energy E_0 is[1]

H. Yakowitz—Institute for Materials Research, Metallurgy Division, National Bureau of Standards, Washington, D.C.

$$n = \frac{N_0 C_A}{A_A} \int_{E_0}^{E_c} \frac{Q}{dE/dX} \, dE \qquad (1)$$

where dE/dX is the mean energy change of an electron in traveling a distance X, N_0 is Avogadro's number, ρ is the density of the material, A_A is the atomic weight of A, C_A is the concentration of element A, E_c is the critical excitation energy for whatever characteristic x-ray line is of interest, and Q is the ionization cross section, defined as the probability per unit path length of an electron of given energy causing ionization of a particular electron shell (K, L, or M) of an atom in the specimen. The effect of backscattering electrons can be taken into account by introducing a factor R equal to the ratio of x-ray intensity actually generated to that which would have been generated if all of the incident electrons had remained within the specimen. Now the intensity I_A of A x-rays generated is proportional to n. Hence

$$I_A = \text{const} \times C_A R \rho \int_{E_c}^{E_0} \frac{Q}{dE/dX} \, dE \qquad (2)$$

To a first approximation, R, ρ, Q, and dE/dX can be assumed equal for both the specimen and pure A. Hence

$$I_A / I^{(A)} = C_A \qquad (3)$$

where $I^{(A)}$ is the intensity generated in pure A. The measured intensities from specimen and standard may, however, need to be corrected for absorption within the solid and for differences in R, ρ, Q, and dE/dX in order to arrive at the ratio of generated intensities as in equation (3) and hence the value of C_A. A publication in which such proposed correction procedures were applied to a great amount of uncorrected data from diverse sources shows a disappointingly wide range of deviation of the calculated results with respect to the expected "true" values.[2] Recognition of the complexity of the problem has led numerous investigators[3–5] to expand the theoretical treatment of quantitative analysis proposed by Castaing.

It is very difficult, however, to check on the correctness of theoretical models unless the effects of uncertainty in the values of input parameters are recognized. Much of the existing confusion concerning the quantitative aspects of electron probe analysis is due to neglect of this important factor. Such input parameters include physical constants—x-ray mass attenuation coefficients, fluorescence yields, backscatter correction coefficients, among others—as well as data concerning the conditions of the measurement—such as the accelerating potential, the x-ray emergence angle, and the detector coincidence losses.

All schemes for carrying out quantitative analysis use a standard of known composition. In many cases, especially for metals, pure elements are suitable. In the case of mineral or petrological samples, homogeneous, compound standards close in average atomic number to the unknown are usually chosen. What the analyst measures is the relative x-ray intensity ratio between the elements of interest in the specimen and the same element in the standard. Both specimen and standard are examined under identical experimental conditions. The measured relative intensity ratio, commonly called k, must be accurately determined or any quantitative analysis scheme will result in the same inaccuracy. X-ray peak intensities can usually be measured more easily and with better precision than the integrated intensity under a particular peak. Using peak intensity values gives results as accurate as integrated intensities[6]; hence in all that follows, peak intensity is used. All k values in order to be accurate must include consideration of the Poisson statistics of the process, coincidence losses (detector dead time),[7] and possible effects of contamination of the sample and/or standard,[8] and they must be corrected for the background arising from the continuum radiation.[9,10]

Once the k values have been obtained, they must be corrected for several effects, including (1) the differences in electron scattering and retardation in the specimen and the standard, i.e., the so-called atomic number effect, expressed by the factor k_Z, (2) absorption of x-rays within the specimen, k_A, (3) fluorescence effects, k_F, and (4) continuum fluorescence, k_c. A common form of the correction equation is

$$C = kk_Zk_Ak_Fk_c \tag{4}$$

where C is the weight fraction of the element of interest. This method is often referred to as the ZAF method. Each of the effects listed above will be discussed in further detail.

In this chapter, quantitative microprobe analysis will be considered from the classical point of view as well as by the so-called empirical approach. The relations used to convert raw data to mass fractions will be discussed and scrutinized with respect to sources of error in the final analysis. Histograms depicting the typical spread of results are used to show what can be expected. By way of illustration, the complete planning and execution of the analysis of a binary silicon–iron alloy will be described in detail. Effects associated with detector dead times, background subtraction and instrumental instabilities will be discussed here. Minicomputer schemes for obtaining analysis results on-line or rapidly off-line will be outlined in the next chapter.

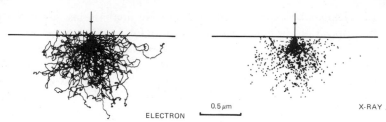

FIGURE 1. Schematic representation of 100 electron paths in a copper specimen; primary beam voltage 20 kV. The corresponding x-rays generated by these electrons are shown as well. (After Curgenven and Duncumb.[43])

II. THE ABSORPTION FACTOR k_A

A. Formulation

Since the x-rays produced by the primary beam are created at some nonzero depth in the specimen (see Figure 1), they must pass through the specimen on their way to the detector. On this journey, the x-rays undergo absorption due to interactions with the atoms in the specimen—both with atoms of the element emitting the radiation of interest as well as with atoms of other elements present. Therefore, the intensity of the x-radiation finally reaching the detector is reduced in magnitude. Castaing[1] has described the intensity dI of characteristic radiation—without absorption—generated in a layer of thickness dz having density ρ at some depth z below the specimen surface as

$$dI = \phi(\rho z) \, d(\rho z) \tag{5}$$

where $\phi(\rho z)$ is defined as the distribution of characteristic x-ray production with depth. The shape of the $\phi(\rho z)$ curve for Cu $K\alpha$ radiation is shown in Figure 2; this shape is typical for all elements.[1]

Thus, in the absence of absorption, the total flux generated and detected by the spectrometer system would be

$$I = \int_0^\infty \phi(\rho z) \, d(\rho z) \tag{6}$$

But absorption of the generated x-rays does occur and so the actual flux recorded, I', is given by

$$I' = \int_0^\infty \phi(\rho z) \, \exp\{-(\mu/\rho)(\rho z) \csc \Psi\} \, d(\rho z) \tag{7}$$

where μ/ρ is the x-ray mass attenuation coefficient of the specimen for the characteristic line of interest, ρz csc Ψ is the absorption path length, and Ψ is the angle between the x-ray detector and the specimen surface, the take-off angle. The quantity $\chi = (\mu/\rho)$ csc Ψ. Philibert[4] referred to I as $F(0)$ when χ was zero, and he referred to I' as $F(\chi)$. The ratio $F(\chi)/F(0)$ was called $f(\chi)$, which is equivalent to I'/I. Then for the determination of an element designated "A" in any composite specimen

$$k_A = f(\chi)_{\text{STD}}/f(\chi)_{\text{SPEC}} \qquad (8)$$

where STD and SPEC refer to standard and specimen, respectively. Hereafter, all quantities referring to the alloy or specimen will be denoted by an asterisk. All quantities referring to the standard will be left unmarked. Thus we write

$$k_A = f(\chi)/f(\chi)^* \qquad (9)$$

The term $f(\chi)$ can be written as f_p in order to distinguish absorption of primary x-radiation.

B. Expressions Used to Calculate f_p

The absorption correction factor f_p depends upon the respective mass absorption coefficient μ/ρ, the x-ray emergence angle Ψ, the operating

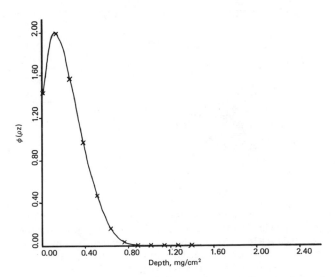

FIGURE 2.   Computed $\phi(\rho Z)$ curve for copper assuming a primary beam voltage of 20 kV. The value of $\phi(\rho Z)$ is the number of $K\alpha$ quanta generated at the indicated depth per incoming electron.

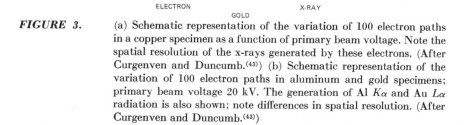

a

10.5KV

15KV

0.5 μm

ELECTRON   20KV   X-RAY

b

1.0 μm

ALUMINUM

0.2 μm

ELECTRON   X-RAY
GOLD

FIGURE 3.   (a) Schematic representation of the variation of 100 electron paths in a copper specimen as a function of primary beam voltage. Note the spatial resolution of the x-rays generated by these electrons. (After Curgenven and Duncumb.[43]) (b) Schematic representation of the variation of 100 electron paths in aluminum and gold specimens; primary beam voltage 20 kV. The generation of Al $K\alpha$ and Au $L\alpha$ radiation is also shown; note differences in spatial resolution. (After Curgenven and Duncumb.[43])

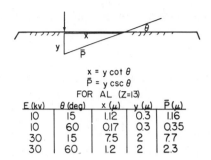

$$x = y \cot \theta$$
$$\bar{P} = y \csc \theta$$

FOR AL (Z=13)

E (kv)	θ (deg)	x (μ)	y (μ)	$\bar{P}$ (μ)
10	15	1.12	0.3	1.16
10	60	0.17	0.3	0.35
30	15	7.5	2	7.7
30	60	1.2	2	2.3

FIGURE 4.   Schematic representation of path lengths for absorption in an aluminum specimen as a function of primary beam voltage E and x-ray emergence angle θ.

voltage E_0, the critical excitation voltage for K, L, or M radiation E_c, and the mean atomic number and mean atomic weight of the specimen Z^* and A^*. Hence we write

$$f_p = f(\chi = f[(\mu/\rho) \csc \Psi, E_0, E_c, Z^*, A^*] \tag{10}$$

There have been some direct determinations of $\phi(\rho z)$. Philibert fitted a relation for f_p to empirical $\phi(\rho z)$ curves,[4]

$$\frac{1}{f_p} = \left(1 + \frac{\chi}{\sigma}\right)\left(1 + \frac{h}{1+h}\frac{\chi}{\sigma}\right) \tag{11}$$

where $h = 1.2A/Z^2$, A and Z being the atomic weight and number of the specimen, respectively. The coefficient 1.2 is a fitting parameter. The absorption parameter is $\chi = (\mu/\rho) \csc \Psi$ and Ψ is the angle between the centerline of the detector and the specimen surface—the takeoff or emergence angle. The parameter σ is a factor to account for the voltage dependence of the absorption process. At higher excitation potentials, electrons penetrate more deeply and the absorption path is lengthened. Figure 3a shows representations of the electron and x-ray distributions in copper as a function of primary beam voltage. Figure 3b shows representations of the electron and x-ray distributions in aluminum and gold. Figure 4 shows schematically the absorption path in an aluminum target for different experimental conditions. The path length increases rapidly with increasing operating voltage and with decreasing emergence angle.

By multiplication and rearrangement, equation (11) can be written in the more generally useful form

$$1/f_p = a_0 + a_1\gamma\chi + a_2\gamma^2\chi^2 \tag{12}$$

If Philibert's original formulation is fitted to equation (12), then one obtains

$$a_0 = 1.0$$

$$a_1 = 2 \times 10^{-6} \frac{1 + 2h}{1 + h}$$

$$a_2 = 4 \times 10^{-12} \frac{h}{1 + h} = \frac{h(1 + h)}{(1 + 2h)^2} a_1^2 \tag{13}$$

$$\gamma = E_0^{1.65}$$

Thus all the parameters in equation (10) were taken into account except for E_c. Duncumb and Shields[5] proposed a relation which yields

$$a_0 = 1.0$$

$$a_1 = 4.18 \times 10^{-6} \frac{1 + 2h}{1 + h}$$

$$a_2 = \frac{h(1 + h)}{(1 + 2h)^2} a_1^2 \tag{14}$$

$$\gamma = E_0^{1.5} - E_c^{1.5}$$

This formulation came to be known as the Philibert–Duncumb equation. Later Heinrich,[11] after critical examination of existing experimental f_p data, suggested

$$a_0 = 1.0$$

$$a_1 = 2.22 \times 10^{-6} \frac{1 + 2h}{1 + h}$$

$$a_2 = \frac{h(1 + h)}{(1 + 2h)^2} a_1^2 \tag{15}$$

$$\gamma = E_0^{1.65} - E_c^{1.65}$$

This grouping is known as the Philibert–Duncumb–Heinrich (PDH) equation and is currently the most popular expression for f_p available. Accordingly, the PDH equation is often used in microprobe correction schemes to calculate f_p.

Shortly before Philibert published equation (11), Birks suggested that the E_c, Z^*, and A^* dependences in f_p could be neglected.[12,13] In other words, if $h = 0$ and the effect of E_c is small, a universal f_p curve for each operating voltage can be postulated. In fact, Birks' concept can be put into the form of equation (12) with only slight modification. Then,

$$a_0 = 1.0$$

$$a_1\gamma = 4 \times 10^{-5} + 8 \times 10^{-7}E_0^2 \tag{16}$$

$$a_2\gamma^2 = 0.07(a_1\gamma)^2$$

Equations (13)–(16) are not, in fact, very much different from one another. Figure 5 shows the value of σ in equation (11) plotted against the operating voltage E_0 in kilovolts. The similarities are apparent. However, the effect of E_c has been neglected in Figure 5.

Equations (13) and (16) are not used; the authors of these expressions never cast them into analytical form. Equation (15) evolved from equation (14) and, as noted, is very widely used. An examination of experimental results is helpful in evaluating the justification of the form of the PDH equation [equation (15)] and its accuracy.

C. *Experimental Results and the f_p Expressions*

There are several difficulties involved in attempting to obtain purely empirical expressions for f_p. First, there is very little direct experimental data for either $\phi(\rho z)$ or f_p available in the literature. What is available

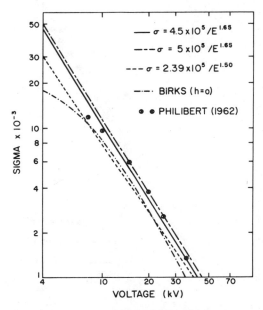

FIGURE 5. Comparison of proposed expressions for the σ term in the absorption correction showing their basic similarities.

must be cast into a unified form, i.e., indirect fluorescence-induced radiation must be taken into account and the same set of mass absorption coefficients applied. The result is a set of 37 usuable curves of $1/f_p$ vs. χ, comprising 11 elements, namely carbon, aluminum, titanium, iron, copper, germanium, molybdenum, neodymium, tantalum, gold, and uranium.

In equation (12), the effect of the $a_2\gamma^2\chi^2$ is small compared to the $a_1\gamma\chi$ term. (In cases of moderate χ, a_2 is often assumed zero with little loss of accuracy.) Hence an f_p expression capable of accurately reproducing experimental $a_1\gamma$ values is desired. Experimental $a_1\gamma$ values can be taken as the slope of the $1/f_p$ vs. χ curve at low χ values.

Figure 6 shows experimental $a_1\gamma$ values plotted against operating voltage. No regard has been paid to atomic number, line type (K, L, M), or critical excitation potential. Three relations for $a_1\gamma$ are also shown in an attempt to infer the most useful coefficient in the expression for γ. Clearly, the value of 1.65 fits the experimental data more closely than does 1.5 or 2. Hence, based on the experimental data available, we may conclude that the

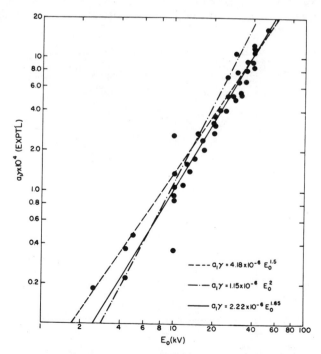

$$a_1\gamma = 4.18 \times 10^{-6}\, E_0^{1.5}$$
$$a_1\gamma = 1.15 \times 10^{-6}\, E_0^{2}$$
$$a_1\gamma = 2.22 \times 10^{-6}\, E_0^{1.65}$$

FIGURE 6.   Leading term $a_1\gamma$ in the absorption correction against primary beam voltage. Proposed expressions for $a_1\gamma$ purporting to fit the experimental results are superimposed.

exponent of 1.65 in the γ expression is justified, i.e., $\gamma = E_0^{1.65} - E_c^{1.65}$ as originally suggested by Heinrich.[11]

Heinrich et al.,[14] in a critical study of the calculation of f_p, concluded that there is no justification at present for retaining the atomic number Z and atomic weight A dependences in the calculation of f_p. This conclusion is based on a critical comparison with presently available experimental f_p vs. χ curves. Hence the term $h \equiv 1.2A^*/Z^{*2}$ can be abandoned in the calculation of f_p. (In any case, h can be shown to depend only on Z^*.) The terms containing h in the calculation of f_p have been replaced by the constant 1.35. This replacement leads to only slightly better analytical results (based on about 500 analyses) than does the Philibert–Duncumb–Heinrich equation but is somewhat simpler. The comparison is shown in Figures 7a, b, and 8a, b. In these figures the k_Z and k_F correction schemes were the same.

The exponent 1.65 in the voltage-dependent terms agrees well with available experimental data. Hence the use of this exponent is justified. The inclusion of the excitation potential in the voltage-dependent terms results in a small but useful gain. Omission of the excitation potential will usually not cause large errors in the absorption correction.

The formulation resulting from the Heinrich et al. study[14] is

$$1/f_p = 1 + 3 \times 10^{-6}[E_0^{1.65} - E_c^{1.65}]\chi + 4.5 \times 10^{-13}[E_0^{1.65} - E_c^{1.65}]^2\chi^2 \quad (17)$$

This expression, called H:Y, has been in use for about two years. Results obtained with it appear to be at least as accurate as those obtained with the PDH expression.

Monte Carlo calculations of $1/f_p$ vs. χ curves do not generally agree exactly with experimental determinations of these curves. At present, the use of the Monte Carlo technique for routine calculations of absorption correction factors is not warranted. However, in cases of special geometries, such as a thin film on a substrate, the Monte Carlo method can be of great value, as will be discussed later.

Comparison of different expressions for the absorption correction with the aid of well-characterized alloys is to some extent disappointing. On balance, it appears that one can expect errors within $\pm 1\%$ relative about one-third of the time, within $\pm 2\frac{1}{2}\%$ relative about half the time, and within $\pm 5\%$ most of the time. It would appear that the most urgent need for quantitative electron probe microanalysis is more experimental f_p data, both on alloys and pure elements.

Consideration of the error propagation in equation (17) leads to the expression

$$\frac{\Delta f_p}{f_p} = (1 - f_p)\left[\cot\psi\,\Delta\psi - \frac{\Delta(\mu/\rho)}{\mu/\rho} + \frac{1.65E_0^{0.65}\,\Delta E_0}{E_0^{1.65} - E_c^{1.65}}\right] \quad (18)$$

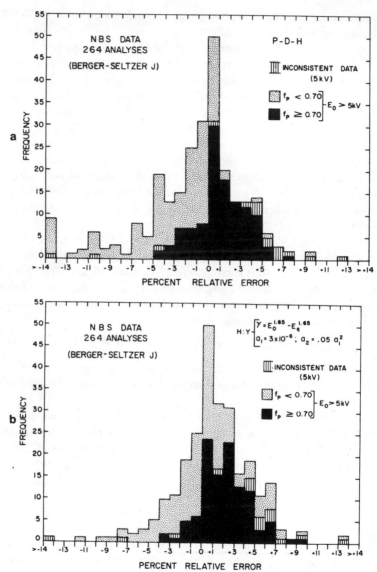

FIGURE 7.   (a) Histogram of results for binary alloy quantitative electron probe microanalysis using the Philibert–Duncumb–Heinrich (PDH) absorption expression. (b) Histogram of results for binary alloy quantitative electron probe microanalysis using the absorption expression proposed by Heinrich *et al.*[14]

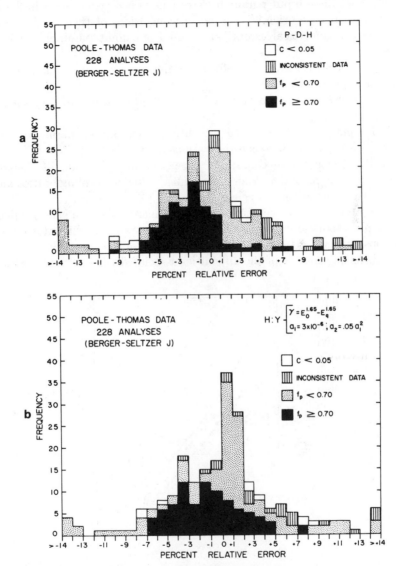

FIGURE 8.   (a) Histogram of results for data taken from Poole[2] using the PDH absorption expression.[14] (b) Histogram of results for data taken from Ref. 2 using the H:Y absorption expression.

The effects of these input parameter errors have been considered in detail by Yakowitz and Heinrich.[15] The major conclusions of their study were:

1. Serious analytical errors can result from input parameter uncertainties.

2. In order to reduce the effects of these errors, the product $a\gamma_1\chi$ should be kept to 0.25 or less.

3. In consequence, the value of the absorption function $f(\chi)$ should be 0.7 or greater.

4. Keeping $f(\chi)$ above 0.7 has the added advantage of minimizing the effects of discrepancies between the various models for $f(\chi)$.

5. To achieve these conditions experimentally, instruments should be capable of giving stable, high count rates at low overvoltage ratios and have high x-ray emergence angles.

6. To further improve the accuracy of microprobe analysis, particularly of low-atomic-number elements, more accurate experimental determinations of mass attenuation coefficients and of the function $f(\chi)$ are required. As an example of the difficulties associated with mass absorption

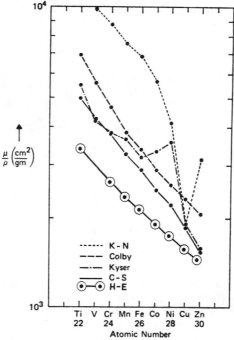

FIGURE 9. Proposed values of $(\mu/\rho)(A, A\ L\alpha)$, where A is an element with $22 \leq Z \leq 30$ (adapted from Kyser[16]). Comparison values from Henke and Ebisu[41] (H–E), Kamen Nuclear[63] (K–N), Colby,[64] and Cooke and Stewardson[65] (C–S).

coefficient uncertainties, Figure 9 shows the value of $M(A, A\ L\alpha)$ where M is μ/ρ and A is an element in the range $22 \leq Z \leq 30$.[16] Huge discrepancies are obvious; soft x-ray effects are discussed in detail in Chapter XII. A generalized $\Delta f_P/f_P$ based on equation (18) is shown plotted against μ/ρ in Figure 10. Clearly, errors due to the effect of μ/ρ, and $\Delta\mu/\rho$, $\Delta\theta(\Delta\Psi)$, and ΔE are reduced if low operating voltages and high takeoff angles are chosen.

Next, the case of analysis with an energy-dispersive detector or crystal spectrometer mounted on a typical SEM must be considered. If the speci-

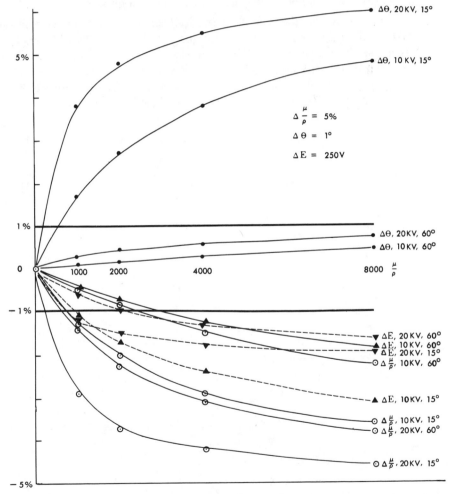

FIGURE 10.   Generalized error curve for the absorption correction assuming errors in input parameters as shown ($\Delta\mu/\rho = 5\%$, $\Delta\theta = 1°$, $\Delta E = 250$ V). The abscissa is in units of μ/ρ.

men is flat and smooth, only a few extra obstacles to quantitative analysis will be encountered. The most important of these are the correction for background, which has already been discussed in detail in Chapter VII, and the exact value of Ψ, the x-ray takeoff angle.

In the SEM the specimen is usually tilted in order to provide a suitable value of the takeoff angle. Uncertainties in Ψ mainly affect the absorption correction calculation; such uncertainties increase as Ψ decreases since the error propagates as cot Ψ.[15] Thus values of Ψ in excess of 30° are needed. Furthermore, it is of crucial importance to ensure that the specimen and standard are measured with identical x-ray takeoff angles.

In many SEM's, the lithium-drifted silicon detector chip is placed relative to the specimen as shown in Figure 11. For the cases illustrated in Figure 11, the specimen is tilted at an angle θ as indicated on the SEM stage control; the center of the detector chip is at a distance X from the impact point of the electron beam on the specimen. The distance of the impact point above (case I) or below (case III) the detector chip centerline is ΔZ. Then

$$\text{case I:} \qquad \Psi = \theta - \arctan(\Delta Z/X) \qquad\qquad (19a)$$

$$\text{case II:} \qquad \Psi = \theta \qquad\qquad\qquad\qquad\qquad\quad (19b)$$

$$\text{case III:} \qquad \Psi = \theta + \arctan(\Delta Z/X) \qquad\qquad (19c)$$

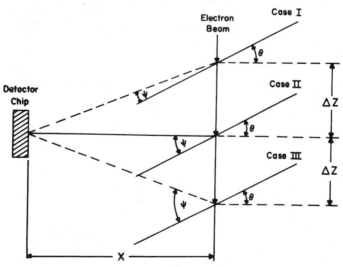

FIGURE 11.   Geometry for deducing x-ray emergence angle Ψ using a tilted specimen and fixed energy-dispersive detector chip.

Hence it is usually advantageous to try to adjust the height of the specimen such that case II is obtained so as to eliminate the need to know ΔZ and X accurately. The closer the detector chip (smaller X), the greater the uncertainty of Ψ. Furthermore, since the solid angle accepted by the detector increases as the distance X decreases, using the position of the center of the detector as a basis for computing Ψ becomes only an approximation.

For crystal spectrometers mounted on the electron probe microanalyzer or SEM, the value of Ψ can be computed from the specific crystal–detector–specimen arrangement available. Small differences in the specimen level can lead to intensity differences increasing with the distance of the x-ray source from the Rowland circle. Hence it is important that the specimen height be the same for all measurements and that specimen and standard be at the same height in order to obtain accurate k values. One way to obtain appropriate specimen heights with respect to the spectrometer is to adjust the specimen height for a maximum x-ray intensity on the line of interest. This procedure can be repeated for the standard as well.

III. ATOMIC NUMBER CORRECTION k_z

The atomic number effect in electron microprobe analysis is determined by two phenomena, namely electron backscattering and electron retardation, both of which depend upon the average atomic number of the target. Therefore if there is a difference between the average atomic number of the specimen given by

$$\bar{Z} = \sum_i C_i Z_i \tag{20}$$

and that of the standard, an atomic number correction is required. For example, in Fe–3 wt % Si, the value of $\bar{Z}$ is 25.64, so a somewhat larger effect for Si ($Z = 14$) analysis would be expected than for Fe analysis ($Z = 26$). In general, unless this effect is corrected for, analyses of heavy elements in a light matrix generally yield erroneously low values, while analyses of light elements in a heavy matrix usually yield erroneously high values. Hence, for Si analysis in Fe–Si, the magnitude of the atomic number correction is in the opposite sense to that of the absorption correction.

At present, the most accurate formulation of k_Z appears to be that given by Duncumb and Reed[17]

$$k_Z = \frac{R \displaystyle\int_{E_c}^{E} (Q/S)\, dE}{R^* \displaystyle\int_{E_c}^{E} (Q/S^*)\, dE} \tag{21}$$

where R and R^* are the backscattering correction factors for standard and specimen, respectively:

$$R = \frac{\text{total number of photons actually generated in the sample}}{\text{total number of photons generated if there were no backscatter}}$$

Q is the ionization cross section, defined as the probability per unit path length of an electron with a given energy causing ionization of a particular inner electron shell of an atom in the target,[18] and S is the electron stopping power (see Chapter III) in the region $1 \le E \le 50$ keV, given by Bethe[19] as

$$S = \text{const} \times \frac{Z}{A} \frac{1}{E} \ln \frac{C_1 E}{J} \tag{22}$$

with C_1 given by 1166 for electrons (E in keV) and J the so-called mean ionization potential in eV.

In order to avoid the indicated integration in equation (21), Thomas proposed that an average energy $\bar{E}$ may be taken as $0.5(E_0 + E_c)$ with little loss in accuracy.[20,21]

There is experimental evidence that

$$S^* = \sum_i C_i S_i \tag{23}$$

and Duncumb and Reed have postulated that

$$R^* = \sum_i C_i R_i \tag{24}$$

Thus far no experimental evidence that refutes equations (23) and (24) has been found.

There are several tabulations of R values as a function of atomic number[17,22,23] but only one experimental study of R.[24] The tabulation given by Duncumb and Reed very nearly agrees with the experimental determinations where comparisons can be made. Other tabulations do not.[21] Thus the R values most often used are those of Duncumb and Reed.

There are several relations for Q in the literature, all of which are in the form

$$Q = \text{const}_1(\ln U)/U^{\text{const}_2}E_c^2 \tag{25}$$

where $U = E_0/E_c$.

Despite the fact that Q values differ by several percent depending on the value of the constants, Heinrich and Yakowitz[21] have shown that discrepancies in Q value have only a negligible effect on the final value of the concentration. A value for Q for K characteristic radiation is often taken from Green and Cosslett[19] as

$$Q = 7.92 \times 10^{-20}(\ln U)/UE_c^2 \tag{26}$$

The value of J to be used in equation (22) is a matter of controversy. The main problem seems to be that J is not measured directly but is a derived value—and at that from experiments done in the MeV range. The most complete discussion of the question is given by Berger and Seltzer.[25] These authors postulate that a "best" J vs. Z curve after weighing all available evidence is given by

$$J = 9.76Z + 58.8Z^{-0.19} \tag{27}$$

After comparing microprobe results and considering the various relations for J, several U.S. microprobe laboratories feel that equation (27) gives the best value of J presently available. The same conclusion has been reached in France. However, most British workers use the J values tabulated by Duncumb and Reed.[17] Studies of the "J problem" are currently underway. Figure 12 shows J vs. Z for several J relations.

Heinrich and Yakowitz have investigated error propagation in the k_Z term.[21] In general, the magnitude of k_Z decreases as U increases, but very slowly (5% for a tenfold increase in U). But the uncertainty in k_Z remains remarkably constant as a function of U. This is because the R and S uncertainties tend to counterbalance one another. Thus no increase in $\Delta k_Z/k_Z$ is to be expected at low U values and hence the choice of low operating voltages is still valid for obtaining the highest accuracy.

The integration indicated by equation (21) has been carried out and a closed form solution obtained.[26] Hence the integration can be carried out readily in the computer without recourse to numerical methods. For the case of minicomputer data reduction, it has been shown that the value of k_Z can be satisfactorily approximated as

$$k_Z = RS^*/R^*S \tag{28}$$

The use of equation (28) almost always leads to very small errors in k_Z.[21]

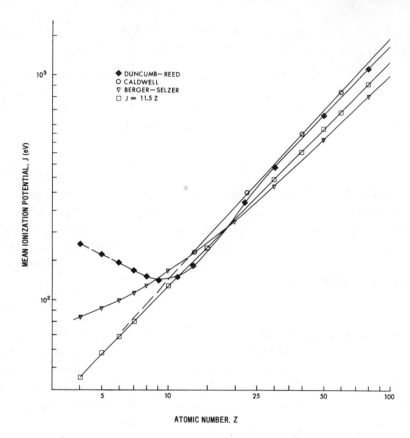

FIGURE 12. Comparison of several expressions for mean ionization potential *J*.

IV. THE CHARACTERISTIC FLUORESCENCE CORRECTION k_F

If the energy of a characteristic x-ray peak E_l, e.g., $E_{K\alpha}$, $E_{K\beta}$, or $E_{L\alpha}$, from element B in a specimen lies such that $0 < (E_l - E_c) < 5\,\text{keV}$, where E_c is the critical excitation potential to eject electrons from the inner K, L, or M shells of element A, then parasitic fluorescence must be accounted for in the correction procedure for element A. Such a correction is necessitated because the energy of the x-ray peak from element B is sufficient to excite x-rays secondarily from element A. Thus, more A x-rays are present than would have been produced by electron excitation alone.

Electrons are attenuated more strongly than x-rays of comparable energy. Thus fluorescent radiation can originate at greater distances from the point of impact of the electron beam than primary radiation. Hence the mean depth of production of fluorescent radiation is greater than that of primary radiation. Therefore the intensity of fluorescent emission relative to that of primary emission increases with increasing x-ray emergence angle. However, the error associated with the term k_F does not increase with this angle so that low-U, high-Ψ operation is entirely satisfactory even in cases requiring large fluorescence corrections.[27]

The form of the correction is

$$k_F = 1/(1 + r_f')$$

where r_f' is the fluorescence correction factor. The most popular version of r_f' was derived by Reed for binary combinations[28]:

$$r_f' = C_B Y_0 Y_1 Y_2 Y_3 P_{ij} \tag{29}$$

C_B is concentration of the element causing the parasitic fluorescence, i.e., the fluorescer; Y_0 is a constant:

$$Y_0 = 0.5 \frac{r_A - 1}{r_A} \omega_B \frac{A_A}{A_B}$$

with r_A the so-called absorption jump ratio for element A—for a K line, $(r_A - 1)/r_A$ is 0.88 and for an L line, $(r_A - 1)/r_A$ is 0.75 with very little error; ω_B is the fluorescent yield for element B (see Chapter III for discussion of ω_B); A_i is the atomic weight of the ith element; $Y_1 = [(U_B - 1)/(U_A - 1)]^{1.67}$, with $U = E_0/E_c$; $Y_2 = M(A, B\ K\alpha)/M(*, B\ K\alpha)$, i.e., the mass attenuation coefficient of A for B $K\alpha$ radiation divided by the mass attenuation coefficient of the alloy for B $K\alpha$ radiation. The other terms are

$$Y_3 = \frac{\ln(1 + u)}{u} + \frac{\ln(1 + v)}{v} \text{ with}$$

$$u = \csc \Psi \frac{M(A, A\ K\alpha)(1 - C_B) + C_B[M(B, A\ K\alpha)]}{M(A, B\ K\alpha)(1 - C_B) + C_B[M(B, B\ K\alpha)]}$$

$$v = \frac{3.3 \times 10^5}{E_0^{1.65} - E_c^{1.65}\{M(A, B\ K\alpha) - C_B[M(A, B\ K\alpha) - M(B, B\ K\alpha)]\}}$$

and finally P_{ij} is a factor for the type of fluorescence occurring. If KK (a K line fluoresces a K line) or LL fluorescence occurs, $P_{ij} = 1$. If LK or KL fluorescence occurs, $P_{ij} = 4.76$ for LK and 0.24 for KL.

The Reed relation is used in most computer-based schemes for correction. This relation has been found to be accurate under real analysis conditions.[29] Heinrich and Yakowitz tested the Reed model for its response to imput parameter uncertainties.[27] They found that ω_B produces the worst uncertainties; the other variables produce negligible errors. The extension to cases other than binary is straightforward. Wherever C_B appears in the expression for r_f', a replacement of the sum of concentrations for all elements causing secondary fluorescence of element A is needed. A more rigorous approach to the calculation of r_f' was made by Henoc et al.[30] In this approach, the number of photons n of the x-ray line generated by ionization of the electron shell m is obtained by multiplying the number of ionizations $n_A R$ by the fluorescence yield ω_m and by the ratio of the intensity of the line of interest to the intensity of all lines originating from this shell (weight of the line, p_{mn}):

$$I'_{Ap} = n_A R \omega_m p_{mn} \tag{30}$$

Then the ratio of intensity from the element causing fluorescence to the intensity of the fluoresced element can be fully calculated using equation (35) for primary intensities, just as in the atomic number correction. This improves the accuracy of calculation of fluorescence of K lines by L lines and vice versa, as well as the calculation of fluorescence of $K\alpha$ lines by $K\beta$ lines, which can be performed separately. As pointed out by Criss,[31] the exponential approximation to primary distribution in depth of x-ray emission used by Castaing to calculate the term v in equation (29) can be replaced by a more accurate model. Henoc et al. use the model contained in Philibert's calculation for the function $f(\chi)$.[4] With the computer facilities presently available to most analysts, this more rigorous approach is possible without an excessive increase in the cost of computation.

V. THE CONTINUUM FLUORESCENCE CORRECTION

Whenever electrons are used to excite x-ray spectra, a band of continuous radiation beginning at a wavelength λ_{SWL} corresponding to $12.398/E_0$ always accompanies the characteristic peaks used as analytical lines. This continuum arises since the incoming electrons interact with the specimen atoms to produce photons of different energies as they proceed through the solid.

The continuum band contains quanta of energy sufficient to excite characteristic radiation according to the relation developed by Henoc[32]:

$$I_c' = 0.5 C_A K \omega_A \left(\frac{r_A - 1}{r_A}\right) P_A \frac{Z^*}{\chi^*} F(\lambda) \tag{31}$$

where I_c' is the intensity of the characteristic lines produced by the continuum, C_A is the concentration of element A, K is Kramer's constant describing the continuum,[33] ω_A is the fluorescence yield of element A, $(r_A - 1)/r_A$ is the absorption jump ratio of element A, P_A is the power of the characteristic line of interest in element A, i.e., the amount of energy under the characteristic peak divided by the total spectral energy, Z^* is the average atomic number of the specimen, χ^* is the average χ value of the specimen, and

$$F(\lambda) = \int_{\lambda_{EDGE}}^{\lambda_{SWL}} M(A, \lambda) \left(\frac{1}{\lambda_{SWL}} - \frac{1}{\lambda}\right) \frac{1}{\lambda} \ln \left\{1 + \csc \Psi \frac{M(*, A\ K\alpha)}{M(*, \lambda)}\right\}$$

(32)

where $M(A, \lambda)$ is the mass attenuation coefficient of element A for continuum wavelength λ; λ_{SWL} is the short-wavelength limit of the continuum, $12.398/E_0$; λ_{EDGE} is the wavelength of the absorption edge for the characteristic emission of interest; and $M(*, A\ K\alpha)$ and $M(*, \lambda)$ are the mass attenuation coefficients of the specimen for A $K\alpha$ and continuum wavelength λ, respectively.

The computation of $F(\lambda)$ requires a computer and even then is expensive in terms of computation time and required storage space.[30] Therefore Myklebust *et al.* investigated the continuum correction in order to pinpoint cases where the correction can be safely ignored.[33] They came to the following conclusions.

1. The magnitude of the continuum fluorescence correction cannot be neglected in cases where $f(\chi)$ of the material is 0.95 or greater. This corresponds to analysis using hard x-ray lines in a light matrix, e.g., analysis of the "heavy" component in an oxide.

2. Voltage variation and errors in the takeoff angle, fluorescence yield factor, and jump ratio have no significant effect on the continuum fluorescence correction.

3. The continuum fluorescence correction can be ignored when $f(\chi)$ is 0.95 or less and when the concentration of the analyzed element is greater than 0.5.

4. All of the results support the choice of the operating parameters suggested for quantitative analysis: low overvoltage and high takeoff angle.[5]

Perhaps the most important conclusion is the first, illustrated in Figure 13. The implication is that uncertainties may result if one employs hard x-ray lines from heavy elements such as Zn, Hg, etc. found as traces in light matrices, such as B (Figure 13). For trace amounts of copper in biological tissues, the continuum radiation contribution provides most of the

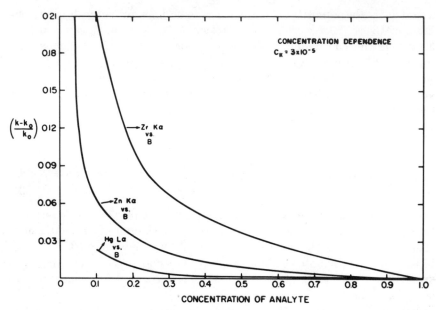

FIGURE 13.   Amount of characteristic intensity produced by continuous radiation for analysis using energetic x-ray lines in a boron matrix. The ordinate is the calculated k value with continuum correction applied, relative to the calculated k value with no continuum correction applied.

observed copper intensity. Hence, when analyzing for small amounts of heavy elements in light matrices, the softest possible characteristic x-ray line should be chosen as the analytical line.

VI. SUMMARY DISCUSSION OF THE ZAF METHOD

We now consider the way in which the individual corrections are utilized to actually calculate the composition of the specimen. In the usual experiment, one determines a set of k values. But the factors k_A, k_Z, and k_F depend upon the true composition of the specimen, which is unknown. This problem is handled by using the k values as the first estimate of composition in order to calculate the correction factors. Then the resulting mass fractions are used as a new estimate of composition, to compute more accurate correction factors. Iteration proceeds until convergence of results occurs.

The iteration procedure most often used was suggested by Criss and Birks[34] and is based on the idea that the curve relating C and k is a hyperbola.[1,35] (This hyperbolic approximation will be discussed in detail in the next section.) For each step in the iterative process, the next estimate for the mass fraction C_m corresponding to the measured intensity ratio k_m can be calculated from

$$C_m = \frac{k_m C(1 - k)}{k_m(C - k) + k(1 - C)} \tag{33}$$

since the product $[(1 - k)/k][C/(1 - C)]$ is a constant if the anaytical curve is indeed a hyperbola. Thus, to apply equation (33), one assumes a reasonable estimate of the value of C. The k corresponding to that C can be calculated since k_A, k_Z, and k_F can also be obtained. Then one measures an actual k_m and solves for C_m. In practice, the k ratio measured is usually taken as the first estimate of C and a k value is calculated from it for use as "C" and "k" in equation (33).

This iteration procedure has been extensively tested. It converges rapidly since the hyperbolic approximation is, in fact, almost always a very good representation of the analytical curve of C against k. Rarely are more than three iterations necessary.

The ZAF method is often used as a means of obtaining quantitative results from measured relative intensity data. This method is supposed to be applicable to any class of specimen. But ZAF may not be successful in the analysis of light elements, due mainly to the lack of knowledge of required input parameters and approximations in the correction models themselves. For these reasons, no other technique is fully reliable, except perhaps when standards of nearly the same composition as the specimen are available.

An example of the accuracy that can be expected when good specimens are available is shown in Figures 7a and 7b, which show the relative error distribution in a group of 264 homogeneous, well-analyzed, binary metal alloys certified as standard reference materials by NBS. These alloys were examined under a variety of operating conditions in a commercial electron probe microanalyzer; crystal spectrometers were used. The relative intensities were corrected by means of a theoretical procedure developed into a computer program called COR2,[30] which is believed to be the best quantitative analysis program available. Some of the experimental measurements were made at higher operating voltages than the optimum or with softer x-ray lines than optimum, i.e., Cu L instead of Cu K, or Au M instead of Au L. The results obtained from cases where nonoptimum experimental conditions were used provide an indication of what occurs when the nature of the specimen makes it difficult for the investigator to ensure that $f(\chi) > 0.7$. The absorption correction factor is the only term in equation (7) that the operator can control easily; fluorescence effects are inherent in the specimen and the atomic number correction is virtually insensitive to operating conditions.[21] The only way to minimize the atomic number effect is to provide a standard whose atomic number is nearly the same as that of the specimen. Providing such a standard is not always easy.[36]

As a further example, some 228 analyses of binary materials carried out by a wide variety of laboratories[21] were corrected using COR2. The results are presented in the histograms of Figures 8a and 8b. One way to sum up the implications of Figures 7a, b and 8a, b is in gambling terms. The odds are 2 to 1 against an analysis being within 1% relative (1% of amount present), even money on the analysis being within $2\frac{1}{2}$% relative, and 7 to 1 for the analysis being within 5% relative. These odds are accuracy estimates based on nearly optimum specimens, i.e., properly prepared[36] and placed normal to the electron beam in a commercial electron probe microanalyzer.

VII. THE EMPIRICAL METHOD FOR QUANTITATIVE ANALYSIS

In the early 1960's, the ZAF method was not nearly as well developed as it is presently. Hence histograms such as those of Figures 7a, b and 8a, b showed a much wider distribution of errors.[20] Furthermore, computer data reduction was much less generally available. In response to this state of affairs, Ziebold and Ogilvie[35] developed what is known as the hyperbolic or empirical correction method, first described in 1964. In fact, Castaing[1] laid the basis for this development in his thesis in what he referred to as "the second approximation." This second approximation was a statement that the true weight fraction C and the measured relative intensity k were related such that a plot of C against k would be a hyperbola.

Ziebold and Ogilvie expressed this relationship in the form

$$\left(\frac{C}{1-C}\right)\frac{1-k}{k} = a \tag{34}$$

Clearly, if a specimen which is homogeneous at the micrometer level and for which C is known can be procured, k can be measured and a so determined. For example, the National Bureau of Standards certifies SRM-480 to be homogeneous and to consist of 0.215 Mo and 0.785 W. The k values determined using an operating voltage of 20 kV were reported as 0.143 for

Mo and 0.772 for W. Solving equation (34) for the respective a values, a is 1.649 for Mo in Mo–W and 1.078 for W in Mo–W. These values are valid only for a 20-kV operating voltage. The solution for C in the case of a binary is

$$C = \frac{ka}{1 + k(a - 1)} \qquad (35)$$

Therefore any composition of W–Mo can be analyzed by carefully determining k with an operating voltage of 20 kV and solving equation (35) with a desk calculator.

There are two major drawbacks, however. First, it is often difficult to obtain the desired standards, and second, the extension to more components than two is not immediately obvious. The extension to more than a binary and the accuracy of the hyperbolic approximation have been considered in detail by Bence and Albee.[37] These authors were concerned with providing a rapid, accurate means to analyze specimens for geological studies. Mineralogical and petrological specimens may be heterogeneous and frequently contain 6–8 elements with concentrations in excess of 1% by weight. For reasons of simplicity and economy, many such specimens are analyzed by means of the ZAF method. Real-time computer reduction of data is desirable since knowledge of the composition and the calculated formula of the phase is often a necessary prerequisite to the operator's deciding what the next stage in the analytical procedure will be. Bence and Albee noted that a plot of C/k against either C or k must be a straight line of slope a for small C in any binary system. Parenthetically, if a series of alloys in a system is analyzed, a plot of k/C vs. k must be a straight line or the data are internally inconsistent.[37] Figure 14 illustrates such curves; data taken with an operating voltage of 5 kV are internally inconsistent. Furthermore, Ziebold and Ogilvie[35] had shown that for a ternary

$$\left(\frac{C_1}{1 - C_1}\right)\frac{1 - k_1}{k_1} = a_{123} \qquad (36)$$

and

$$a_{123} = \frac{a_{12}C_2 + a_{13}C_3}{C_2 + C_3} \qquad (37)$$

In other words, the empircal coefficient for the ternary can be determined from the individual binaries, i.e., standards for each binary are needed. Extending this, it was shown for a system of n components, that for the nth component[37]

$$C_n = k_n \beta_n$$

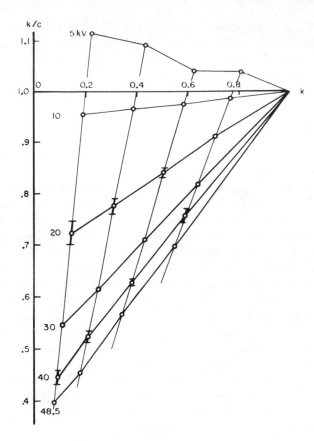

FIGURE 14.   Plot of k/C vs. k for silver in alloys of silver-gold. The Ag $L\alpha$ line
was used for analysis; primary beam voltages were 5, 10, 20, 30, 40,
and 48.5 kV, respectively, as noted on the figure. The data taken at
5 kV are internally inconsistent, as indicated by the deviation from
linearity.

where

$$\beta_n = \frac{k_1 a_{n1} + k_2 a_{n2} + k_3 a_{n3} + \cdots + k_n a_{nn}}{k_1 + k_2 + k_3 + \cdots + k_n} \tag{38}$$

Here a_{n1} is the a value for the $n{:}1$ binary, i.e., for the determination of
element n in a binary consisting of element n and of element 1. Thus a_{n2}
is the a value for a binary consisting of element n and of element 2, and so
on to a_{nn}, which is for n in n and so is unity. Similarly, the notation a_{1n}
would represent the a value for element 1 in a binary of 1 and n.

The value of k_1 is the relative intensity ratio found for element 1 in the specimen to the standard used for element 1. This notation can be carried along so that k_n is the relative intensity ratio for element n to the standard used for element n.

To solve completely a system of n components, one needs $n(n - 1)$ values for a (the a value for each of the n pure elements is unity, as noted above). Thus $n(n - 1)$ standards are required. Bence and Albee procured a number of suitable standards of interest in geology and determined their respective a values.[37] Given the necessary matrix of a values, the analyst chooses his analytical standards and determines $k_1, k_2, \ldots, k_n$. Then, $\beta_1, \beta_2, \ldots, \beta_n$ can be calculated. From these, a first approximation to $C_1', C_2', \ldots, C_n'$ is found. Then a new set of β values is calculated as $\beta_1', \beta_2', \ldots, \beta_n'$,

$$\beta_n' = \frac{C_1' a_{n1} + C_2' a_{n2} + \cdots + C_n' a_{nn}}{C_1' + C_2' + \cdots + C_n'} \tag{39}$$

Reiteration can continue until differences between calculated values of β are made arbitrarily small. In practice, the result really depends on how good the standard is; in geological work, standards of nearly the same atomic number as the specimen are usually chosen. Hence atomic number effects are reduced. Fluorescence effects are small in silicate rocks and similar materials. Hence most of the final correction is for absorption.

In 1972, Japanese geologists from twenty laboratories participated in a cooperative study of ten petrologic specimens, including natural basalt glasses, Bornite and Bournonite, synthetic silicates, and synthetic sulfides.* Data were corrected by the method of Bence–Albee[37] and by the ZAF-based method of Sweatman and Long.[40] The spread of measured k values for a given specimen was about $\pm 5\%$ relative; but for a given k value, the results of correcting by the two schemes were within $\pm 2\%$. Other tests of the validity of this method indicate that it yields results comparable to those obtainable with the ZAF method. The chief obstacle to its general use is the need for many homogeneous, well-characterized standard materials. For this reason, the a method is sometimes combined with the ZAF method. In this case, one computes the necessary a matrix by assuming a C value and using the ZAF method to compute the corresponding k. Then, of course, a can be obtained from equation (34). Such an a value is subject to all of the uncertainties associated with the ZAF method as outlined previously. However, as will be shown in detail in the chapter on rapid analysis computation methods, using the ZAF method to infer an a matrix is one method of arranging for on-line analysis with very small computational devices.

---

*Report available from Hirokawa.[39]

VIII. COMMENTS ON ANALYSIS INVOLVING ELEMENTS OF ATOMIC NUMBER OF 11 OR LESS

Next, the question of analysis of samples with $Z \leq 11$ (Na) can be considered briefly. Major difficulties are obtaining enough x-ray intensity and suitable standards for the light elements. Until now, energy-dispersive analysis for the light elements has been very difficult. Special crystals are needed in spectrometers in order to diffract the soft (low-energy) x-rays produced by these elements. Once the intensity problem has been overcome, the chief obstacle is that x-ray mass attenuation coefficients for the light elements are both large and poorly known (see Figure 9). There has been recent work on determining these coefficients.[16,41] But a great deal of uncertainty concerning the correct values of the coefficients persists and more work is needed in this area. In light element analysis, the choice of operating voltage may be crucial. As the primary beam voltage increases, x-ray production increases. But the depth from which emission occurs also is greater (see Figure 3). Hence, with light elements, there is an optimum operating voltage at which increased x-ray production is just balanced by increased absorption, i.e., intensity as a function of operating voltage passes through a definite maximum in the case of light elements[16] (see Chapter XII).

The atomic number correction uncertainty can play a large role when light elements comprise a portion of the specimen even though these elements are not analytes. For example, uncertainties of about 2% relative occur in the analysis of Ti in TiC performed using an operating voltage of 10 kV.[21]

Chemical binding effects can cause a shift of the peak position for a given element.[42] These shifts become larger as the atomic number decreases. Hence care must be taken to assure that the intensity is taken from the actual peak position and not from some other point on the peak sides, i.e., spectrometers must be repeaked from specimen to standard.

In any analysis involving light elements, the carbonaceous contamination layer built up by electron beam interaction with oil or dirt in the electron column must be considered since a thick layer may strongly absorb the radiation of interest. Some means of reducing or eliminating the contamination layer, such as a cold finger, is useful (see Chapter XII).

The analysis of biological material, whether it be calcified material or soft tissue, could be attempted with the foregoing methods, ZAF or empirical, provided that the specimen is infinitely thick to the electron beam and that the beam does not alter the specimen. The latter is of course crucial to any analysis. Care must be taken to ensure that contamination layer buildup is minimized. Reducing voltage and beam current reduces intensity and so longer dwell times may be needed to obtain statistically valid data. The investigator must balance between increased exposure time and higher beam intensity in order to optimize x-ray analytical experimental conditions (see Chapter XIII).

IX. QUANTITATIVE ANALYSIS WITH NONNORMAL ELECTRON-BEAM INCIDENCE

There is the difficulty that all the correction models presume that the electron beam is normal to the specimen. However, in the SEM the specimen is usually tilted in order to provide a suitable value of the takeoff angle. The effect of specimen tilt on the accuracy of quantitative analysis was investigated by Monte Carlo calculations of electron–solid interactions, which clearly indicated changes in x-ray distribution and emission as a function of specimen tilt[43,44] (see Figure 15). However, experience with tilted specimens indicates that if specimen and standard are handled identically and tilts are not extreme (>70°), large quantitative analysis errors do not occur.[45,46]

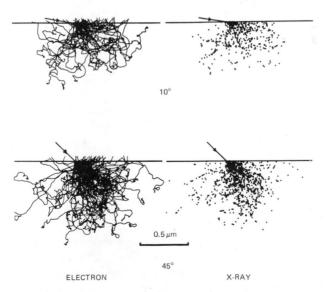

10°

0.5 μm

45°

ELECTRON   X-RAY

FIGURE 15.   Schematic representation of 100 electron paths in copper; primary beam voltage was 20 kV. The effects of tilting the specimen on the paths and on the corresponding x-ray generation are apparent. (Compare with Figure 1.)

The chief difficulty lies in the fact that the ZAF correction model is predicated on normal beam incidence. The effect of a tilted specimen on the parameters of the ZAF method has not been extensively dealt with. Reed has determined that backscattering factors R change considerably when the sample is tilted 45° with respect to the beam.[47] The absorption correction may well be satisfactory provided the correct emergence angle Ψ is obtained for the geometry available. It is difficult to assess stopping power and fluorescence effect behavior as a function of tilt *a priori*. Probably, extensive Monte Carlo calculations and more experimental work will be needed to quantitatively assess the effect of a tilted specimen on the analytical results. However, long experience would seem to indicate that errors introduced by tilting specimen and standard are not fatal.

X. ANALYSIS INVOLVING SPECIAL SPECIMEN GEOMETRIES

The analysis of thin films, either free standing or supported on a substrate, or small spheres, cylinders, and inclusions requires special treatment in order to account for effects of electrons (1) passing through the specimen, (2) scattering out of its side, or (3) backscattering from a substrate into the specimen (see also Chapter VIII for a discussion of this topic). Monte Carlo techniques are perhaps the most obvious means in such cases for attempting to obtain quantitative results. However, the use of Monte Carlo methods does not guarantee accuracy but merely provides a way to handle the geometry problem. Monte Carlo programs are expensive to run, even on large, fast computers. Hence Monte Carlo methods are usually employed to establish a few calibration curves for systems of interest. It is impractical to use Monte Carlo methods for routine analysis.

In any Monte Carlo program, an electron scattering model must be chosen. A mathematical model is constructed and a number of electron trajectories are simulated so that electrons are allowed to follow a random walk through the solid. Scattering angles and the coordinates of the electron are selected by a random number process—hence the name Monte Carlo. The production of an x-ray photon at any stage is calculated probabilistically from an x-ray cross section. Any such program requires input parameters including x-ray mass attenuation coefficients, mean ionization potentials,[14] etc. Thus the accuracy of the calculation should not be expected to be any better than, say, the ZAF method. Nevertheless, Monte Carlo programs can be very versatile in that a wide variety of analytical conditions, e.g., tilted specimens, thin film on a substrate, micrometer size spheres, can be handled. Recently, a great deal of work involving Monte Carlo methods has appeared in the literature.[43,44,48-50]*

A Monte Carlo program based on Curgenven and Duncumb's original formulation[43] has been developed for use at the National Bureau of Standards. The Monte Carlo procedure used in this work is based upon a simple Rutherford scattering model for the primary electron–solid interaction. Since inelastic collisions cause the electrons to lose energy with little angular deflection, inelastic scattering can be described by Bethe's energy loss law without consideration of concomitant changes of direction of the electrons. The number of elastic interactions has been reduced to 60 per trajectory in order to save computer time. It should be noted that the Monte Carlo procedure is, for the most part, empirical in nature, and broad assumptions are involved in the scattering and energy loss models.

A variable step length between collisions in which the step length is related to the current value of the energy of the electron is used. The sum of the step lengths for 60 steps is equal to the Bethe range for the incident energy. The effects of a finite beam size and the distribution of electron impacts within the beam have been taken into account by using a Gaussian-distribution random number generator to model the beam (Figure 16), for which a circle containing 80% of the electrons is defined as the beam diameter.[51] For the calculation, parameters such as the electron beam diameter, the operating potential, the position of the beam center on the specimen, and the composition, size, and orientation of the specimen can be selected. The program gives results in good agreement with experimental findings for such entities as electron backscatter coefficients, electron transmission through films, $\phi(\rho z)$ curves, $f(\chi)$ curves, and photon generation.

Unsupported thin specimens or those mounted on a grid and hence unsupported near where the electron beam passes through are of special interest since the bulk of biological analysis is performed on such samples. Therefore various other schemes have been developed in order to analyze this important class of specimens. The main assumption underlying these schemes is that the x-rays produced by the beam can escape the specimen surface after suffering only negligible absorption and fluorescence effects. Hence the relationships between observed intensity and actual concentration are linear. If this assumption is valid, then formulating the desired equation or calibration curve becomes an empirical process depending on (1) the elements present, (2) the way in which they are bound in the specimen, (3) the section thickness, and (4) the experimental conditions chosen.[52-55] Hall and co-workers[56] have derived a method based on nor-

---

*Reference 50 contains several Monte Carlo papers.

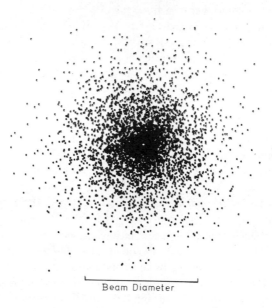

Beam Diameter

FIGURE 16. Gaussian representation of the electron probe; a circle containing 80% of the electrons is defined as the probe diameter.

malization of the observed x-ray continuum. With it, neither variation of specimen thickness nor the actual thickness of the thin standards need be known (see Chapter XIII on biological analysis).

Many specimens to be analyzed in the SEM do not have flat, smooth surfaces. For qualitative analysis, various specimen tilts should ensure that the elements present are identified (see Figure 17). For quantitative analysis, local specimen orientation must be known in order to obtain a correct value of Ψ. Furthermore, features outside the area of interest may intervene in the x-ray path and affect the observed intensity. Bomback has described a stereoscopic-computer method for reorienting planar regions on rough surfaces to some standard position. He claims a relative error of no more than 10% for intensities measured this way. Stereoscopic techniques reveal blocking by remote obstructions. However, extraneous radiation produced remotely could not be taken into account.[57]

An example of a nonflat surface is shown schematically in Figure 17; the sources of extraneous radiation in the case of energy-dispersive detectors can cause considerable difficulty. For example, the blocking effect could cause confusion in qualitative analysis with an energy-dispersive detector if radiation from the electron-bombarded area excited radiation from

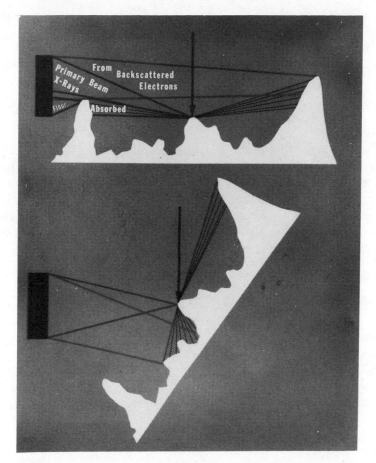

FIGURE 17.   Schematic representation of the processes that may occur when fracture surfaces are examined. See text for explanation.

remote obstructions on the surface of a rough sample or anywhere else in the line of sight of the detector. Changing the tilt or rotation would aid in this regard. The difficulties are less serious if a crystal spectrometer is used since radiation sources must be on the Rowland circle in order to be detected.

The chief problem is to determine the x-ray takeoff angle Ψ. Perhaps a means of attacking this problem is to measure the ratio of specimen current i_s to beam current i_B. Now the backscatter coefficient η is given by

$$\eta = 1 - (i_s/i_B) \tag{40}$$

The value of η as a function of tilt was measured for Al, Fe, and Pt. As Figure 18 shows, η is a linear function of tilt θ in the range $40° \leq \theta \leq 70°$. Thus with a and b as empirical constants

$$i_s/i_B = 1 - \eta = a + b\theta \tag{41a}$$

and

$$\theta = \frac{1}{b}\left(a - \frac{i_s}{i_B}\right) \tag{41b}$$

For Fe, $a = 0.97$ and $b = 0.008$ with $E_0 = 30$ keV.

From data such as these, the tilt of a given facet can be determined provided the specimen is biased to remove secondary electrons and care is taken to minimize rebackscattering of primary electrons from extraneous sources. Given the specimen tilt, the value of the takeoff angle can be computed. However, analysis from rough surfaces should be regarded as semiquantitative at best.

XI. DISCUSSION

The bases currently used for carrying out quantitative elemental analysis in electron probe microanalysis or scanning electron microscopy have been outlined. The actual computation of results is almost invariably

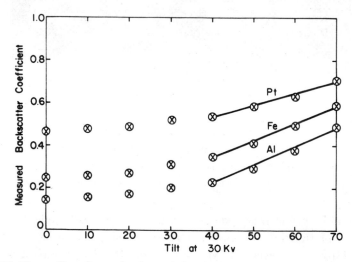

FIGURE 18.   Experimental values of the electron backscatter coefficient for Al, Fe, and Pt as a function of specimen tilt.

carried out with the aid of a computer. In the next chapter, we discuss computational methods with strong emphasis on minicomputers (about 8K of word memory). The uncertainties produced by the ZAF model rest with input parameters such as x-ray mass attenuation coefficients, ionization potentials, and backscattering factors. Experimental difficulties such as the need to tilt the specimen, uncertain x-ray takeoff angle, or incorrect determination of operating voltage also play a role. The models themselves have not been proven rigorously correct. Nevertheless, the method almost always yields results within 5% relative and about once in three times gives results within 1% relative.

The empirical method is predicated on the analytical curve being accurately represented by a hyperbola. Experience has shown that this is often the case. However, procuring appropriate standards to apply the method is not always easy. Therefore this method is often combined with the ZAF method in order to obtain the necessary a matrix. In this form, the method is then subject to the difficulties associated with the ZAF method.

Special methods for special geometries have been developed, some based on Monte Carlo methods, others empirical. The Monte Carlo method is subject to uncertainties introduced by input parameters in the same way as the ZAF method. However, the Monte Carlo method can handle specimens having difficult geometry, while the ZAF method cannot.

In any case, the ultimate accuracy of the analysis depends not only on uncertainties in the correction schemes but also on the accuracy with which the relative intensity ratio k is determined. Therefore care is needed in measuring and in planning the strategy of analysis in order to obtain good k values.

The appendix describes the entire procedure used for the analysis of SRM-484, a binary iron–silicon alloy, from planning to final data reduction.

APPENDIX. THE ANALYSIS OF AN IRON–SILICON ALLOY

The specimen is NBS-SRM-483, which is certified to contain 3.22 ± 0.01 wt % silicon, balance iron. It is referred to as Fe–3 Si. The analysis was carried out at NBS using an instrument having a 52.5° takeoff angle. What follows is a typical quantitative analysis strategy.

The microprobe experiment consists of measuring the x-ray flux emitted from the specimen. Both the Fe $K\alpha$ and Si $K\alpha$ peaks can be investigated simultaneously using two spectrometer channels. Scalers are used to obtain the counts per unit collection interval. Then the x-ray flux from reference standards for Si and Fe is monitored on the channel used for each of these elements. The ratio of the fluxes from specimen and standard is called k by convention among microprobe analysts. Raw data must be corrected for instrumental effects, such as background, to obtain k.

The conversion of k values to C values will be discussed in detail later. What is of concern now are the steps needed to obtain the best (most accurate) value of k possible. As these steps are described, the strategy of a typical microprobe analysis will be delineated. At the end, the experimental conditions will be completely fixed for the Fe–3 Si investigation. The analyst hopes that all factors will be optimized so that the most accurate analysis possible can be made.

The steps are:

1. Deciding on the experimental parameters.
2. Specimen preparation.
3. Choosing appropriate reference standards.
4. Data acquisition, including corrections for instrumental errors and x-ray background effects.
5. Conversion of the data to the quantitative composition of the analyzed region, i.e., the Fe and Si concentrations present.

The way the first of these steps is carried out depends almost entirely on how much is known about the specimen beforehand. In the case of Fe–3 Si, the elements of interest are Fe and Si and the approximate Si content is 3% by weight. As shown, the accuracy of quantitative electron probe microanalysis improves as the operating voltage is lowered and the angle Ψ at which the x-rays emerge from the specimen is raised. However, the operating voltage E_0 cannot be decreased below about 1.5 times the critical excitation potential E_c of the characteristic line of interest. Below this value of $U = E_0/E_c = 1.5$, x-ray emission drops off rapidly. Furthermore, microprobes may become unstable when voltages of 6 kV or less are used. For Fe, the value of E_c is 7.1 kV for the $K\alpha$ peak; for Si $K\alpha$, E_c is 1.84 kV. Thus 10 kV is about as low as one can safely go. The $K\alpha$ lines are the only suitable choices for Fe and Si; all other lines have too low an energy for practical use.

The choice of crystals used to diffract the x-rays emitted from the sample is limited since one needs suitable crystals for Fe $K\alpha$ (1.936 Å) and Si $K\alpha$ (7.110 Å). The only practical crystal for the former is LiF; for Si $K\alpha$ either ammonium dihydrogen phosphate (ADP) or ethylene diamine-*d*-tartrate will be satisfactory.[58]

Specimen preparation for microprobe analysis has been investigated.[36] For the Fe–3 Si case, an ordinary electropolished metallographic finish is entirely suitable. Electropolishing is satisfactory since no smearing of constituents occurs with this technique, inclusions are not of interest, and small surface blemishes are of no consequence.[36]

The only appropriate standards for analysis are pure Si and pure Fe. Standards are needed because it is impossible to accurately determine absolute x-ray intensities. Therefore all measurements made with the microprobe must refer the intensity of a characteristic x-ray line from the specimen to the intensity of the same characteristic line from a reference standard of known composition. Pure elements are usually chosen for several reasons: (1) it is very difficult to obtain alloys or compounds for use as standards which are truly homogeneous at the submicrometer level,[59,60] (2) it is possible to convert the raw data to concentrations by means of fairly rigorous physical models when using elemental standards in microprobe analysis, and (3) pure elements are often easy to obtain. The subject of standards has been treated elsewhere.[36] Characterization of standards is tedious and time-consuming.[38] Further discussion of standards and specimen preparation is given in Chapter XI.

There are two chief sources of instrumental errors: dead-time (coincidence loss) effects and instrument drift or instability. Dead time is the time interval after a pulse is recorded during which the system cannot respond to another pulse. Dead time is caused by the detectors used (sealed proportional counters in the case of Fe–3 Si) and by the readout electronics. The question of correction for dead time in proportional counter systems in electron probe microanalysis has been investigated by Heinrich, Vieth, and Yakowitz.[7] These authors showed that the Ruark–Brammer relation,

$$N = N'/(1 - \tau N') \qquad (42)$$

where N' is the measured count rate, N is the true count rate, and τ is the dead time in seconds, is valid at least up to observed count rates $N' \simeq 5 \times 10^4$ counts/sec. The value of τ has been determined for the two spectrometer channels used in the Fe–3 Si study on many occasions. (Reference 7 describes the methods and gives the first results.) The value of τ is 2.3 ± 0.05 μsec (1σ) for spectrometer channel one (SC-1), and τ is 2.3 ± 0.13 μsec (1σ) for spectrometer channel two (SC-2). Thus if N' is 20,000 counts/sec, N is 21,000 counts/sec. The actual microprobe experimental work consists in determining the ratio of the true count rate from the specimen N^* and the true count rate from the standard N:

$$k = \frac{N^*}{N} = \frac{N'^*}{N'} \frac{(1 - \tau N')}{(1 - \tau N'^*)} \qquad (43)$$

From equation (43) we obtain by differentiation

$$\Delta k/k = (N^* - N)\, \Delta \tau \tag{44}$$

which represents the expected error in k due to dead time error $\Delta \tau$, e.g., 0.05 μsec for SC-1. Thus, the absolute count rate difference between specimen and standard should be made small in order to minimize errors due to τ.

However, for a Si content of 3 wt % in the specimen referred to a pure Si standard, there is only one way to minimize $\Delta k/k$: reduce the count rate for the standard. Reducing the value of N also reduces the total number of counts accumulated N_T, and the statistical counting fluctuation given by $\sqrt{N_T}$ is increased.

A good compromise is to arrange matters so that N is about 20,000 counts/sec. For $\Delta \tau$ of about 0.1 μsec, we get a relative error $\Delta k/k$ of 0.2% for Si. Since $k_{Si} \simeq 0.03$, we have then $k_{Si} \simeq 0.03 \pm 0.00006$ due to errors in τ. This uncertainty in k is negligible. The value of N_T can be set to about 25,000 so that $3\sqrt{N_T}$ is about 500. Taking $3\sqrt{N_T}$ is equivalent to stating a 99% confidence interval for N_T. The value $3\sqrt{N_T}/N_T$ gives the relative spread in the N_T determination; for the conditions stated this value is 2% relative. This contribution makes the projected k_{Si} uncertainty $k_{Si} \simeq 0.03 \pm 0.0007$, or $3 \pm 0.07\%$.

For the iron, the difference between N^* and N will be small because specimen and standard have nearly the same composition. With the microprobe set to give N for Si of 20,000 counts/sec, the value of N for the Fe will be similar. Hence $3\sqrt{N_T}/N_T$ will be very small for the Fe measurement.

The other major source of experimental error is drift of the probe during the entire data-taking process. Drift is caused primarily by the warping of the electron-gun filament during the course of the analysis. Long-term drift in the instrument used for the Fe–3 Si analyses caused x-ray count rates to vary by about 0.5% per hour. However, the problem of drift can be overcome by using a fixed increment of monitor current i_m to control the dwell time at an analysis point. Monitor current is obtained by measuring the beam-induced current on the electron objective lens aperture. For the particular instrument in question, the constancy of the ratio of actual electron beam current i_B to monitor current has been established experimentally.[14] The ratio i_B/i_M is 3.5 $\pm$ 0.1 in the region $10^{-9} \leq i_B \leq 10^{-6}$ A.

The experimental arrangement is to convert the monitor current to a voltage, go through a voltage-to-frequency converter, and then to scale the frequency. This scaler can be made the master controller of the scalers recording the Fe $K\alpha$ and Si $K\alpha$ x-ray signals; time can also be recorded. Thus the same number of incident electrons will always be responsible for the observed Fe and Si counts. The time variation, if any, is a measure of the drift effect.

The next problem is that of background contribution to the observed counts. The most important background source in electron microprobe analysis is the continuous radiation that originates when high-energy electrons strike the specimen and lose their energy by deceleration (Bremsstrahlung). Continuous radiation is always produced when electrons are used to excite characteristic x-rays. The overall intensity of the continuous radiation is a function of the accelerating potential, electron beam current, and atomic number of the specimen; it increases with increasing operating voltage, beam current, and atomic number of the specimen. Further contributions to the overall background originate from cosmic rays, scattered x-rays, and electrons, circuit noise, and fluorescence radiation produced in the diffracting crystal.

For the Fe–3 Si study, the background correction method outlined by Ziebold[9] was used. In this method, the background is determined by leaving the two spectrometers set for the Fe $K\alpha$ and Si $K\alpha$ peaks. The x-ray flux from the element next lower in the periodic table is found. For Fe $K\alpha$, the background is determined on Mn and for Si $K\alpha$, on Al.

Ziebold states that in the study of alloys, it is sufficiently accurate to assume that the specimen background level can be found from weight fraction averages of the background measured from each element individually.[9]

For example, the Si background in Fe–3 Si would be

$$B(\text{Si } K\alpha) = C_{\text{Si}}B_{\text{Al}}(\text{Si } K\alpha) + C_{\text{Fe}}B_{\text{Fe}}(\text{Si } K\alpha) \tag{45}$$

where $B_{\text{Al}}(\text{Si } K\alpha)$ and $B_{\text{Fe}}(\text{Si } K\alpha)$ are the x-ray fluxes at the Si $K\alpha$ wavelength measured on pure Al and pure Fe, respectively. Similarly, the Fe background in Fe–3 Si is

$$B(\text{Fe } K\alpha) = C_{\text{Fe}}B_{\text{Mn}}(\text{Fe } K\alpha) + C_{\text{Si}}B_{\text{Si}}(\text{Fe } K\alpha) \tag{46}$$

An alternate background correction method is to displace the spectrometer (above and below the peak) and measure off-peak background values on the actual sample. The average of these two values is taken to be the background. Unless the exact shape of the continuous spectral distribution is known, such a procedure can lead to erroneous background values. The final value of k is now computed as

$$k = (N^* - B^*)/(N - B) \tag{47}$$

The entire experimental procedure can now be summarized as follows:

1. Electropolish (and lightly etch) specimen.
2. Operating voltage: 10 kV. This is as low as we care to go in order

to get enough Fe $K\alpha$ intensity but is sufficiently low to minimize quantitative analysis errors for Si determination.

3. Monitor current adjusted on pure Si standard so that about 20,000 counts/sec are obtained. This condition gives about 600 counts/sec from the Si in the specimen.

4. Accumulate about 25,000 counts from Si in Fe–3 Si. The dwell time per point is then about 40–45 sec.

5. Monitor current controls the dwell time to minimize possible drift effects.

6. Use pure Si and pure Fe as reference standards.

7. Use adjacent element method of Ziebold for background correction. Pure Al and pure Mn are needed.

Based on the previous discussion using these experimental conditions, we can reasonably expect a k value for Si to be accurate to within 3% relative, which is about 0.001 absolute units. The Fe k value will also be uncertain by about 0.001 units. Now the conversion of k to C values can be considered.

First, one inquires whether a fluorescence correction will be necessary. Specifically, will Fe K lines excite Si $K\alpha_{1,2}$ and will Si K lines excite Fe $K\alpha_{1,2}$? The Si K spectrum has most of its power at about 2 kV and will not excite any Fe lines. The Fe K spectrum power is centered near 7 kV and so some excitation of Si $K\alpha_{1,2}$ would be expected. However, Heinrich and Yakowitz have shown for an atomic number difference of eight or greater ($Z_{\mathrm{Fe}} - Z_{\mathrm{Si}} = 12$) that omitting the fluorescence correction causes a negligible error in the C value.[27] Thus k_F can be taken as unity and $\Delta k_F/k_F$ as zero for Fe–3 Si. Fluorescence of characteristic lines by the continuum can also be ignored in the case of Fe–3 Si with very little loss of accuracy.

Next, the absorption correction term k_A can be considered. The value of f_p should be 0.7 or more to minimize errors in k_A. Hence the experimental conditions chosen before must be tested to see if this criterion is met.

A value of 10 kV was picked as operating voltage; Ψ is 52.5°, so csc $\Psi = 1.260$. The values of (μ/ρ) given by Heinrich were chosen.[61] These are listed in Table I. The computed value of $f(\chi)$ for Si in Fe–3 Si is 0.757 and that for Fe is 0.996. At 15 kV, $f(\chi)$ for Si in Fe–3 Si drops to 0.602. Thus the choice of 10 kV essentially satisfies the 1% accuracy criterion for Si and certainly does so for Fe.

The estimated uncertainty of k_Z for Si in Fe–3 is 3% relative, due mainly to uncertainties in J and R. In sum, there is an error $\Delta k/k$ of 3%,

TABLE I. X-Ray Mass Absorption Coefficients for Silicon and Iron

Absorber	Emitter Si $K\alpha$ $\lambda = 7.13$ Å	Emitter Fe $K\alpha$ $\lambda = 1.936$ Å
Si	328	115.5
Fe	2502	71.4

$\Delta k_F/k_F$ of zero, $\Delta k_A/k_A$ of 1%, and $\Delta k_Z/k_Z$ of 3% as estimates in the analysis of Si in Fe–3 Si. Then $\Delta C/C$ is 7% or about 0.2 weight %. Consequently, any two analyses for Si differing by 0.3 weight % can be considered to be representative of real differences in the Si content between the two analyzed regions.

No significant variation in Si content was found in tests of homogeneity in several samples.[6] Over 100 separate analyses were carried out for Fe and Si simultaneously while traversing several grains in a systematic fashion. Special positions such as grain boundaries or triple points were also tested. Within the ability of the electron microprobe to measure a variation, no significant variation of Si content exists as a function of position in the specimens.[62]

The results for Fe and Si can be summarized as follows:

1. The silicon content is 3.14% (105 determinations).
2. The observed range of Si content was 2.89–3.29%.
3. We can be 99% confident that the mean Si concentration lies between 3.12 and 3.16%.
4. The percent coefficient of variation for Si is 2.74% ($\%CV = 100s/C_{Si}$).
5. The iron content is 96.87% (106 determinations).
6. The observed range of Fe content was 94.4–99.4%.
7. We can be 99% confident that the mean Fe concentration lies between 96.6 and 97.2%.
8. The percent coefficient of variation for Fe is 1.20%.

The true silicon percentage is 3.22%, so that the microprobe experiment gave a result too low by 2.5% relative; the iron result is 1% too high. As a final note, it is interesting to compute the empirical values for a_{Si} and a_{Fe} in iron–silicon so that they can be compared with those inferred from the ZAF method. One obtains for a_{Si} a value of 1.149 as compared to a calculated value of 1.134; the corresponding calculation for Fe gives 1.158 as compared to a ZAF-based value of 1.196.

REFERENCES

1. R. Castaing, Thesis, University of Paris (1951), ONERA Publication No. 55.
2. D. M. Poole, in *Quantitative Electron Probe Microanalysis*, (K. F. J. Heinrich, ed.), NSB Special Publication 298 (1968), p. 93.
3. D. B. Wittry, ASTM Special Technical Publication No. 349 (1963), p. 128.
4. J. Philibert, in *X-Ray Optics and X-Ray Microanalysis, Proceedings of the Third International Symposium, Stanford University* (H. H. Pattee, V. E. Cosslett, and A. Engstrom, eds.), Academic Press, New York (1963), p. 379.
5. P. Duncumb and P. K. Shields, in *The Electron Microprobe, Proc. Symp. Electrochem. Soc., Washington, D.C. 1964* (T. D. McKinley, K. F. J. Heinrich, and D. B. Wittry, eds.), Wiley, New York (1966), p. 284.
6. R. L. Myklebust and K. F. J. Heinrich, ASTM-STP 485 (1971), p. 232.
7. K. F. J. Heinrich, D. L. Vieth, and H. Yakowitz, in *Advances in X-Ray Analysis*, Vol. 9 (1966), p. 208.
8. P. S. Ong, in *Advances in Electronics and Electron Physics*, Supplement 6, *Electron Probe Microanalysis*, (A. J. Tousimis and L. Marton, eds.), (1969), p. 137.
9. T. O. Ziebold, *The Electron Microanalyzer and Its Applications*, MIT Press, Boston, Massachusetts (1965), p. S-5.
10. E. Lifshin, M. F. Ciccarelli, and R. B. Bolon, in *Proceedings of the 8th National Conference on Electron Probe Analysis, EPASA, New Orleans* (1973), Paper 29.
11. K. F. J. Heinrich, NBS Technical Note 521 (1969).
12. L. S. Birks, *J. Appl. Phys.*, **32**, 387 (1961).
13. L. S. Birks, *J. Appl. Phys.*, **33**, 233 (1962).
14. K. F. J. Heinrich, H. Yakowitz, and D. L. Vieth, in *Proceedings of the 7th National Conference on Electron Probe Analysis, EPASA, San Francisco* (1972), Paper 3.
15. H. Yakowitz and K. F. J. Heinrich, *Mikrochim. Acta*, **1968**, 183.
16. D. F. Kyser, in *Proceedings of the 6th International Conference on X-Ray Optics and Microanalysis* (G. Shinoda, K. Kohra, and T. Ichinokawa, eds.), (1972), p. 147.
17. P. Duncumb and S. J. B. Reed, in *Quantitative Electron Probe Microanalysis* (K. F. J. Heinrich, ed.), NBS Special Publication 298 (1968), p. 133.
18. M. Green and V. E. Cosslett, *Proc. Phys. Soc. (London)*, **78**, 1206 (1961).
19. H. Bethe, *Ann. Phys.*, **5**, 325 (1930).
20. P. M. Thomas, U.K. Atomic Energy Authority Report AERE-R 4593 (1964).
21. K. F. J. Heinrich and H. Yakowitz, *Mikrochim. Acta*, **1970**(1), 123.
22. M. Green, Thesis, Cambridge University (1964).
23. G. Springer, *Mikrochim. Acta*, **1966**(3), 587.
24. J. D. Derian and R. Castaing, in *Optique des Rayons X et Microanalyse* (R. Castaing, P. Deschamps, and J. Philibert, eds.), Hermann, Paris (1966), p. 193.
25. M. J. Berger and S. M. Seltzer, National Academy of Science, National Research Council Publication 1133, Washington, D.C. (1964), p. 205.
26. J. Philibert and R. Tixier, in *Quantitative Electron Probe Microanalysis* (K. F. J. Heinrich, ed.), NBS Special Publication 298 (1968), p. 13.
27. K. F. J. Heinrich and H. Yakowitz, *Mikrochim. Acta*, **1968**(5), 905.
28. S. J. B. Reed, *Brit. J. Appl. Phys.*, **16**, 913 (1965).
29. J. W. Colby, National Lead Co., Ohio Report NLCO-969 (1965).
30. J. Henoc, K. F. J. Heinrich, and R. L. Myklebust, NBS Technical Note 769 (1973), p. 127.

31. J. W. Criss, in *Quantitative Electron Probe Microanalysis* (K. F. J. Heinrich, ed.), NBS Special Publication 298 (1968), p. 52.
32. J. Henoc, in *Quantitative Electron Probe Microanalysis* (K. F. J. Heinrich, ed.), NBS Special Publication 298 (1968), p. 197.
33. R. L. Myklebust, H. Yakowitz, and K. F. J. Heinrich, in *Proceedings of the 5th National Conference on Electron Probe Analysis, EPASA, New York* (1970), Paper 11.
34. J. W. Criss and L. S. Birks, in *The Electron Microprobe* (T. D. McKinley, K. F. J. Heinrich, and D. B. Wittry, eds.), Wiley, New York (1966), p. 217.
35. T. O. Ziebold and R. E. Ogilvie, *Anal. Chem.*, **36**, 322 (1964).
36. H. Yakowitz, ASTM-STP 430 (1968), p. 383.
37. A. E. Bence and A. Albee, *J. Geol.* **76**, 382 (1968).
38. K. F. J. Heinrich, R. L. Myklebust, S. D. Rasberry, and R. E. Michaelis, NBS Special Publication 260-28 (1971).
39. K. Hirokawa, Tohoku Univ., Sendai, Japan.
40. J. Sweatman and J. V. P. Long, *J. Petrology*, **10**, 332 (1969).
41. B. L. Henke and E. S. Ebisu, in *Advances in X-Ray Analysis*, Vol. 17, Plenum, New York (1974), p. 150.
42. E. W. White, in *Microprobe Analysis* (C. A. Andersen, ed.), Wiley–Interscience, New York (1973), p. 349.
43. L. Curgenven and P. Duncumb, Tube Investments Research Labs Report 303 (1971).
44. P. Duncumb, in *Proc. EMAG*, Conference Series No. 10, Institute of Physics, London (1971), p. 132.
45. R. B. Bolon and E. Lifshin, in *Proceedings of the 8th National Conference on Electron Probe Analysis, EPASA, New Orleans* (1973), Paper 31.
46. J. W. Colby, D. R. Wonsidler, and D. K. Conley, in *Proceedings of the 4th National Conference on Electron Probe Analysis, EPASA* (1969), Paper 9.
47. S. J. B. Reed, *J. Phys. D (Appl. Phys.)*, **4**, 1910 (1971).
48. R. B. Bolon and E. Lifshin, *SEM/1973 Proceedings of the 6th Annual SEM Symposium*, IITRI, Chicago, Illinois (1973), p. 285.
49. D. F. Kyser and K. Murata, in *Proceedings of the 8th National Conference on Electron Probe Analysis, EPASA, New Orleans* (1973), Paper 28.
50. E. Preuss (ed.), *Quantitative Analysis with Electron Microprobes and Secondary Ion Mass Spectroscopy*, Kernforschungsanlage, Jülich, Germany (1972).
51. A. N. Broers, in *Microprobe Analysis* (C. A. Andersen, ed.), Wiley, New York, (1973), p. 83.
52. J. C. Russ, in *SEM/1973 Proceedings of the 6th Annual SEM Symposium*, IITRI, Chicago, Illinois (1973), p. 113.
53. P. Duncumb, *J. de Microscopie*, **7**, 581 (1968).
54. M. H. Jacobs and J. Baborovska, in *Proceedings of the 5th European Conference on Electron Microscopy*, Institute of Physics, London (1972), p. 136.
55. G. Cliff and G. W. Lorimer, in *Proceedings of the 5th European Conference on Electron Microscopy*, Institute of Physics, London (1972), p. 140.
56. T. A. Hall, in *Proc. EMAG*, Conference Series No. 10, Institute of Physics, London (1971), p. 146.
57. J. L. Bomback, in *SEM/1973 Proceedings of the 6th Annual SEM Symposium*, IITRI, Chicago, Illinois (1973), p. 97.
58. L. S. Birks, *Electron Probe Microanalysis*, 2nd ed., Wiley—Interscience, New York (1971), p. 41.
59. R. E. Michaelis, H. Yakowitz, and G. A. Moore, *J. Res. NBS* **68A**, 343 (1964).

60. J. I. Goldstein, F. J. Majeske, and H. Yakowitz, in *Advances in X-Ray Analysis*, Vol. 10, Plenum Press, New York (1967), p. 431.
61. K. F. J. Heinrich, in *The Electron Microprobe* (T. D. McKinley, K. F. J. Heinrich, and D. B. Wittry, eds.), Wiley, New York, (1966), p. 296.
62. H. Yakowitz, C. E. Fiori, and R. E. Michaelis, NBS Special Publication 260-22 (1971).
63. B. L. Bracewell and W. J. Veigele, Developments in Applied spectroscopy, Vol. 9, Plenum Press, New York (1971), p. 357.
64. J. W. Colby, Advances in X-Ray Analysis, Vol. 11, Plenum Press, New York (1968), p. 287.
65. B. A. Cooke and E. A. Stewardson, *Brit. J. Appl. Phys.* **15**, 1315 (1964).

COMPUTATIONAL SCHEMES FOR QUANTITATIVE X-RAY ANALYSIS: ON-LINE ANALYSIS WITH SMALL COMPUTERS

H. Yakowitz

I. INTRODUCTION

This chapter will deal primarily with rapid means to perform data reduction for quantitative electron probe microanalysis with the aid of small computational devices (desk calculator to 8K machine). Since the two basic correctional procedures—ZAF and hyperbolic approximation—have already been discussed in Chapter IX, the computational schemes will be emphasized. Included will be a complete discussion of a ZAF program packaged for on-line data reduction and the basis for use of the hyperbolic approximation on-line. For the laboratory that only does occasional quantitative analysis, a brief discussion is given of how to obtain an analysis with a desk calculator in a reasonable time—about 2 hr for a six-component system or 30 min for a binary is needed, starting from scratch.

H. YAKOWITZ—Institute for Materials Research, Metallurgy Division, National Bureau of Standards, Washington, D.C.

II. SUMMARY OF COMPUTATIONAL SCHEMES FOR QUANTITATIVE ANALYSIS

Methods of data reduction for quantitative x-ray microanalysis by computers began proliferating in about 1963. Beaman and Isasi critically reviewed and compared more than 50 such programs in 1970.[1] These programs were by and large batch types which could be used only after the data were taken. Hence the analyst was able to have the tedious arithmetic of the corrections carried out, but usually somewhat after the fact of the analytical procedure. Easily the most popular of this generation of programs is the one known as MAGIC (Microprobe Analysis, General Intensity Corrections) promulgated by Colby.[2] This program is a ZAF-based number cruncher capable of providing a statistical evaluation of the input data as well as the calculated concentrations. MAGIC has been periodically updated to include the input parameters, such as x-ray mass attenuation coefficients, currently considered most accurate. Furthermore, the ZAF models themselves have been altered in MAGIC to correspond to those currently considered to best represent physical fact. MAGIC is still the most widely used batch program; it has been giving results considered satisfactory by its users for several years. MAGIC represents the kind of program that can be built on the classical ZAF method outlined in Chapter IX.

Another type of program in which all of the "classical corrections" are handled as rigorously as possible from first principles of x-ray intensity generation from solids bombarded by electrons was developed by Henoc *et al.*[3] Called COR2 (second version CORrection scheme), this program includes all the calculations for contributions to the observed intensity from both specimen and standard, including continuum fluorescence. This program is considered to be the most accurate data reduction scheme available for x-ray microanalysis. It is a batch-type program requiring a large ($\sim$50K) computational facility. If batch data reduction is satisfactory for all of a laboratory's purposes, then an appropriate program can be selected and used.

However, for many investigations, large batch-type programs are not entirely satisfactory. For example, in the analysis of mineralogical and petrological specimens, real-time computer reduction of data is desirable since knowledge of the composition and the calculated formula of the phase is often a necessary prerequisite to the analyst's deciding what the next stage in the analytical procedure must be. Until recently, no ZAF program would permit such rapid on-line analysis to be performed on an economically practical basis. Therefore the empirical method originated by Ziebold and Ogilvie was used in order to simplify matters.

Recently a procedure using a minicomputer for on-line correction of x-ray data from electron microprobe analysis was developed by Yakowitz, Myklebust, and Heinrich. This program, called FRAME, allows mass fractions to be computed on-line by the ZAF approach. Besides x-ray intensities, the only input data required are (1) the atomic numbers of the elements present, (2) the analytical line being used, and (3) the operating voltage. Other required parameters such as atomic weights and x-ray mass attenuation coefficients are stored or calculated by the program. FRAME need not be used on-line. If desired, the program can be run as batch or time-shared program for all analyses. The program is flexible, short, simple to use, and accurate. FRAME is in FORTRAN IV, and requires about 4K of computer core. Results from FRAME are very close to those of COR2.

The way in which the empirical approach is used and a complete description of the FRAME program will be given. Because the parameters for use in the empirical or hyperbolic approach are often calculated from a ZAF or other classical source, FRAME will be described first.

III. THE FRAME PROGRAM*

In the program FRAME, which is described here, there is incorporated into a simple computer model[4] the decision making and the internal calculation of parameters which were first developed for COR2. In this manner, a program simple enough to be used on-line in a small computer, including procedures to correct the observed x-ray intensities for dead time and background, was obtained so that the scaler readings can be directly used as input. The simplifications used in FRAME, do not observably degrade the accuracy of the procedure if certain working conditions are adopted. The working rules concerning the choice of lines and operating voltage are discussed at the end of the chapter. A comparison of

---

*The FRAME program was developed by R. L. Myklebust, K. F. J. Heinrich, and H. Yakowitz. The author thanks RLM and KFJH for permission to use here the material originally appearing as NBS Technical Note 796.

important features of the programs mentioned above is given in Tables IA–IC.

A. *Fundamentals of FRAME*

The program (see Figure 1) operates on the assumption that the relationship between concentration and background dead-time corrected intensity ratios can be expressed as in Chapter IX by

$$C = kk_Zk_Ak_F \tag{1}$$

where C is the mass fraction of the element being measured, k is the x-ray intensity ratio obtained from the microprobe corrected for background and

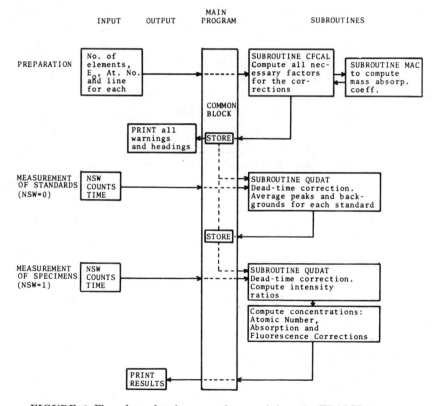

FIGURE 1. Flow sheet showing operations carried out by FRAME program.

TABLE IA. Comparison of NBS Programs for Quantitative Microprobe Analysis: Mechanical Aspects

	COR2	FRAME3
Memory	50K	4K
Intended mode of operation	Batch	Time-share or dedicated
Input	Cards and magnetic tape	Teletype or on-line
Output	Line printer	Teletype

TABLE IB. Comparison of NBS Programs for Quantitative Microprobe Analysis: Input

	COR2	FRAME3
Number of elements	Yes	Yes
Atomic numbers	Yes	Yes
Atomic weights	Data file	Stored
Operating voltage	Yes	Yes
Line	Yes	Yes
Critical excitation potential	Data file	Calculated
Fluorescence yield	Data file	Calculated
μ/ρ	Calculated	Calculated
Fluorescence decisions	Made internally	Made internally
Valences	Yes (if used)	Yes (if used)
X-ray intensities	Raw or k values	Raw or k values

TABLE IC. Comparison of NBS Programs for Quantitative Microprobe Analysis: Contents and Options

	COR2	FRAME3
Background	Yes	Yes
Dead time	Yes	Yes
Atomic number correction	Philibert–Tixier	Thomas modified
Absorption correction	Philibert–Duncumb Heinrich sigma	Heinrich–Yakowitz
Characteristic fluorescence	Hénoc	Reed
Continuum fluorescence	Hénoc	No
k values	Yes	Yes
Element by difference	Yes	Yes
Element by stoichiometry	Yes	Yes
Element with fixed concentration	Yes	Not yet

deadtime, k_Z is the atomic number correction, k_A is the absorption correction, and k_F is the characteristic fluorescence correction. There is no provision for fluorescence due to the continuum.

Specific features of FRAME are as follows:

1. As an input, FRAME can use x-ray photon counts collected in fixed time periods on scalers. A background correction is performed by subtracting background counts, which must also be accumulated in scalers, from the line intensities. The operator must determine from where the background readings are to be collected. In the present version, the background readings are made on the standards, and FRAME distinguishes background from standard line readings by the intensities of the observed radiation.

2. A correction for counter coincidence (dead time), according to the equation of Ruark and Brammer,[5] is applied to the observed count rates for specimens and standards. A method for determining the dead-time constant τ has been described elsewhere.[6] The relative x-ray intensity ratio k is calculated for each measurement by dividing the corrected intensity from the specimen by that of the standard.

3. The energies of the K, L, and M absorption edges are calculated by means of fits based on Moseley's law. The energies of the $K\alpha$, $L\alpha$, and $M\alpha$ lines are also calculated in the same way. One of these lines is almost always chosen as the analytical line. If conditions unfavorable for analysis result from choice of line by the operator, a warning in printed by the program. The operator can then use another line or readjust the operating voltage, depending on the nature of the difficulty.

4. The correction for absorption in the specimen of primary x-rays is carried out with use of the absorption correction factor f_p proposed by Heinrich et al.,[7]†

$$1/f_p = 1 + 3 \times 10^{-6}(E_0^{1.65} - E_c^{1.65})\chi + 4.5 \times 10^{-13}(E_0^{1.65} - E_c^{1.65})^2\chi^2 \quad (2)$$

where $\chi = (\mu/\rho) \csc \Psi$, with Ψ the x-ray emergence angle; and E_0 and E_c are the operating voltage and critical excitation potential for the line of interest, in kilovolts.

The absorption correction is

$$k_A = f_p/f_p^* \quad (3)$$

(Here and in what follows terms with an asterisk refer to the unknown, and terms without superscript refer to the standard.)

The mass attenuation coefficients are calculated as proposed by

---

† The exponent -13 in the third term of equation (2) is correct. This value was erroneously given as -14 in Ref. 7.

Heinrich.[8] To reduce requirements for storage, least squares fits were made to the logarithm of the coefficient C and the exponent n in the general relation

$$\mu/\rho = C\lambda^n$$

in which λ is the wavelength in angstroms.

A comparison of the calculated values with those given by Heinrich[8] showed agreement within a few percent. To avoid uncertainties that arise where lines close to edges are used, a warning is printed if the analytical line falls within 100 eV of an absorption edge of another element present in the specimen.

5. The correction for atomic number used in FRAME is

$$k_Z = (R^*/R)(\bar{S}/\bar{S}^*) \tag{4}$$

The electron backscattering correction factors R and R^* are those obtained by Duncumb and Reed from Monte Carlo calculations,[9] fitted by us with respect to the overvoltage U and the mean atomic number Z as follows:

$$R = R_1' - R_2' \ln(R_3'\bar{Z} + 25) \tag{5}$$

where

$$R_1' = 8.73 \times 10^{-3}U^3 - 0.1669U^2 + 0.9662U + 0.4523$$
$$R_2' = 2.703 \times 10^{-3}U^3 - 5.182 \times 10^{-2}U^2 + 0.302U + 0.1836$$
$$R_3' = (0.887U^3 - 3.44U^2 + 9.33U - 6.43)/U^3$$

with $U = E_0/E_c$, $\bar{Z} = \sum C_iZ_i$, and C_i the weight fraction of element i. If U is greater than 10, then U is set equal to 10.

Figure 2 shows the values given by Duncumb and Reed and the curves predicted by equation (5). Note that in Figure 2 W is equal to $1/U$. The value of R^* is obtained from

$$R^* = \sum_i C_iR_i \tag{6}$$

Equation (5) is simpler than the fifth-order fit in voltage and atomic number[10] often used and gives very similar results.

The term $\bar{S}$ in equation (4) is an average stopping power according to a form of Bethe's equation[11] [see equation (22), Chapter IX]. In the stopping power calculation the mean electron energy is obtained, as previously proposed by Thomas,[12] as follows:

$$\bar{E} = (E_0 - E_c)/2$$

This approximation is sufficiently accurate except at extreme overvoltage values, which should be avoided for accurate analysis.[13]

The atomic weights required for the stopping power calculation are

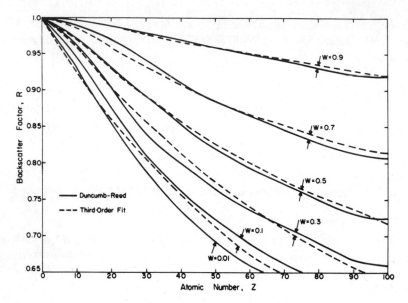

FIGURE 2. Analytical fit to backscatter factor R.

stored in the program. The characteristic fluorescence correction is performed with the aid of Reed's equation.[14] In FRAME, fluorescence yields for K and L shells are calculated by means of empirical fits. No fluorescence correction is carried out if fluorescence is excited by M lines. The absorption jump ratio is assumed to be 0.88 for the K edge and 0.75 for the L_3 edge.

FRAME contains the following provisions, which decide whether fluorescence correction is required. Call E_l the energy of the exciting line (K or L line), and E_K and E_L the critical excitation potentials for the excited K and L lines.

If $0 < (E_l - E_K) \leq 5$, then the K line is indirectly excited.

If $0 < (E_l - E_L) \leq 3.5$, then the L line is indirectly excited.

If fluorescent excitation occurs, the program prints a message to this effect. Then it calculates the new mass attenuation coefficients required for Reed's equation[14] and performs the necessary calculation for the fluorescence effect. This automatic selection of corrections for fluorescent excitation where necessary eliminates the need for the operator to decide if such corrections must be applied.

As mentioned in Chapter IX, the hyperbolic iteration procedure used in COR2—and again in FRAME—converges very rapidly. For this reason, the iteration is ended after three cycles in order to reduce the time of computation. The resulting errors are smaller than the errors of measurement. The sectors for input and output will require modifications for use on other computers. These adaptations should be easy to carry out.

The interaction of the main program with the subprograms can be observed on the flow diagram in Figure 1. Subroutines CFCAL and MAC, which perform all preliminary calculations that are independent of the composition of the specimen and standard, require 2.8K words of storage (see Figure 1). Subroutine QUDAT (1.5K words) carries out the conversion of intensity ratios to concentrations.*

B. Limitations of FRAME

As mentioned above, the limitations of available memory have required several simplifications (e.g., omission of the correction for fluorescence due to the continuum). These simplifications will not cause significant increases in the error of the analyses, as long as the following operational conditions are maintained:

1. The overvoltage E_0/E_c should not be smaller than 1.5.
2. The primary x-ray absorption factor f_p should not be smaller than 0.75.
3. If an x-ray line falls within 100 eV on the high-energy side of an absorption edge, or 30 eV on the low-energy side of any element present, then this line should not be used because of the uncertainties concerning mass absorption coefficients near the edge.

The restrictions imposed by these conditions are not serious or specific to FRAME, since working outside these specifications is unwise, even when a more detailed computer program is used. FRAME produces a printed warning when an analysis is computed with disregard of the above rules. In the case of f_p, the warning criterion is applied to the absorption factor of the standard. For further control of conditions, the values of f_p for the specimen are printed in the output of FRAME.

In the NBS version of FRAME, it is assumed that the x-ray emergence angle is equal to 52.5°. The value of this angle can be changed by a single

*Program listings for FRAME are available on request from the National Bureau of Standards, Washington, D.C. 20234.

statement, and even a change to different angles for each spectrometer could be introduced without much effort.

The number of elements in the specimen is presently limited to six. Since the program is primarily intended for on-line use, it is assumed that the six intensities are read simultaneously. However, the number of x-ray lines that can be read simultaneously in an instrument cannot exceed that of the spectrometers, unless part of the information is provided by an energy-dispersive spectrometer. Since this limitation may be significant in the analysis of minerals, modifications can be made that would permit the reading of x-ray intensities in two steps in time, so that spectrometers could be realigned to read more than one line per analysis.

All measurements for a case are assumed to have been performed at the same operating voltage. If a two-step readout is considered, this criterion will be modified so that, if desired, some lines can be read at a lower voltage than the rest. This experimental procedure is, however, awkward and is not used frequently. Therefore the limitation to a single operating voltage is not stringent.

All versions of FRAME permit the determination of one element by difference. The use of stoichiometric relations to determine the concentration of one element is provided in FRAME4 and can easily be incorporated into other versions.

At present, it is not possible to replace the mass absorption coefficients calculated in the program by others provided externally. Such a feature will be incorporated later, as well as an option to assume that the concentration of certain elements is known. The number of iterations, which presently is limited to three, could be changed by a simple statement. However, as indicated by the data presented in Tables II and III, computation beyond three iterations does not provide any significant improvement in accuracy.

C. Comparison of Results from FRAME and COR2

We have compared the FRAME and COR correction programs, using the measurements on 228 binary alloys listed by Poole[15] as well as 16 analyses on the binary alloys Au–Ag and Au–Cu designated NBS Standard Reference Materials 481 and 482, respectively.[16] The analyses listed by Poole were performed in several laboratories from all over the world, and they cover a variety of systems and experimental conditions.

A comparison of the results by FRAME with those by COR2 for the Poole-Thomas compilation is shown in Figure 3a and 3b. We note that the disagreement between COR2 and FRAME is small compared with the

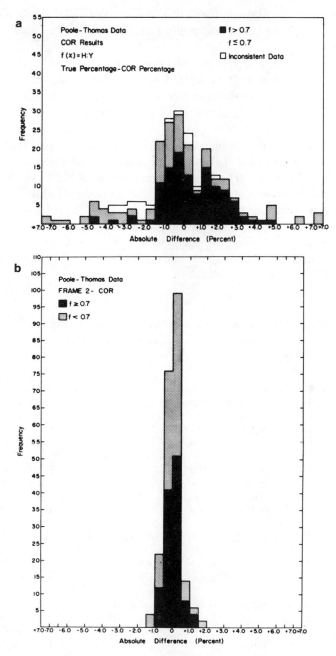

FIGURE 3.   (a) Results of analysis on alloy data given by Poole[15] as computed by COR using the Heinrich–Yakowitz $f(\chi)$ expression. (b) Comparison of FRAME2 and COR for the alloy data shown in (a).

˙errors shown in the diagram for COR2. We also note that larger discrepancies occur when the operating conditions are outside the suggested $f(\chi)$ boundaries.

Analyses performed at NBS are presented in Tables II, III, IVA, and IVB. The results show that FRAME is suitable for a wide variety of analytical studies. They also show, in the case of the two minerals analyzed, that the method of determining the background level was not critically important.

An example of the input and output for the analysis of the Diopside 65–Jadeite 35 is shown in Table V. All required input statements follow a colon or question mark. The program version is FRAME4. Table V shows that the concentrations obtained by calculating oxygen by difference or by stoichiometric relations are the same. The FRAME4 routine was run on a time-shared computer with background dead-time corrected relative intensity ratios (k values) already calculated. The k values shown are referred to pure element standards as calculated from the mineral standards listed in Table IVA.

TABLE II. Analysis of Alloys: SRM's 481 and 482 (Au–Ag and Cu–Au)

Element and line	Operating voltage, kV	k_{meas}	C_{FRAME} [a]			C_{COR2}	C Certified	$f(\chi)$
			I	II	III			
Au $M\alpha$	10	0.201	0.227	0.227	0.227	0.234	0.2243	0.84
Au $M\alpha$	10	0.362	0.399	0.399	0.399	0.409	0.4003	0.84
Au $M\alpha$	10	0.559	0.598	0.598	0.598	0.608	0.6005	0.84
Au $M\alpha$	10	0.771	0.799	0.799	0.799	0.806	0.8005	0.85
Ag $L\alpha$	10	0.764	0.778	0.776	0.776	0.764	0.7758	0.89
Ag $L\alpha$	10	0.584	0.600	0.598	0.598	0.582	0.5993	0.86
Ag $L\alpha$	10	0.386	0.399	0.398	0.398	0.382	0.3992	0.83
Ag $L\alpha$	10	0.191	0.198	0.197	0.197	0.187	0.1996	0.80
Cu $K\alpha$	25	0.240	0.206	0.205	0.205	0.203	0.1983	0.90
Cu $K\alpha$	25	0.453	0.413	0.411	0.411	0.402	0.3964	0.91
Cu $K\alpha$	25	0.651	0.619	0.618	0.617	0.604	0.5992	0.93
Cu $K\alpha$	25	0.834	0.818	0.817	0.817	0.805	0.7985	0.95
Au $L\alpha$	25	0.745	0.788	0.795	0.795	0.794	0.8015	0.92
Au $L\alpha$	25	0.529	0.592	0.598	0.598	0.598	0.6036	0.91
Au $L\alpha$	25	0.331	0.391	0.395	0.395	0.396	0.4010	0.90
Au $L\alpha$	25	0.154	0.192	0.193	0.193	0.194	0.2012	0.89

[a] I–III refer to the iterations. III is the final one.

TABLE III. Analysis of the Silicate Diopside 65–Jadeite 35 with Oxygen by Stoichiometry Using FRAME4 [a]

	Na	Mg	Al	Si	Ca
k_m	0.0281	0.0593	0.0363	0.236	0.111
C, Iteration I	0.0392	0.0730	0.0439	0.270	0.121
C, Iteration II	0.0384	0.0718	0.0433	0.268	0.121
C, Iteration III	0.0384	0.0718	0.0433	0.268	0.121
C, Chemical analysis	0.0398	0.073	0.0467	0.266	0.120
f_p	0.69	0.77	0.81	0.86	0.97

[a] $E_0 = 10$ kV. Here k_m is the measured intensity ratio corrected for background and dead-time effects and C is the mass fraction.

D. Example Showing On-Line Analysis Using FRAME

An iron–silicon alloy containing 3.22% Si and 96.8% Fe, designated SRM-483, was analyzed. This material is certified to be homogeneous with respect to iron and silicon.[77] Five individual analyses were carried out. The results, as printed on a teletype display during analysis, are shown in Table VI. The average silicon content is 3.34% and the average iron content is 96.2%.

TABLE IVA. Analysis of the Silicates Johnstown Hypersthene and Diopside 65–Jadeite 35 Using FRAME4: Chemical Analysis of Minerals Used [a]

Mineral	Na	Mg	Al	Si	Ca	Ti	Fe
NaCl	0.393	—	—	—	—	—	—
SiO$_2$	—	—	—	0.467	—	—	—
Fayalite	—	—	—	0.137	—	—	0.525
Diopside·2%TiO$_2$	—	0.110	—	0.254	0.181	0.012	—
Garnet 110752	—	0.043	0.120	0.188	0.129$_5$	0.002	0.088
Diopside 65–Jadeite 35	0.0398	0.073	0.0467	0.266	0.120	—	—
Johnstown Hypersthene	—	0.164	0.002	0.251	0.0099	0.0013	0.121

[a] Standards used:
NaCl for Na in Diopside 65–Jadeite 35.
SiO$_2$ for Si in both.
Fayalite for Fe in Johnstown Hypersthene.
Diopside·2%TiO$_2$ for Mg, Ca, Ti in Johnstown Hypersthene.
Garnet 110752 for Mg, Al, Ca in Diopside 65–Jadeite 35.
Al$_2$O$_3$ used as background specimen for Fe, Si, Mg, Ca, Ti; MgO used as background specimen for Na, Al, Ca: Scheme I.
Off-peak (above and below) used for all elements: Scheme II.

TABLE IVB [a]

Analysis of Johnstown Hypersthene [b]
$E_0 = 15$ kV

	Mg	Si	Ca	Ti	Fe
k_m, S1	0.114	0.202	0.011	0.0008	0.104
k_m, S2	0.117	0.202	0.010	0.0007	0.104
C_{CALC}, S1	0.165	0.260	0.0110	0.0009	0.121
C_{CALC}, S2	0.169	0.260	0.0110	0.0008	0.121
C, Chemical analysis	0.164	0.251	0.0099	0.0013	0.121

Analysis of Diopside 65–Jadeite 35
$E_0 = 10$ kV

	Na	Mg	Al	Si	Ca
k_m, S1	0.0281	0.0593	0.0363	0.236	0.111
k_m, S2	0.0275	0.0595	0.0363	0.236	0.111
C_{CALC}, S1	0.0384	0.0716	0.0433	0.268	0.121
C_{CALC}, S2	0.0375	0.0718	0.0433	0.268	0.121
C, Chemical analysis	0.0398	0.073	0.0467	0.266	0.120

[a] S1, background taken on Al_2O_3 or MgO. S2, background taken off-peak. k_m, measured intensity ratio corrected for background and dead-time effects, C, mass fraction.

[b] Analysis carried out calculating oxygen by stoichiometry was exactly the same as that carried out calculating oxygen by difference.

E. Conclusions

The FRAME program is entirely suitable for on-line or off-line use to reduce data from the electron probe microanalyzer to concentrations of the elements present. FRAME employs the analytical ZAF approach, but only occupies about 4K of computer memory. The FRAME program should be readily adaptable to most minicomputers for use in many microprobe laboratories.

Comparisons have shown that FRAME gives virtually the same results as the much larger batch-type program COR, provided that fluorescence effects due to the continuum can be safely neglected. All of the data discussed in this chapter, including those reported by Poole, fall into this category. The input requirements for FRAME are virtually the same, however (see Table IB). Results on two minerals, Johnstown Hypersthene and Diopside 65–Jadeite 35, give results close to the chemical analysis (Tables III, IVA, IVB, and V). Furthermore, the results were only slightly

TABLE V. Input–Output for FRAME4 Analysis of Diopside 65–Jadeite 35

(a) Analysis with oxygen calculated by difference

NUMBER OF ELEMENTS AND E_0: 6 10

ATOMIC NUMBER AND LINE FOR EACH: 11 1 12 1 13 1 14 1 20 1 8 0

WARNING — THE LINE OF 11 IS CLOSE TO AN EDGE OF 11

ELEMENT	11	IS EXCITED BY ELEMENT	12
ELEMENT	11	IS EXCITED BY ELEMENT	13
ELEMENT	11	IS EXCITED BY ELEMENT	14
ELEMENT	11	IS EXCITED BY ELEMENT	20
ELEMENT	12	IS EXCITED BY ELEMENT	13
ELEMENT	12	IS EXCITED BY ELEMENT	14
ELEMENT	12	IS EXCITED BY ELEMENT	20
ELEMENT	13	IS EXCITED BY ELEMENT	14
ELEMENT	13	IS EXCITED BY ELEMENT	20
ELEMENT	14	IS EXCITED BY ELEMENT	20

VALENCES: 0 0 0 0 0 0

STANDARDS DATA? NO

k-VALUES: .0281 .0593 .0363 .2364 .111

AT. NO.	k-VALUE	CONCENTRATION	$f(\chi)$
11.	.0281	.0384	.6867
12.	.0593	.0718	.7690
13.	.0363	.0433	.8121
14.	.2364	.2685	.8555
20.	.1110	.1207	.9670
8.	0.0000	.4573	1.0000

(b) Analysis with oxygen calculated stoichiometrically

NUMBER OF ELEMENTS AND E_0 6 10

ATOMIC NUMBER AND LINE FOR EACH: 11 1 12 1 13 1 14 1 20 1 8 0

WARNING — THE LINE OF 11 IS CLOSE TO AN EDGE OF 11

ELEMENT	11	IS EXCITED BY ELEMENT	12
ELEMENT	11	IS EXCITED BY ELEMENT	13
ELEMENT	11	IS EXCITED BY ELEMENT	14
ELEMENT	11	IS EXCITED BY ELEMENT	20
ELEMENT	12	IS EXCITED BY ELEMENT	13
ELEMENT	12	IS EXCITED BY ELEMENT	14
ELEMENT	12	IS EXCITED BY ELEMENT	20
ELEMENT	13	IS EXCITED BY ELEMENT	14
ELEMENT	13	IS EXCITED BY ELEMENT	20
ELEMENT	14	IS EXCITED BY ELEMENT	20

VALENCES: 1 2 3 4 2 2

STANDARDS DATA? NO

k-VALUES: .0281 .0593 .0363 .2364 .111

AT. NO.	k-VALUE	CONCENTRATION	$f(\chi)$
11.	.0281	.0384	.6871
12.	.0593	.0718	.7693
13.	.0363	.0433	.8123
14.	.2364	.2685	.8557
20.	.1110	.1207	.9671
8.	0.0000	.4531	1.0000

TOTAL C = .9957

TABLE VI. Results of On-Line Analysis Performed at NBS of SRM 483 Containing 0.0322 Si–0.968 Fe

FRAME3 1118HRS 187 DAY, 1973
ELECTRON PROBE ANALYSIS 15.0 KV

ELEMENT 14 IS EXCITED BY ELEMENT 26

Z-LINE	Z-ABSORBER	MASS ABS. COEFF.
14	14	330.0
14	26	2495.8
26	14	113.8
26	26	70.9

AT. NO.	k-VALUE	CONC	$f(\chi)$
14	.0237	.0337	.5544
26	.9532	.9595	.9834
	TOTAL C = .9932		
14	.0233	.0332	.5543
26	.9583	.9640	.9834
	TOTAL C = .9972		
14	.0238	.0339	.5544
26	.9530	.9593	.9834
	TOTAL C = .9932		
14	.0234	.0333	.5543
26	.9578	.9635	.9834
	TOTAL C = .9968		
14	.0236	.0337	.5544
26	.9552	.9612	.9834
	TOTAL C = .9949		

affected by the method of determining background level or by using FRAME to calculate the results obtaining the oxygen content by difference or by stoichiometric relations.

IV. DATA REDUCTION BASED ON THE HYPERBOLIC METHOD

The hyperbolic or empirical method first described by Ziebold and Ogilvie has been discussed in Chapter IX. For purposes of preparing a small computer program, this method has much to recommend it. For a binary system, the hyperbolic relation can be expressed as

$$C_1/k_1 = a_{12} + (1 - a_{12})C_1 \tag{7}$$

Here C_1 is the concentration of element 1 in a binary of elements 1 and 2,

k_1 is the relative intensity ratio corrected for background and dead-time effects for element 1 in the specimen to the standard for element 1, and a_{12} is the a factor for the determination of element 1 in the binary constituted of elements 1 and 2.

From equation (7), it is clear that a_{12} can be determined as C_1 goes to zero. Investigations have shown that the correction term for a multi-component system of n components can be expressed for the nth component as[18]

$$C_n = k_n \beta_n \tag{8}$$

$$\beta_n = \frac{k_1 a_{n1} + k_2 a_{n2} + \cdots + k_n a_{nn}}{k_1 + k_2 + \cdots + k_n} \tag{9}$$

In equation (9), a_{n1} is the a value for the determination of element n in a binary of elements n and 1, a_{n2} is the a value for determining element n in a binary of n and 2, and so on to a_{nn}, which is unity. Thus the notation a_{1n} represents the a value for the determination of element 1 in the binary n and 1. The $k_1, k_2, \ldots, k_n$ represent the measured relative intensities of element $1, 2, \ldots, n$ in the specimen with respect to the respective standards for these elements. In order to compute the final value of C_n, one iteratively computes a set of β_n values until differences between successive calculations are made arbitrarily small. Usually, no more than three or four such iterations are necessary.

The main requirement for successful application of the hyperbolic technique is to have an appropriate set of a values. Bence and Albee determined such a set for each of ten elements (Na, Mg, Al, Si, K, Ca, Ti, Cr, Mn, Fe) in oxides (H_2O, Na_2O, MgO, Al_2O_3, SiO_2, K_2O, CaO, TiO_2, Cr_2O_3, MnO, and FeO) of interest in geological investigations.[18] This matrix of 110 values for a is useful if one always employs experimental conditions identical to those used to obtain the matrix. Furthermore, the number of empirical a values available for other systems is small, nor is it usually easy to prepare one's own a values due to a lack of suitable, homogeneous standards for this purpose. Therefore it is often advantageous to use a calculational approach to obtain the desired matrix of a values.

For a four-component system, the necessary a matrix is

$$\begin{pmatrix} a_{11} = 1 & a_{12} & a_{13} & a_{14} \\ a_{21} & a_{22} = 1 & a_{23} & a_{24} \\ a_{31} & a_{32} & a_{33} = 1 & a_{34} \\ a_{41} & a_{42} & a_{43} & a_{44} = 1 \end{pmatrix}$$

The appropriate set of a values is chosen to compute the β values needed, e.g.,

$$\beta_3 = \frac{k_1 a_{31} + k_2 a_{32} + k_3 + k_4 a_{34}}{k_1 + k_2 + k_3 + k_4}$$

In other words, a computational scheme such as MAGIC, COR2, or FRAME is given a C value and computes the corresponding k and a values. At first, the advantage of this procedure may not be apparent. The advantage is that, given a qualitative analysis of the specimen to be subjected to quantitative analysis, a large computer can be used to provide the a matrix quickly. Then a minicomputer can be used to store this matrix and carry out corrections on-line. The on-line computer needs only to:

1. Receive and store contents of scalers for the unknown, standards, and background, and analysis time.
2. Reduce the contents of the scalers to counts per unit time.
3. Perform dead-time correction on the count rate data.
4. Subtract background counts appropriately.
5. Compute statistical data on counting strategy.
6. Compute k values, $k_1, k_2, \ldots, k_n$.
7. Enter a matrix for appropriate a value.
8. Calculate $\beta_1, \beta_2, \ldots, \beta_n$ from equation (9).
9. Calculate $C_1, C_2, \ldots, C_n$ from equation (8).
10. Reiterate until β calculations, and hence C calculations, converge.
11. Print out k, C, and statistical data for each element.

If desired, provisions for analysis of one element by difference or by assumed stoichiometry can be included.

Obviously, the size of the computer needed once an a matrix is available can be very small. In fact, it is possible to hard-wire a device to carry out the appropriate steps. The calculations can even be done on a desk calculator at the instrument by the operator. Having the a matrix eliminates the need for computer calculation of such parameters as mass attenuation coefficients, stopping powers, etc. The computer does not even need to be given the operating voltage or the analytical line given the a matrix.

Therefore a laboratory unable or unwilling to provide a computational system capable of using FRAME on-line can still provide a means of rapid analysis. A batch program such as MAGIC or even FRAME or COR2 can be available on a time-sharing basis off-line. Then routine off-line analysis can be done with the batch program. For on-line analysis, the batch program is made to calculate the a matrix appropriate to the specimen and chosen experimental conditions. The analyst then inserts this a matrix into a small computational device capable of carrying out the functions listed previously. Or, given a programmable desk calculator, the analyst can carry them out in about 3 min per analytical point for a six-component system. Hence both batch-type and rapid analysis facilities can be available to the laboratory.

Consider briefly now the laboratory that has no ready access to any computational scheme. This may be the case in an SEM laboratory equipped for energy-dispersive analysis, which rarely needs to perform quantitative analysis. The question is, how can this laboratory carry out occasional analysis? One way is to prepare a master tape of some program such as MAGIC or FRAME and load a time-sharing device with it when quantitative analysis is needed, or to contract with a time-sharing firm already having such a program in its library. But suppose the computer is down and so is the client—down in the analysis laboratory screaming for results. What then?

In this extreme, it is possible to compute the appropriate a matrix for the experiment by hand, quickly if no fluorescence corrections are needed and less so if they are. For this purpose, then,

$$\frac{C_1}{k_1} = \frac{f_1}{f_2}\frac{R_1}{R_2}\frac{S_2}{S_1}\frac{1}{1 + r'_{f,2}} \tag{10}$$

while

$$\frac{C_2}{k_2} = \frac{f_2}{f_1}\frac{R_2}{R_1}\frac{S_1}{S_2} \tag{11}$$

The possibilities concerning r_f' are paired as follows:

$$
\begin{aligned}
r'_{f,1} &= 0, & r'_{f,2} &= 0 \\
r'_{f,1} &> 0, & r'_{f,2} &= 0 \\
r'_{f,1} &= 0, & r'_{f,2} &> 0
\end{aligned} \tag{12}
$$

The necessary R values can be obtained from equation (5). Table VII lists the fitting parameters R_1', R_2', and R_3' for equation (5) as a function of the overvoltage U. The stopping power values S can be obtained from

$$S = \frac{Z}{A(E_0 + E_c)} \ln \frac{583(E_0 + E_c)}{J} \tag{13}$$

Table VIII lists $583/J$ computed from the Berger–Seltzer relation. In equation (13) note that E_0 and E_c are in keV, but J is expressed in eV.

The absorption correction parameter f can be obtained from equation (2). If U or χ (or both) is small, then it is satisfactory to take

$$1/f = 1 + 3 \times 10^{-6}(E_0^{1.65} - E_c^{1.65})\chi \tag{14}$$

Finally, r_f' must be computed from equation (31) of Chapter IX. Table VIII lists ω_K and ω_L values as a function of atomic number.

For an unknown of n elements, one needs to obtain $n(n - 1)$ values

TABLE VII. Fitting Parameters for Duncumb–Reed Backscattering Correction Factor R.[a]

U	R_1'	R_2'	R_3'
1	1.26	.069	.347
1.1	1.32	.089	.639
1.2	1.38	.108	.778
1.3	1.44	.127	.834
1.4	1.50	.145	.846
1.5	1.55	.161	.835
1.6	1.60	.178	.811
1.7	1.65	.193	.783
1.8	1.70	.207	.752
1.9	1.74	.221	.723
2.	1.78	.234	.695
2.1	1.82	.247	.670
2.2	1.86	.258	.647
2.3	1.89	.269	.626
2.4	1.93	.280	.608
2.5	1.96	.289	.592
2.6	1.98	.298	.578
2.7	2.01	.307	.566
2.8	2.04	.315	.555
2.9	2.06	.322	.546
3.	2.08	.329	.538
3.1	2.10	.335	.532
3.2	2.12	.340	.526
3.3	2.13	.345	.522
3.4	2.15	.350	.518
3.5	2.16	.354	.515
3.6	2.17	.358	.513
3.7	2.18	.361	.511
3.8	2.19	.364	.510
3.9	2.19	.366	.509
4.	2.20	.368	.509
4.1	2.20	.369	.509
4.2	2.21	.370	.510
4.3	2.21	.371	.510
4.4	2.21	.372	.511
4.5	2.21	.372	.512
4.6	2.21	.372	.514
4.7	2.21	.371	.515
4.8	2.21	.370	.517
4.9	2.20	.370	.518
5.	2.20	.368	.520
5.1	2.19	.367	.522
5.2	2.19	.365	.524
5.3	2.18	.363	.526
5.4	2.17	.361	.529
5.5	2.17	.359	.531

TABLE VII. (cont'd)

U	R_1'	R_2'	R_3'
5.6	2.16	.354	.533
5.7	2.15	.357	.535
5.8	2.14	.352	.538
5.9	2.13	.349	.540
6.	2.12	.346	.543
6.1	2.11	.343	.545
6.2	2.10	.341	.547
6.3	2.09	.338	.550
6.4	2.08	.335	.552
6.5	2.07	.332	.555
6.6	2.06	.329	.557
6.7	2.05	.326	.560
6.8	2.05	.323	.562
6.9	2.04	.321	.564
7.	2.03	.318	.567
7.1	2.02	.315	.569
7.2	2.01	.313	.571
7.3	2.00	.311	.574
7.4	2.00	.308	.576
7.5	1.99	.306	.578
7.6	1.98	.305	.581
7.7	1.98	.303	.583
7.8	1.97	.301	.585
7.9	1.97	.300	.588
8.	1.97	.299	.590
8.1	1.96	.299	.592
8.2	1.96	.298	.594
8.3	1.96	.298	.596
8.4	1.96	.298	.598
8.5	1.96	.299	.600
8.6	1.97	.300	.603
8.7	1.97	.301	.605
8.8	1.97	.303	.607
8.9	1.98	.305	.609
9.	1.99	.307	.611
9.1	2.00	.310	.613
9.2	2.01	.313	.615
9.3	2.02	.317	.616
9.4	2.03	.321	.618
9.5	2.05	.326	.620
9.6	2.07	.331	.622
9.7	2.08	.337	.624
9.8	2.10	.343	.626
9.9	2.13	.350	.628
10.	2.15	.357	.629

[a] See equation (5).

TABLE VIII. Berger–Seltzer J Values,[20] Factor $583/J$, and Fluorescent Yield Values[21]

Z	J (eV)	$583/J$	ω_k	ω_{L_3}
1	68.2	8.54	—	—
2	70.8	8.23	—	—
3	76.7	7.59	—	—
4	83.9	6.94	—	—
5	91.8	6.34	—	—
6	100.	5.81	0.0009	—
7	108.	5.36	0.0015	—
8	117.	4.96	0.0022	—
9	126.	4.61	0.004	—
10	135.	4.30	0.008	—
11	144.	4.03	0.01	—
12	153.	3.79	0.028	—
13	162.	3.58	0.038	—
14	172.	3.38	0.038	—
15	181.	3.21	0.084	—
16	190.	3.05	0.098	—
17	200.	2.91	0.113	—
18	209.	2.78	0.130	—
19	218.	2.66	0.147	—
20	228.	2.55	0.166	—
21	237.	2.45	0.187	—
22	247.	2.35	0.209	—
23	256.	2.27	0.232	—
24	266.	2.18	0.257	—
25	275.	2.11	0.283	—
26	285.	2.04	0.310	—
27	294.	1.97	0.339	—
28	304.	1.91	0.370	—
29	313.	1.85	0.402	—
30	323.	1.80	0.436	—
31	333.	1.75	0.471	—
32	342.	1.70	0.508	—
33	352.	1.65	0.546	—
34	361.	1.61	0.586	—
35	371.	1.56	0.628	—
36	380.	1.53	0.660	—
37	390.	1.49	0.680	0.01
38	400.	1.45	0.702	—
39	409.	1.42	0.719	—
40	419.	1.39	0.737	—
41	429.	1.35	0.754	—
42	438.	1.32	0.770	—
43	448.	1.30	0.785	—

TABLE VIII. (cont'd)

Z	J (eV)	$583/J$	ω_K	ω_{L3}
44	457.	1.27	0.799	—
45	467.	1.24	0.812	—
46	477.	1.22	0.822	—
47	486.	1.19	0.833	0.05
48	496.	1.17	0.843	—
49	506.	1.15	—	—
50	515.	1.13	—	0.089
51	525.	1.10	—	0.095
52	535.	1.08	—	0.100
53	544.	1.07	—	0.106
54	554.	1.05	—	0.112
55	564.	1.03	—	0.118
56	573.	1.01	—	0.125
57	583.	0.999	—	0.131
58	593.	0.982	—	0.138
59	602.	0.967	—	0.146
60	612.	0.951	—	0.153
61	622.	0.937	—	0.160
62	631.	0.922	—	0.168
63	641.	0.908	—	0.177
64	651.	0.895	—	0.185
65	660.	0.882	—	0.194
66	670.	0.869	—	0.202
67	680.	0.857	—	0.212
68	689.	0.845	—	0.221
69	699.	0.833	—	0.231
70	709.	0.821	—	0.241
71	718.	0.810	—	0.251
72	728.	0.800	—	0.262
73	738.	0.789	—	0.272
74	748.	0.779	—	0.284
75	757.	0.769	—	0.295
76	767.	0.759	—	0.307
77	777.	0.750	—	0.319
78	786.	0.740	—	0.331
79	796.	0.731	—	0.344
80	806.	0.723	—	0.357
81	815.	0.714	—	0.370
82	825.	0.706	—	0.384
83	835.	0.697	—	0.398
84	845.	0.689	—	0.412
85	854.	0.682	—	0.427

TABLE VIII. (cont'd)

Z	$J(\mathrm{eV})$	$583/J$	ω_K	ω_{L3}
86	864.	0.674	—	0.442
87	874.	0.666	—	0.457
88	883.	0.659	—	0.472
89	893.	0.652	—	0.488
90	903.	0.645	—	0.505
91	912.	0.638	—	0.522
92	922.	0.631	—	0.539
93	932.	0.625	—	0.556
94	942.	0.618	—	0.574
95	951.	0.612	—	—
96	961.	0.606	—	—
97	971.	0.600	—	—
98	980.	0.594	—	—
99	990.	0.588	—	—
100	1000.	0.583	—	—

for a, n values for U, n^2 values for R, n^2 values for S, n^2 values for f, n^2 values for μ/ρ, n values for E_c, and perhaps as many as n values of r_f'.

The example of SRM-480, which is certified to contain 0.215 Mo and 0.785 W, is given in Table IX to illustrate the procedure. Experimental conditions were $E_0 = 20$ kV, $\Psi = 52.5°$, and the $L\alpha$ line of both W and Mo was used to obtain the data. The measured intensity ratios gave $k_{\mathrm{Mo}} = 0.143$ and $k_{\mathrm{W}} = 0.772$. Fluorescence effects do not require calculation; the criteria listed in Chapter IX are not met.[19]

The final values are calculated as

$$C_{\mathrm{Mo}} = 0.226; \qquad C_{\mathrm{TRUE}} = 0.215$$
$$C_{\mathrm{W}} = 0.797; \qquad C_{\mathrm{TRUE}} = 0.785$$

The calculated value of C_{Mo} is 5.1% too high and C_{W} is 1.5% too high. These results can be compared with those obtained by computer reduction and reported in Ref. 19 as $C_{\mathrm{Mo}} = 0.202$ and $C_{\mathrm{W}} = 0.812$.

The computation for the binary took about 40 min starting from the k values and having Tables VII and VIII and tables of μ/ρ and E_c available along with a desk calculator. Clearly, the same job on a computer would take seconds. Nevertheless, the hand procedure can be used in unusual circumstances with reasonable confidence. The example is useful to illustrate precisely how a computer scheme actually works in practice in that the same arithmetic computations are carried out in the same way. The only differences are that atomic weights are stored; μ/ρ is computed as are J and R_1', R_2', R_3'. E_c is also computed as a function of atomic number by means of Moseley's law.

TABLE IX. Tabulation of Necessary Constants

Mo	W
$E_c = 2.525$ kV; $U = 7.92$	$E_c = 10.2$ kV; $U = 1.96$
$M(\text{Mo, Mo } L\alpha) = 728$; $\chi_1 = 918$	$M(\text{W, W } L\alpha) = 151$; $\chi_2 = 190.3$
$M(\text{W, Mo } L\alpha) = 3145$; $\chi_2 = 3964$	$M(\text{Mo, W } L\alpha) = 140$; $\chi_1 = 176.5$
$Z = 42$	$Z = 74$
$A = 95.94$	$A = 183.85$
$583/J_{\text{Mo}} = 1.329$	$583/J_{\text{W}} = 0.779$

REQUIRED CALCULATIONS

$$1/f = 1 + 3 \times 10^{-6}(E_0^{1.65} - E_c^{1.65})\chi + 4.5 \times 10^{-13}(E_0^{1.65} - E_c^{1.65})^2\chi^2$$

For Mo:

$E_0^{1.65} - E_c^{1.65} = 135.6$
$1/f_{\text{Mo}} = 1/f_1 = 1.380$; $\quad f_1 = 0.724$
$1/f_{\text{W}} = 1/f_2 = 2.743$; $\quad f_2 = 0.365$

Stopping Power

$E_0 + E_c = 22.525$
$(Z/A)_{\text{Mo}} = 0.4377$, $\quad (Z/A)_{\text{W}} = 0.4025$
$S = [Z/A(E_0 + E_c)] \ln[583(E_0 + E_c)/J]$
$S_1 = S_{\text{Mo}} = 0.0660$, $\quad S_2 = S_{\text{W}} = 0.0512$

Backscattering Factor, At $U = 7.92$, Table VII gives

$R_1' = 1.97$, $\quad R_2' = 0.300$, $\quad R_3' = 0.589$
$R = R_1' - R_2' \ln(R_3'Z + 25)$
$R_1 = R_{\text{Mo}} = 0.798$, $\quad R_2 = R_{\text{W}} = 0.702$
$a_{12} = \dfrac{f_1}{f_2}\dfrac{R_1}{R_2}\dfrac{S_2}{S_1} = 1.749$

For W:

Absorption Calculation

$E_0^{1.65} - E_c^{1.65} = 140.2 - 46.2 = 94$
$1/f_{\text{Mo}} = 1/f_1 = 1.050$; $\quad f_1 = 0.952$
$1/f_{\text{W}} = 1/f_2 = 1.054$; $\quad f_2 = 0.949$

TABLE IX. (cont'd)

Stopping Power

$$E_0 + E_c = 30.2$$
$$(Z/A)_{Mo} = 0.4377, \qquad (Z/A)_W = 0.4025$$
$$S_1 = S_{Mo} = 0.0535, \qquad S_2 = S_W = 0.0421$$

Backscattering Factor. At $U = 1.96$, Table VII gives

$$R_1' = 1.76, \qquad R_2' = 0.228, \qquad R_3' = 0.706$$
$$R_1 = R_{Mo} = 0.848, \qquad R_2 = R_W = 0.769$$
$$a_{21} = \frac{f_2 R_2 S_1}{f_1 R_1 S_2} = 1.149$$
$$\beta_1 = \frac{k_1 + k_2 a_{12}}{k_1 + k_2} = \frac{0.143 + (0.772)(1.749)}{0.143 + 0.772} = 1.632$$
$$C_1' = C'_{Mo} = (0.143)(1.632) = 0.233$$
$$\beta_2 = \frac{k_1 a_{21} + k_2}{k_1 + k_2} = \frac{(0.143)(1.149) + 0.772}{0.143 + 0.772} = 1.023$$
$$C_2' = C_W' = (0.772)(1.023) = 0.790$$

First Iteration

$$\beta_1' = \frac{0.23 + (0.790)(1.749)}{0.790 + 0.233} = 1.578$$
$$C_1'' = (1.578)(0.143) = 0.226$$
$$\beta_2' = \frac{0.233(1.149) + 0.790}{0.790 + 0.233} = 1.034$$
$$C_2'' = 0.798$$

Second Iteration

$$\beta_1'' = \frac{(0.226) + (0.798)(1.749)}{(0.226)\ 0.798} = 1.584$$
$$C_{Mo} = (1.584)(0.143) = 0.226$$
$$\beta_2'' = \frac{(0.226)(1.149) + 0.798}{0.226 + 0.798} = 1.033$$
$$C_W = 0.797$$

V. SUMMARY

Minicomputer methods for data reduction in x-ray microanalysis have been described in detail. Any laboratory doing or planning to do such analysis requires some computer data reduction scheme. The minicomputer methods appear to provide calculated compositions very nearly equal to those calculated by much larger batch-type programs such as MAGIC or COR2. Furthermore, input to these minicomputer programs has been reduced to a minimum (see Table I). Therefore programs based on FRAME or a combination of FRAME and the hyperbolic method have much to recommend them for the average laboratory.

REFERENCES

1. D. R. Beaman and J. A. Isasi, *Anal. Chem* **42**, 1540 (1970).
2. J. W. Colby, in *Advances in X-Ray Analysis* (J. B. Newkirk, G. R. Mallett, and H. G. Pfeiffer, eds.), Vol. 11, Plenum Press, New York (1968), p. 287.
3. J. Henoc, K. F. J. Heinrich, and R. L. Myklebust, NBS Technical Note 769 (1973).
4. K. F. J. Heinrich, R. L. Myklebust, H. Yakowitz, and S. D. Rasberry, "A Simple Correction Procedure for Quantitative Electron Probe Microanalysis," NBS Technical Note 719, U.S. Government Printing Office, Washington, D.C. (1972).
5. A. Ruark and F. E. Brammer, *Phys. Rev.*, **52**, 322 (1937).
6. K. F. J. Heinrich, D. L. Vieth, and H. Yakowitz, in *Advances in X-Ray Analysis* (G. R. Mallett, M. J. Fay, and W. M. Mueller, eds.), Vol. 9, Plenum Press, New York (1966), p. 208.
7. K.F.J. Heinrich, H. Yakowitz, and D. L. Vieth, in *Proceedings of the 7th National Conference on Electron Probe Microanalysis, San Francisco* (1972), Paper 3.
8. K. F. J. Heinrich, in *The Electron Microprobe* (T. D. McKinley, K. F. J. Heinrich, and D. B. Wittry, eds.) Wiley, New York (1966), p. 296.
9. P. Duncumb and S. J. B. Reed, in *Quantitative Electron Probe Microanalysis* (K. F. J. Heinrich, ed.), NBS Special Publication 298, U.S. Government Printing Office, Washington, D.C. (1968), p. 133.
10. K. F. J. Heinrich, in *Present State of the Classical Theory of Quantitative Electron Probe Microanalysis*, NBS Technical Note 521, U.S. Government Printing Office, Washington, D.C. (1970), p. 5.
11. K. F. J. Heinrich, *Anal. Chem.* **44**, 350 (1972).
12. P. M. Thomas, U.K. Atomic Energy Authority, AERE Report 4593, (1964).
13. K. F. J. Heinrich and H. Yakowitz, *Mikrochim. Acta*, **1970**, 123.
14. S. J. B. Reed, *Brit. J. Appl. Phys.*, **16**, 913 (1965).
15. D. M. Poole, in *Quantitative Electron Probe Microanalysis* (K. F. J. Heinrich, ed.), NBS Special Publication 298, U.S. Government Printing Office, Washington, D.C. (1968), p. 93.
16. K. F. J. Heinrich, R. L. Myklebust, and S. D. Rasberry, *Preparation and Evaluation of SRM's 481 and 482 Gold–Silver and Gold–Copper Alloys for Microanalysis*, NBS

Special Publication 260-28, U.S. Government Printing Office, Washington, D.C. (1971).

17. H. Yakowitz, C. E. Fiori, and R. E. Michaelis, *Homogeneity Characterization of Fe-3 Si Alloy*, NBS Special Publication 260-22, U.S. Government Printing Office, Washington, D.C. (1971).

18. A. E. Bence and A. Albee, *J. Geol.* **76**, 382 (1968).

19. H. Yakowitz, R. E. Michaelis, and D. L. Vieth, in *Advances in X-Ray Analysis*, Vol. 12, Plenum Press, New York (1969), p. 418.

20. M. J. Berger and S. M. Seltzer, National Academy of Science, National Research Council Publication 1133, Washington, D.C. (1964), p. 205.

21. R. W. Fink, R. C. Jopson, H. Mark and C. D. Swift, *Rev. Mod. Phys.* **38**, 513 (1966).

PRACTICAL ASPECTS OF X-RAY MICROANALYSIS

H. Yakowitz and J. I. Goldstein

For the solution of many practical problems, identification of microconstituents is all that is required. For example, foreign matter may have somehow incorporated itself into the grain boundary of a material, with deleterious effect. Then the identification of the foreign substance, in order to locate its source, completes the analyst's job.

The next stage of complication is when the distribution of the identified constituents is needed. This task is usually carried out by means of x-ray area scanning. The net result is a means to view the specimen's microstructure in terms of composition. Such elemental maps represent the major contribution to the ultimate solution of many problems in science and industry.

Finally, quantitative x-ray microanalysis may be required to pinpoint the source of certain materials problems or for research applications such as diffusion and phase diagram determination. For the highest accuracy to be attained, appropriate reference standards are needed, and both specimen and standard must be properly prepared. Usually, computer methods are used to convert the raw data to compositional information.

This chapter will describe methods to obtain identification and elemental distribution; selected applications of these methods will then be

H. YAKOWITZ—Institute for Materials Research, Metallurgy Division, National Bureau of Standards, Washington, D.C.
J. I. GOLDSTEIN—Department of Metallurgy and Materials Science, Lehigh University, Bethlehem, Pennsylvania.

outlined to indicate the way in which problems can be attacked. Next, specimen preparation and standards will be discussed in the context of carrying out quantitative analysis. Finally, a variety of examples for which quantitative analysis was required in the solution of a problem will be discussed. In all instances, applications were picked selectively; no effort is made to exhaustively review the literature.

1. GRAPPLING WITH THE UNKNOWN

A. Elemental Identification

Most scanning electron microscopes and electron microprobes are equipped with energy-dispersive spectrometer (EDS) facilities. All microprobes and many SEM's have wavelength-dispersive spectrometers (WDS), also called crystal spectrometers, as well. The EDS system is capable of identifying elements with atomic number $Z \geq 11$ in a few minutes.[1,2] WDS units capable of identifying elements with $Z < 11$ are available as well as WDS units meant for investigating the region of $Z \geq 11$. (Chapter VII has dealt with EDS and WDS systems in detail.) Accordingly, only a brief review of element identification with EDS and WDS facilities will be given here.

The EDS detector collects the entire x-ray spectrum and a multichannel analyzer calibrated so that each channel corresponds to some convenient energy, e.g., 25 or 50 eV per channel, is used to divide the spectrum into energy packets. Moseley's law relates the characteristic energy of an x-ray peak and the atomic number of the element responsible for the peak. Hence, by determining the channel numbers (energy) of peaks in the spectrum, appropriate atomic numbers can be assigned for the elements present in the electron-irradiated region. Because of x-ray absorption in the detector window (usually beryllium sheet about 10 μm thick) and in the lithium-drifted silicon chip itself, elements with atomic number Z less than ten are not usually identified or analyzed energy-dispersively. In addition, the characteristic energies of these elements are only about 100–150 eV apart. The vast majority of present-day energy-dispersive analysis equipment cannot cleanly resolve peaks of adjacent elements in the range $4 \leq Z \leq 11$. Therefore energy-dispersive analysis is useful for identification and analysis of elements with $Z \geq 11$. Figure 1a shows a typical spectrum of a mineral obtained with an energy-dispersive detector system.

Instrumentation can be made to indicate what elements are detected in a given spectrum and to display the information, thus saving the investi-

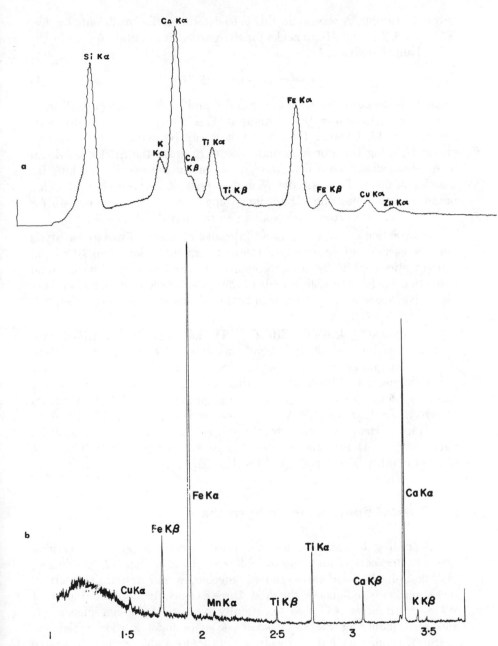

FIGURE 1.   (a) Typical energy-dispersive spectrum (EDS); all elements present of atomic number greater than 11 are indicated. Sample—Kakanui Hornblende. (b) Typical wavelength-dispersive spectrum (WDS) of same specimen as that shown in Figure 1a, obtained with an LiF diffraction crystal; 20 keV.

gator the trouble. A way to do this is to determine the coefficients for $K\alpha$, $K\beta$, $L\alpha$, $L\beta$, $L\gamma$, and $M\alpha$, β peaks for the particular calibration used in the multichannel analyzer, i.e.,

$$Z = \exp[A(\ln C_n + B)] \tag{1}$$

where C_n is the peak channel number and A and B are numerical coefficients depending on the calibration. Equation (1) is a way of stating Moseley's law (Figure 15, Chapter III). A parabola can be fitted to the peaks to obtain C_n using three or more data points to make the fit. Hence Z can be determined automatically. Ambiguities can be reduced by checking for β series peaks or for both the K and L lines of a given element being present in the spectrum. Data smoothing and very sensitive tests for emergence of peaks from background can be carried out as well.[3]

Subtraction of the background intensity in energy-dispersive analysis must be carried out accurately if elements present in less than 1000 ppm concentrations are to be properly identified. The question of background correction has been troubling analysts almost since the inception of energy-dispersive analysis. Recently, significant advances have been made[3–5] (sec Chapter VII).

In summary, elements with $Z \geq 11$ can be readily identified with energy-dispersive analysis equipment, usually in a few minutes or even less. For elements present in trace amounts, longer counting times are necessary but instrument stability limits counting time to 10 or 15 min in practice. Lifshin and Ciccarelli have discussed difficulties in identifying elements present in small amounts.[2] Among these are the fact that detector dead time is fixed. Hence a limiting total count rate is implied to prevent pulse pileup or loss. At high count rates, the peaks may broaden and/or shift channel number (see Chapter VII for details).

B. Crystal Spectrometer Systems

According to Birks,[6] the four most important points concerning spectrometer performance are as follows: (1) intensity, (2) resolution, (3) line-to-background ratio, and (4) mechanics and design. The last of these markedly influences the first three. Hence the potential user of WDS spectrometers will need to carefully evaluate the design, placement, and performance of the spectrometer system under a wide variety of operating conditions. Even if spectrometers are available, energy-dispersive x-ray analytical capability is virtually indispensable.

For qualitative analysis with spectrometers, one merely seeks solutions to Bragg's law by driving the detector through a range of angles. Whenever the angle and the wavelength λ of the x-rays emitted from the specimen are such that Bragg's law is satisfied, a peak results. Since the values of wavelength λ are a characteristic function of atomic number, qualitative identification can be carried out by comparing each λ found with tabulated values for each element (note Chapter III). This process is more time-consuming than the energy-dispersive process. However, higher beam intensities can be used so that weak peaks can perhaps be more easily identified. The resolution of the spectrometer is less than 5 eV, and so ambiguities due to peak overlap can be reduced. Methods of analyzing spectra so as to more positively identify the elements present have been detailed by Heinrich and Giles.[7] A typical spectrum of the same mineral as in Figure 1a is shown in Figure 1b.

It should be noted that Figure 1a is the full spectrum obtainable with the present energy-dispersive system while Figure 1b is a partial spectrum obtained with a LiF diffraction crystal. Other diffractive crystals such as EDDT, ADP, or RAP would be needed to provide the entire spectrum. On an instrument equipped with more than one spectrometer, simultaneous spectra could be prepared in order to obtain the complete picture. However, a SEM may be equipped with only one crystal spectrometer plus an energy-dispersive system. Then, using the example of the Hornblende of Figures 1a and 1b, the energy-dispersive data clearly suggest that another crystal is needed in order to study the silicon or zinc distribution in detail. Furthermore, a crystal such as RAP could be used to determine what elements of atomic number less than 11 are present. Note that Mn is indicated in Figure 1b, but in Figure 1a, the Mn peak is lost in the Fe $K\alpha$ tail. For reasons such as this, a crystal spectrometer combined with energy-dispersive capabilities is virtually a necessity for a laboratory intending to carry out all but the most routine x-ray analytical work. For instance, if elemental distribution maps of both Ca and K are desired, the crystal spectrometer would be needed to separate the two satisfactorily. (The resolution of the energy-dispersive detector used to prepare Figure 1a was rated as 211 eV at Fe $K\alpha$.) Furthermore, with a crystal spectrometer, higher beam current could be used to prepare the distribution maps, provided the specimen can stand such treatment. Higher current may overload the EDS system and therefore simultaneous analysis using WDS and EDS may not be feasible. To overcome this problem, an absorber could be placed before the EDS system to reduce total incoming counts.

The abscissa of Figure 1b is in angstrom units. The "hill" between 1 and 1.5 Å is caused in part by the proximity of the detector to the x-ray source and in part by the fact that the characteristic peak-to-continuum level decreases with increasing energy, i.e., decreasing wavelength. These problems cause the Zn $K\alpha$ peak to be lost in the noise, but the energy-dispersive detector clearly indicates the presence of zinc.

Note the relative peak height ratios for iron and calcium in Figures 1a and 1b. These differences are caused by the efficiencies of the detector systems. The actual ratio in the specimen is 1.15 Fe:1.0 Ca. Misleading data can be obtained by comparing peaks within a given spectrum; changes can be monitored this way, however.

Light element identification, i.e., $Z \leq 11$, is possible with appropriate diffracting media such as rubidium acid phthalate (RAP) or lead stearate dodecanoate (LSD). The detector in such cases is usually a flow proportional counter having an ultrathin (~ 1 μm) window. However, intensities are usually low due to absorption and the physical fact that the x-ray fluorescence yield for light elements is small. In addition, peak overlap can be a problem. Nevertheless, if light element identification is to be a major task of the x-ray analytical capability, an appropriately equipped crystal spectrometer is virtually indispensable.

Peak-to-background ratio is usually higher with the crystal spectrometer, all other experimental conditions being equal. Lifshin and Ciccarelli have presented data comparing energy-dispersive systems and crystal spectrometers for a variety of pure elements and experimental conditions.[2] Identification of elements with $Z \geq 11$ is readily possible with energy-dispersive systems. However, weak peaks due to trace or very low concentrations may be difficult to identify unambiguously. Peak overlap may occur. Special precautions may be necessary in accounting for all observed peaks, e.g., extraneous peaks due to pulse pileup.[2,5] Crystal spectrometers are virtually a necessity if extensive studies involving elements with $Z < 11$ are to be undertaken. For unambiguous identification of all elements, a 30-kV accelerating potential is sufficient. In energy-dispersive systems, the multichannel analyzer can be done away with, and its operations carried out by a small, programmed computer. Such a procedure, which is gaining popularity, allows both greater operator control and more flexibility for both identification of elements and quantitative analysis.

C. Elemental Distribution (X-Ray Area Scanning)

The technique of x-ray area scanning provides the investigator with what amounts to a scanning x-ray microscope. The amplified signal from the detector system—energy-dispersive or crystal spectrometer—is made to modulate the brightness of a cathode ray tube (CRT) scanned in synchronism with the electron probe. Thus, on the CRT, a picture is obtained by the variation of x-ray emission from the surface. The x-ray scanning system is shown schematically in Figure 2. The same magnification controls sweep system and amplifier as in the scanning electron microscope are used here (see Chapter VI). An x-ray area scan can show tones ranging from black to white, depending on experimental conditions. In places of high concentration of the element in the scanned area, the picture will be nearly photographically white; it will be gray where the element's concentration is lower and black where the element is absent. An example showing results for a galena ore is illustrated in Figure 3.

When an energy-dispersive system is used to prepare the area scan photograph, care must be taken to ensure that no other peak interferes with the signal of the desired element. The desired peak should be carefully isolated by means of a single-channel analyzer; this analyzer is often built into multichannel analyzers. In a programmable system, the desired peak can be similarly isolated. The amplified output of the single-channel analyzer is displayed on a CRT and so provides the elemental distribution desired. Magnification can be varied in the usual way. Separate micrographs of all elements of interest in the scanned area can be built up to give the complete elemental distribution. A detailed discussion of x-ray area scanning has been given by Heinrich.[8]

In the case of an energy-dispersive detector, the time required to build up a satisfactory x-ray area scan may be very long. This hindrance results from the requirements imposed by Poisson statistics to provide at least 20,000 photons per micrograph; frequently 200,000 are needed and occasionally as many as 500,000 photons must be used. For the SEM often

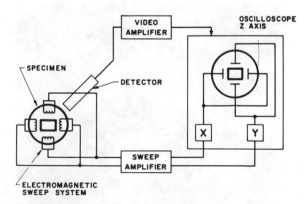

FIGURE 2.   Schematic of x-ray scanning system. The interaction between cathode ray tube (CRT) and the electron beam scanning system is shown.

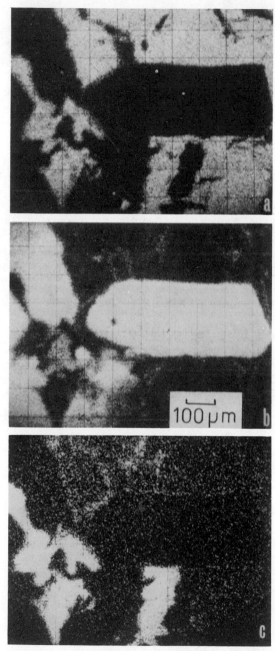

FIGURE 3. X-ray area scans showing distribution of three elements in a galena ore. (a) S $K\alpha$; (b) Si $K\alpha$; (c) Ag $L\alpha$.

20μm

FIGURE 4. X-ray area scans of an Al–W composite material photographed on the SEM with an EDS system. Top: W $L\alpha$; middle: Al $K\alpha$; bottom; secondary electron micrograph. See text (p. 417) for full details.

run with beam currents on the order of 10^{-10} or 10^{-11} A, the total number of photons produced is fairly low ($\sim 10^4$ per second). Since one peak obtained from the total spectrum is of interest, the time to obtain a satisfactory micrograph is correspondingly long. Times in excess of 15 min are not uncommon to prepare x-ray area scans. Figure 4 shows EDS x-ray area scans.

If the specimen will not be harmed, higher beam currents are useful not only for elemental mapping but for all phases of x-ray analysis. The loss of resolution resulting from increasing beam current matters little since the x-ray emission volume is largely determined by electron diffusion and not by probe size.[9] Since it is difficult to decrease the effective volume of x-ray emission below 1 μm, x-ray scanning pictures are limited to magnifications of about 3000 diameters.

Because of the requirement for the x-ray source to be precisely on the Rowland focusing circle of a WDS system, large scan areas may cause intensity dropoff at the edges of the x-ray area scan. This problem becomes more pronounced as the resolution of the WDS system increases. No such difficulty exists in the case of an EDS system. One way to monitor the seriousness of the intensity drop off is to insert a pure element and to prepare x-ray photographs as a function of raster size. This can be done for each crystal in each spectrometer.

The eye can observe contrast differences of about 2% as a minimum. Hence the composition difference responsible for a 2% contrast variation on an x-ray area scan may be of interest. Figure 5 shows kamacite containing about 93 wt % Fe and taenite containing 50–85% Fe in the Tazewell, Tennessee meteorite. The interface separating these kamacite and taenite phases is evident. This case represents a compositional ratio of about 1.5 Fe (kamacite) to 1.0 Fe (taenite). For compositional ratios of abou 1.1 or 1.2:1.0, it will probably be very difficult to unambiguously separa phases on the ordinary x-ray area scan. Heinrich has described method improve on this situation by electronic enhancement methods.[8]

D. Applications of Elemental Identification and Distribution

One of the most important materials for commercial use as a cutting tool is high-speed steel, a complex ferrous alloy containing C, Cr, W, Mo, V, Mn, and Fe. The important characteristic of high-speed steel is its ability to retain hardness and wear resistance at the elevated temperatures produced by high-speed machining. Essentially this feature is achieved by including major quantities of carbon and carbide forming elements such that the microstructure contains an appreciable volume fraction (typically 0.1–0.3) of excess alloy carbides whose inherent hardness and high-temperature stability are responsible for the desired properties of the steel.

The distribution of carbides in the eutectic, or position of the last liquid to solidify, has a large influence on the properties of the high-speed steel after further processing. In a study of the solidification of M2 tool steel, Barkalow *et al.*[10] described the metallography of these carbides and the distribution of elements between eutectic and matrix. Figure 6 shows scanning x-ray pictures of W, Mo, Cr, V, and Fe obtained from a typical eutectic region. All the eutectic carbides contain a substantial concentration of W and Mo. The isolated particles of carbide are high in V and much lower in W and Fe. Chromium, in comparison with W and Mo, is rather uniformly distributed between the matrix and the eutectic, but pronounced compositional differences are shown in the area scans for V and Fe. Further quantitative x-ray analysis can establish the compositions of the carbide.

Black and white photographs showing the topographic distribution of a single signal are widely used. It is often difficult, however, to show the correlation of signals from two or more x-ray lines without the use of color. Composite color photographs using x-ray images from the electron microprobe have been produced. One such method uses the black and white scanning images as color separation positives and with appropriate filters, color prints are made.[11] Figure 7 shows an optical micrograph of a crack in a nickel-base superalloy that occurred during high-temperature service. Figure 7 also shows x-ray area scans for Ni, Cr, and Al; it is difficult to determine the relationships of these elements exactly. Furthermore, the separation of the structure into three component parts may be confusing. If, however, the three black and white area scans are combined in color, the structure in terms of composition can be represented (Figure 8*).

This color scan is obtained by adding (combining) the three photographs of Figure 7 with primary color filters, i.e., red, green, and blue. Red and green add to give yellow; green and blue add to give cyan; red and blue add to give magenta and red; green and blue add to give white. The black and white prints are used as color separation positives; they must be registered accurately for adding onto a Polaroid color film. Table I lists the filters and exposure times needed.

The first requirement for a satisfactory color composite is a set of

---

*Figures 8–11 are color illustrations and will be found following p. 206.

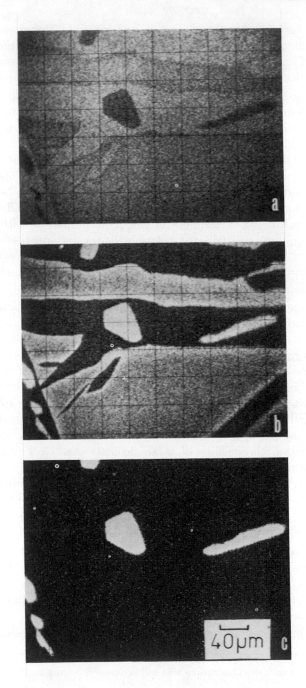

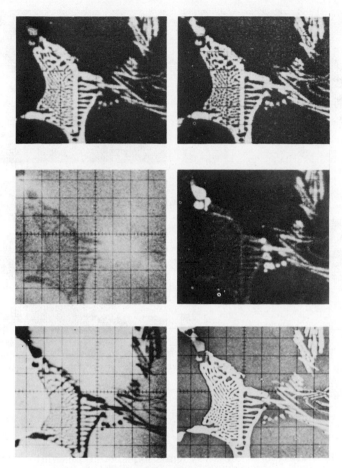

FIGURE 6. X-ray scanning pictures from a solidified section of an M2 tool steel. The scanned area is 90 × 80 μm in size. The various signals are W *Lα* (upper left), Mo *Lα* (upper right), Cr *Kα* (center left), V *Kα* (center right), Fe *Kα* (bottom left), and backscattered electrons BSE (bottom right).

FIGURE 5. X-ray area scans showing distribution of constituents in the Tazewell, Tennessee iron meteorite. Note the nickel enrichment at the phase boundaries; the iron content is about 93% in the kamacite (dark in the Ni picture) and 50-85% in the taenite (light in the Ni picture). Differences in iron content of these phases are apparent in the Fe picture. (a) Fe *Kα*; (b) Ni *Kα*; (c) P *Kα*.

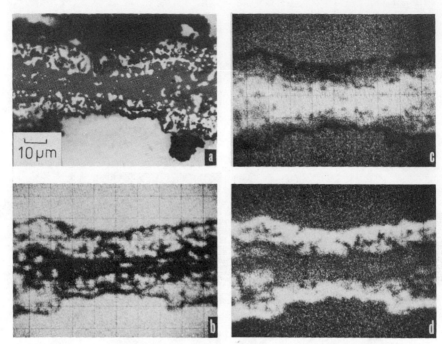

FIGURE 7.   (a) Optical micrograph of a crack in a nickel-base superalloy that occurred during high-temperature service; (b) Ni $K\alpha$; (c) Cr $K\alpha$; (d) Al $K\alpha$.

x-ray black and white images of good quality. The main points to observe are:

1. The areas of high x-ray emission should be white, and those with no exposure, black. The effects of a gray background (e.g., through too short development time of the black and white picture) cannot be easily corrected in the making of the composite.
2. The interpretation of the color picture is often aided if equal lightness of the separation positives corresponds to equal concentrations of the elements represented.
3. A sufficient amount of x-ray counts should be recorded.
4. Care should be taken that the specimen position with respect to the scan does not change between exposures due to electrostatic charges or mechanical displacement.

Returning to Figure 7, it is apparent that working with only the x-ray area scans, a good deal of patience and intuition is required to reconstruct the exact phase positions in the microstructure. Furthermore,

TABLE I. Exposure Times for Color Composites [a]

Kodak Wratten filter set number	Color	Time, sec
Conditions I[b]		
92	Red	1/4
93	Green	12
94	Blue	60[d]
25	Red	1/8
58	Green	5/8
47	Blue	5
33	Red	1/5
93	Green	10
45	Blue	3[e]
Conditions II[c]		
33	—	1.5
93	—	40
45	—	7.5[e]

[a] Intended as a guide only. These values give approximately equal lightness.

[b] Light source: Four reflector flood lamps spaced 2 ft apart, 3 ft from subject. Camera $20\frac{1}{4}$ in. from subject. Lens setting $f5.6$.

[c] Light source: Two cold white fluorescent bulbs 15 in. long placed 18 in. from subject. Camera $3\frac{1}{2}$ in. from subject. Lens setting $f5.6$.

[d] Add 1/30 white.

[e] Add 1/60 white.

the locations of superposed elements are difficult to discern. The color composite with Al red, Cr green, and Ni blue (Figure 8) shows exactly the phase positions in the microstructure. Mixed colors indicate the presence of superposed elements. The composite is topographically analogous to the optical micrograph, but the major constituents responsible for the structure are identified. The conclusion of the metallurgists regarding the specimen was that oxides of Cr and Al formed during service. These oxides are larger in volume than the host lattice and the Ni lattice failed. Crack propagation then proceeded as more oxide formed. The composite indicates that Cr oxidizes rapidly and that after most of the available Cr is oxidized, Al oxidizes. This mechanism is illustrated by the presence of the Al shell around the Cr.

The next application involves composite materials. The availability of sapphire (α-Al_2O_3) fibers (whiskers) in the form of single crystals having very high strength has created extensive interest in the incorporation of whiskers in metallic matrices. Hopefully such composites will combine the desirable properties of ductile metals with those of the ultrahigh-strength whiskers, which would be oriented so as to provide reinforcement in the direction of the applied stresses. In one study, 15 vol % of 1–30-μm-diameter sapphire whiskers added to an aluminum–10% silicon matrix by liquid-phase hot-pressing at 580°C resulted in some cases in an increase of 60% in tensile strength coupled with a 50% increase in elastic modulus. The beneficial effects were still present up to 800°F (426°C); some specimens were stronger than any wrought commerical aluminum alloy at this temperature. However, strengthening from sample to sample occurred with less than satisfactory consistency. It was thought that this lack of consistency might be due to poor bonding of the whisker to the matrix.

In an effort to improve this bond between the matrix and the sapphire, the specimens were prepared using nickel-coated fibers. Nickel was chosen because it could be deposited with relative ease by the thermal decomposition of nickel tetracarbonyl $Ni(CO)_4$. Hence a matte of nickel-coated whiskers could be prepared; the matte could then be inserted into the matrix by liquid-phase hot-pressing. Apparently, a $NiO–Al_2O_3$ spinel forms which serves to bond the Ni to Al_2O_3. In the absence of this spinel, adhesion of Ni to Al_2O_3 may be poor or not occur at all. The presence of the spinel is taken to indicate the probability of good Ni-to-fiber adhesion. In addition, bonding of the fiber matte to the matrix by diffusion during liquid-phase hot-pressing is expected. Hence better bonding of the nickel-coated whiskers to the matrix was expected as contrasted to uncoated whiskers.[12]

However, the mechanical test results were still inconsistent even after the introduction of nickel. Electron probe microanalysis was used in an effort to determine the reason. The area scanning micrograph as given by Yakowitz *et al.*,[13] shown in Figure 9 tells the story. The nickel (red) reacted with the aluminum and is distributed throughout the matrix. The sapphire whiskers were delineated by the cathodoluminescence signal (see Chapter V) and were photographed using the green filter. The silicon in the 10% Si–Al alloy was photographed using the blue filter.

The spotty appearance (yellow and green of the whiskers) is indicative of the process taking place. If complete bonding had occurred, these whiskers would be totally yellow (red plus green equals yellow). Because the Ni bonding agent reacted with the Al-base matrix, incomplete bonding occurred. Hence the whiskers are randomly "spot-welded" into the matrix; this largely explains the inconsistencies found in the mechanical test results. The silicon is inhomogeneously distributed in the matrix as well, as shown by Figure 9. Furthermore, the figure shows that the whisker distribution is poor. Therefore the chief sources of difficulty with the composite could be deduced from Figure 9.

Figure 10 shows a portion of a galena ore. Little phase mixing has occurred; the economically desirable lead is combined as lead sulfide and is delineated by the cyan (light blue) field. Thus the location of the desired constituent can be shown; note also that some of the sulfur is not combined with lead. Further study revealed that this sulfur was, in fact, combined with silver (see Figure 3).

In recent years several studies have been performed on explosively bonded metals. It has been found that the primary requirement for establishing a proper metallurgical bond is the absolute cleanliness of the surfaces of the metals to be bonded together. In order to achieve a good bond, the plate collision velocity divided by the size of the collision angle should not greatly exceed the sonic velocity of the metals being welded. Under such conditions a pressure wave is generated ahead of the collision front, and the surface layers of both interacting metals, if covered with oxides and other contaminants, are jetted into the space between the plates. The existence of a jet during explosive bonding was established experimentally by Bergman et al.[14] This high-velocity impact technique is used to achieve metallurgical bonding on an extremely wide variety of metals with greatly differing physical and mechanical properties. Many practical applications of such composite materials have been found in several fields of engineering.

Problems occurring at the interface of explosively bonded titanium and 1008 steel were found to be responsible for poor mechanical properties of the composite. Figure 11 shows one "wave" of the jetted interface: Fe is green, Ti is red. The topographic image (specimen current signal) is superimposed as well. Note that the steel grains are pulled out like spaghetti in the wave, itself indicating high local stresses. The main feature is the yellow (Fe plus Ti) coating in the interface. Quantitative electron probe studies indicate that the yellow band is TiFe and Ti_2Fe, brittle intermetallic compounds. These intermetallics formed because the bonding speed was too low. The presence of the intermetallics is largely responsible for the observed difficulties with the composite.[15]

An example of x-ray area scanning information obtainable with the EDS system is shown in Figure 4. The specimen is a composite of tungsten wires which was plunged into molten aluminum held at 1100°C under a vacuum of 10^{-4} Torr. The area scans show no Al in W and no W in Al. The micrographs were prepared in an SEM operating at 20 kV and a specimen current of 10^{-10} A; exposure time was 12 min for each. The gamma control was set at 4 and maximum contrast set on the record CRT. Figure 4 compares favorably with similar data prepared with a WDS system; however, the time and effort needed to prepare Figure 4 would be less in the case of WDS facilities.

II. SPECIMEN PREPARATION FOR QUANTITATIVE ANALYSIS

A. Surface Roughness and Polishing

Since the x-ray analysis performed is essentially an analysis of the prepared surface, it is requisite that the prepared surface be truly representative of the specimen. Over the years, a number of qualitative criteria for a properly prepared surface have evolved. These are that the specimen should be polished as flat and scratch-free as possible and be analyzed in the unetched condition so as not to alter the topography or surface chemistry. Such criteria were set forth primarily for metallurgical specimens; they can be applied almost directly to petrographic specimens. However, for biological work such criteria are virtually meaningless since it is rare that a "polished" specimen is used in such work (Chapter XIII).

Yakowitz and Heinrich[16] showed that the relative error in the absorption correction term for quantitative microanalysis is proportional to $\cot \Psi \, \Delta\Psi$, where Ψ is the x-ray emergence angle (see Chapter IX for further details). Hence one wishes to use high values of Ψ in order to minimize effects of local surface inclinations $\Delta\Psi$. This treatment assumes that the measured relative intensity ratio from the specimen to the standard commonly called k, is measured without error, i.e., Δk is zero. However, the effects of differences in surface preparation between specimen and standard will certainly contribute to errors in k. The differences in local inclinations affect the precision error in x-ray microanalysis and are superimposed on the x-ray statistical uncertainty. Instrumental effects, such as beam drift, are assumed to be negligible.

In order to estimate the magnitude of surface effect errors on precision, specimens of pure gold and NBS C1102 brass casting, which is homogeneous to within 0.8% of the amount present for copper and zinc, were used.[17] Presumably, scratches produced by 600-grit silicon carbide (SiC) papers represent the worst surface that might be introduced into an electron microprobe. Both specimens were examined after being polished on 600-grit, first and final polishing stages; the brass was also etched. The $L\alpha$ and $M\alpha$ lines of gold were monitored simultaneously, using a probe voltage of 20 kV. The $K\alpha$ lines of copper and zinc from brass were both monitored using beam voltages of 15 and 30 kV, respectively. Two conclusions were drawn from the experimental results[17]: (a) the softer the x-ray line of interest, the worse the possible effect of a nonflat specimen examined at a given voltage will be, and (b) the effect is worsened by lowering the nominal x-ray emergence angle Ψ.

Picklesimer and Hallerman[18] performed a similar experiment on a 50–50 copper–nickel alloy in the heavily etched, lightly etched, and as-polished conditions and after treatment with 16-μm grit Al_2O_3 papers. They had instruments with x-ray takeoff angles of 15.5° and 35° and found that there was little effect on the results for the first three conditions of surface preparation for both values of Ψ. However, after 16-μm grit paper, no change was observed with 35°, while a 2–3% variation occurred using a 15.5° emergence angle. After 16-μm grit paper there were 2–4-μm steps between scratches. Only the relatively high-energy $K\alpha$ lines of copper and nickel were used; the $f(\chi)$ values for the alloy are 0.96 and 0.91 at 35° and 15.5°, respectively. Therefore, even for cases where $f(\chi) \geq 0.9$, that is, very low absorption of the line of interest, low x-ray emergence angles may lead to difficulty if 2–4-μm steps are encountered.

The existence of scratches is not in itself particularly detrimental to the surface away from the scratches or to the analysis. In order to determine the effect on the surface flatness in the vicinity of a scratch, a corner of the brass sample was deeply cut with a needle. An interference micrograph of the area showed that the surface is essentially undisturbed approximately 50 μm away from the deep cut. Therefore, if scratches can be avoided by the beam, their presence is not harmful.

Flatness of both specimen and standard is a prime requisite. For pure elements and homogeneous materials, it is feasible to prepare relatively flat surfaces since the hardness will not vary greatly over the specimen. This results in fairly uniform stock removal. Unfortunately, most specimens submitted fall outside these two categories. In cases where phases of different hardness coexist, sharp steps may occur at the phase boundaries. Such effects must be taken into account if it is necessary to carry out analysis of regions near the boundary. One way to do this is to rotate the specimen 180° and remeasure the intensities.

A somewhat more insidious and more general problem than sharp steps is that of relief due to polishing, etching, and repolishing. This results in variable surface contours over the specimen face, that is, a hilly surface. These contours can occur in the same phase as well as across phase boundaries. Although grain boundaries show the effect most markedly, variation within the grain may be large. However, variability in the surface within grains is often as large as or larger than at grain boundaries. Therefore quantitative analysis of such a specimen may be difficult; the possibility of errors due to topography cannot be precluded.

Relief polishing is often minimized by using low nap polishing cloths and higher wheel speeds. If a polish–etch–polish procedure is being used, an etchant giving the lowest relief possible is the most desirable. In view of the possible effects of relief polishing and hardness variation steps on the x-ray intensity recorded by a spectrometer, some procedure to identify or minimize such effects or both would be desirable. One procedure is to monitor the same x-ray line with two spectrometers at 180° to each other and then to rotate the specimen 180°. The effects of surface topography can then be sorted out by comparison of the two sets of data. Unfortunately, some commercially available microprobes do not have the suitable spectrometer–specimen geometry to do this.

Another problem is the analysis of inclusions in various matrices. In effect, this is a repeat of the problem of two phases of variable hardness lying adjacent to one another. The major difference is that the inclusion is small in size and hence may be pulled out during specimen preparation. The standard technique to retain inclusions is to use as little lubricant as possible during the polishing procedure or to electropolish the specimen. In the case of inclusions, the entire junction between matrix and inclusion usually can be examined at high magnification on a metallograph. Both matrix and inclusion are sharply in focus throughout if the specimen is properly prepared. Detailed treatment of the subject of inclusion polishing and identification has been given elsewhere. [19, 48]

The counter-problem is the introduction of polishing abrasive into the material of interest. For example, NBS-C1102 brass was polished on SiC papers and a large number of "inclusions" were observed. These contained primarily silicon. On repolishing with Al_2O_3 papers, fewer inclusions were observed; these contained primarily aluminum. Finally, the specimen was electropolished directly from the "as-received" condition. The few inclusions found contained lead, sulfur, silicon, aluminum, and zinc. One should be suspicious of inclusions containing elements used in the polishing preparation. This is particularly true when specimens containing cracks or porosity are to be examined. In such cases, a different preparation technique is often warranted in order to determine whether or not such inclusions are artifacts.

In extremely soft materials such as lead- or indium-based alloys where local smearing of constituents could lead to erroneous microprobe results, electropolishing may be considered. Again, care and patience should be exercised in order to obtain the flattest surface possible. Furthermore, if more than one phase is present, the anodic (less noble) phase will be preferentially attacked[20]; in certain cases this may be an advantage in qualitative phase identification procedures. Thus the matrix could be attacked preferentially to a grain boundary phase, thereby disclosing more of the smaller phase. Occasionally, metallic ions of the anode are added into the bath to increase conductivity—care must be taken so as not to have plating of these or other bath constituents onto the specimen surface. For the same reason, chemical polishing techniques should be used with caution.[18]

Picklesimer investigated the effect of smeared layers and found that films on the order of 100–1000 Å can definitely alter the results of analysis. His feeling is that one should lightly etch the specimen in such cases in order to remove the smeared layer. If one is dealing with materials where $f(\chi)$ is greater than 0.9 and has a high x-ray emergence angle, such a procedure may be justified. It is necessary to prove that the etching causes no detrimental chemical alteration of the surface.

Results on smearing such as these led Picklesimer and Hallerman to consider the effect of an anodic film purposely placed on the specimen surface. Such an anodic film causes interference colors to appear, which under proper control, and in a given alloy system, are representative of the composition and the voltage applied to the specimen during anodizing.[18] The technique has been used on alloys containing titanium, hafnium, zirconium, uranium, tungsten, tantalum, and columbium; with it phase boundaries, segregation gradients, and phases can be delineated by color.[21] Pickelsimer and Hallerman concluded that anodized films can be used satisfactorily in a microprobe provided the standard is also anodized to the same film thickness; in addition, it is also desirable to use the thinnest anodic film that is adequate for color differentiation.[18]

As for specific techniques for the preparation of metallographic and petrographic sections, there are numerous references in the literature.[22–27] Suffice it to say here that there are virtually as many techniques for the preparation of a microsection as there are microsections. Given a properly prepared surface, a few useful instrumental adjuncts are worth mentioning. Polarized-light-equipped microscopes can serve very useful purposes in phase differentiations or determining orientation differences or both. Knowing that grain or phase orientations are different, rotation of the specimen can tell the investigator if possible intensity differences are due to orientation effects or to compositional differences. Finally, it is necessary to have some feature on which to visually focus. This is extremely important since the height of the specimen with respect to the x-ray spectrometer is critical. Unless the spectrometer is perfectly aligned and there are no geometric specimen–spectrometer effects, small differences in the specimen level can lead to intensity differences increasing with the distance of the x-ray source from the Rowland circle. Thus it is imperative that the focal level be the same for all work. Therefore it seems desirable to have a microscope capable of giving a high magnification and having a high numerical aperture lens as well as a lower, searching magnification. Since depth of focus decreases with increase in numerical aperture, an objective having a high numerical aperture will minimize the possibility of error due to the focal level of the specimen.

The need for truly flat specimens is reemphasized for all line or step scanning operations, or both, in which the specimen is mechanically driven with respect to the electron beam. In such cases, the x-ray flux recorded may vary both with specimen focal level and as a result of nonflatness. With certain spectrometer–specimen geometries, this effect may lead to anomalous x-ray flux variations for the elements of interest, resulting in data interpretation difficulties.

B. Choice of Coating Material

The question of specimen coating to ensure electrical and thermal conductivity certainly must be considered since coating is required for most nonmetallic specimens. In addition, for metallic specimens containing small nonmetallic inclusions, sample coating is also advisable. The most frequently employed coatings are carbon, aluminum, gold, and gold–palladium alloy. The usual "rule of thumb" for x-ray microanalysis is that the thinnest coating yielding stable specimen current and x-ray flux is desired. This is true since the thinner the coating, the less x-ray absorption within it and the less the energy loss of the primary electron beam entering the specimen. Furthermore, the thinner the coating, the smaller will be the excitation of x-rays from the coating itself. For gold and gold–palladium coatings, which are often used in scanning electron microscopy to provide enough secondary electron flux, the characteristic and/or continuum radiation produced could interfere with the x-rays lines of interest. Particular problems can occur if the element of interest is present in small or trace amounts.

The coating material may be deposited by standard vacuum evaporator techniques, as described in Chapter VI; the usual thicknesses used range between 50- and 500-Å. For 50–100-Å films of carbon, aluminum, gold, and gold–palladium, the energy loss of the primary beam appears to be of small consequence even at low nominal voltages.[28] However, backscattered electrons obtained from the specimen could excite x-ray radiation from the film. This process may be particularly serious for gold and gold–palladium coatings on specimens with average atomic numbers greater than ten.

Examination of mass attenuation coefficients (μ/ρ) shows that for x-ray lines from 8 to 40 Å, aluminum is lowest of all four, followed in order by carbon, gold, and gold–palladium. In the region below 8 Å, the carbon (μ/ρ) value is lower than those for aluminum, gold, and gold–palladium. However, gold is best when electrical and thermal conductivities are considered, with aluminum about one-third as good, and carbon poor. It would seem that, for general purposes, aluminum is favored by its physical properties for use with x-ray lines of 8–40 Å, while carbon is favored outside this region.

Occasionally, a specific property of the specimen will preclude such a choice. Adler[29] observed that coating fluorite with aluminum was unsatisfactory in that stable specimen currents and x-ray intensities could not be obtained. He assumed that the fluorite decomposed under the electron beam and that the evolving fluorine reacted with the aluminum to form a nonconducting fluoride. The effect was eliminated by using a carbon coating.

Smith has shown experimentally that standards and unknowns should be coated simultaneously and as closely together as possible. Furthermore, his considerations of correction procedures indicate that it may be necessary to use the same coating thickness for all specimens.[30] Smith asserts correctly that the thickness of the carbon coatings that he used was of extreme importance in the consideration of data correction procedures for rock-forming minerals.

C. Standards in Microprobe Analysis

As discussed in Chapter IX, quantitative x-ray microanalysis is based on determining the x-ray flux emitted by an unknown relative to that of a suitable standard. The most easily obtainable standards are those of the pure elements. Data correction procedures can then be applied to reduce the measured intensity ratio to composition. Models for correction, however, are open to question. Input parameters, such as mass attenuation coefficients and fluorescence yield values, are sometimes only poorly known. Furthermore, even if perfect correction procedures were available, some elements, such as sulfur, chlorine, potassium, gallium, etc., cannot be obtained in suitable pure form for use in x-ray microanalysis.

For these reasons, the use of intermediate compositional standards has become widespread. There are three basic approaches: (1) to prepare an entire series of standards so that an empirical calibration curve can be established for the system of interest, (2) to obtain a single standard in order to characterize a particular constituent, and (3) to obtain single standards to monitor instrumental performance or as "anchor points" for correction procedures to be applied to similar systems.

The basic requirements for all such standards are that they be homogeneous at micrometer levels of spatial resolution, stable with respect to time and environment, properly prepared for use in the microprobe, and carefully analyzed by independent techniques   It is preferable that a piece taken from the actual mounted standard be used for such chemical analysis.

Usually, the most stringent requirement is that of homogeneity. This factor should be carefully checked by means of line scans, area scans, and point counting. In this fashion, tolerance limits for the standard and hence for the analysis can be set directly. The effect of instrumental drift can be determined by using the pure elements of which the standard is composed as control specimens.

Preparation of entire sets of standards for a single system is usually confined to metallurgical applications. It is warranted when a system is to be studied in great deail, for example, in the case of constitution diagram preparation. With proper standards, the accuracy of analysis can be made to approach the limits imposed by the x-ray statistics for such systems. It does not cost very much more to produce and characterize six standards of a given system than it does to produce one, especially since only a gram or so of standard is required. Figure 12 shows an empirically determined calibration for the analysis of Fe and Ni in binary Fe–Ni alloys assuming a 30-kV operating potential and a takeoff angle Ψ of 52.5°. The curve was established with the aid of nine well-characterized, homogeneous standards.[31] The Ni $K\alpha$ radiation is heavily absorbed by the nickel and the Fe $K\alpha$ radiation is increased due to x-ray fluorescence by the Ni $K\alpha$ radiation.

Obtaining a single standard to characterize a particular constituent may be adopted when no suitable elemental form is available. Before use, the material in hand should be characterized as to homogeneity and composition. Usually, data taken using such a standard must be corrected in the same fashion as those obtained with elemental standards.

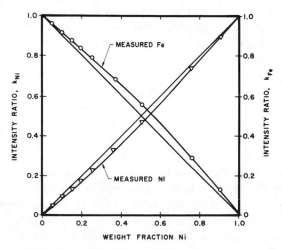

FIGURE 12.   Calibration k vs. C curves for the Fe–Ni system. The data points were obtained from nine well-characterized, homogeneous standards.

For example, pyrite (FeS_2) is often used as a sulfur standard. There is microprobe evidence that it is stoichiometric and homogeneous. Using iron sulfide (FeS_2) as a standard for sulfur in zinc sulfide (ZnS), one finds that the average atomic number is 22 for FeS_2 and 25.5 for ZnS; hence it would seem that atomic number effects would be small. Parasitic fluorescence is also no problem. But there is a 27% correction for absorption at an operating voltage of 20 kV; at 10 kV, this correction is 10%. These values are for an x-ray emergence angle of 41°. Hence such an analysis is subject to accuracy errors that may be somewhat larger than those imposed by x-ray statistics. The superimposed accuracy error with a 20-kV operating voltage for analysis of sulfur in ZnS can be shown to be about 50% of any relative error in the mass absorption coefficient of ZnS for sulfur.[16]

Single standards to check instrumental capabilities are very useful. Brass is especially good for this purpose since no correction is required for zinc, and only a small correction is necessary for copper. Tiepoint standards of systems of more than two elements, for example, iron–nickel–chromium alloys, have been prepared. For trace analysis, standards are also very helpful.

In choosing standards, a number of materials have been proposed. Among these are intermetallic compounds, fully ordered alloys, supposedly stoichiometric inorganic compounds, natural and synthetic minerals, organometallics, and homogenized solid solutions. Such materials as intermetallic compounds may not be homogeneous; a single vertical line on a constitution diagram is not a sufficient criterion. Such materials must be carefully checked, preferably in three dimensions, for homogeneity. Careful chemical analysis is also a prime requisite. Fully ordered alloys such as Cu_3Au may have composition ranges of ±5% of the amount present. All compounds and minerals should be carefully checked optically, by microprobe, and by chemistry. Just observing with the microscope may be enough to show the unsuitability of some materials. Splat-cooling, in which a molten drop of liquid alloy is quenched so rapidly that the "liquid structure" is frozen in place, has been suggested to prepare microprobe specimens. Goldstein, Majeske, and Yakowitz described a relatively simple, inexpensive apparatus in which splat-cooled specimens of any substances having melting points below about 1000°C and which are miscible in the liquid state could be prepared. The success of the method was demonstrated with gold–silicon and with aluminum–magnesium alloys.[32]

The National Bureau of Standards has prepared a few binary and one ternary metal alloy systems suitable for use as standards. At present, the standards available are a low alloy steel,[33] Au–Ag and Au–Cu alloys,[34] W–20% Mo alloy,[35] Fe–3.22% Si alloy,[36] two cartridge brasses,[37] and an Fe–Cr–Ni alloy.[38]

III. APPLICATIONS INVOLVING COMPOSITIONAL ANALYSIS

Several typical examples of problems involving compositional analysis will be given here. These examples illustrate the types of problems that can be investigated. Phosphides and sulfides are common nonmetallic inclusions found in metallic and nonmetallic (stony) meteorites as well as in the lunar samples. Phosphides, $(FeNiCo)_3P$, form by nucleation during cooling from the solid state as the P solubility limit in the metal is exceeded. The growth of the phosphide is controlled by the rate of mass transport of the major elements Fe, Ni, Co, and P. Figure 13 shows the distribution of

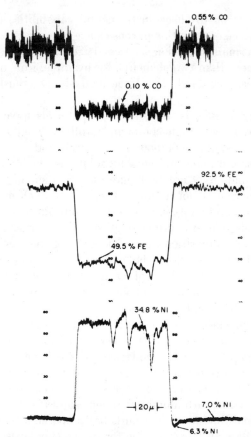

FIGURE 13.   Distribution of Co, Fe, and Ni across a phosphide inclusion in the kamacite (α) phase of the Breece iron meteorite. (From Goldstein and Ogilvie.[39])

Co, Fe, and Ni across a phosphide that is surrounded by the matrix kamacite (α-FeNi) phase of the Breece iron meteorite.[39] The data were taken by attaching a motor to the X or Y drive of the specimen stage and recording the x-ray output for Co, Fe, and Ni on a chart recorder; data were taken in a matter of minutes. Even though the analysis is qualitative in nature, it shows the relative homogeneity of the phosphide and the effect of the cracks that were present in the phosphide. The Ni gradient in the surrounding kamacite forms during the phosphide growth process as Ni diffuses to the growing precipitate. A detailed analysis of the data can lead to an estimate of the rate of growth of the phosphide, its final cooling temperature, and even to the size of the parent body in which it formed.

The structures of iron meteorites were developed during slow cooling of these samples in their parent meteorite bodies. The strikingly regular octahedral pattern that is observed is called the Widmanstatten figure and is found in hundreds of iron meteorites. This octahedral pattern is most commonly encountered in iron meteorites that contain 7–11% Ni as the major solute element in Fe. This two-dimensional pattern is composed of relatively broad, oriented bands of kamacite, a bcc (α) solid solution containing up to $\sim$7.5 wt % Ni in Fe, separated by geometrically regular fields of taenite, a compositionally zoned fcc (γ) solid solution of variable Ni content in the range 20–50 wt %. Microprobe traces taken on a surface of a section at right angles to the kamacite growth front provide true concentration–distance relationships. Such a concentration distance profile across a portion of the Widmanstatten structure is illustrated in Figure 14 for the Grant meteorite. The sample was moved in micrometer-sized steps while the electron beam was held in a static, point position. The Ni vs. distance profile can be obtained by measuring the appropriate k ratios for Ni across the sample, and using the Fe–Ni calibration curve, as shown in Figure 12, to obtain the correct compositions. Quantitative analysis can therefore be obtained directly without using the quantitative correction calculation schemes. The observed diffusion profiles can be explained by the nucleation and diffusion-controlled growth of kamacite-α when originally homogeneous parent taenite-γ is slowly cooled through the two-phase $\alpha + \gamma$ region of the Fe–Ni phase diagram. It is possible to deduce a cooling history appropriate to the observed structure of a particular specimen.[40]

In the course of the electron microprobe study of the Jajh deh kot Lalu, a stony meteorite of the enstatite ($MgSiO_3$) chondrite class, a new mineral containing silicon, nitrogen, and oxygen was discovered.[41] The occurrence of the new mineral sinoite (Si_2N_2O) seems to be of particular significance for the specific and extraordinary environment in which this rock was formed. Apparently this type of meteorite is unusual in that it was

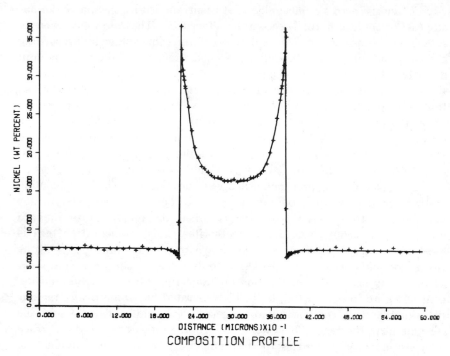

COMPOSITION PROFILE

FIGURE 14.   Concentration gradient of Ni across a kamacite-taenite-kamacite area in the Grant meteorite. Note the Ni depletion in kamacite at the α/γ boundary, and the Ni buildup above the bulk composition $C_0 = 9.4$ wt % Ni in the taenite. (From Goldstein and Axon.[40])

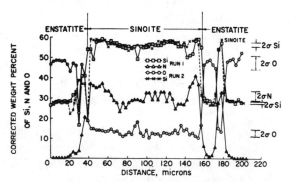

FIGURE 15.   Concentration-distance plot for Si, N, and O across a sinoite (Si_2N_2O) grain in the Jajh deh kot Lalu enstatite chondrite. (From Keil and Andersen.[41])

formed under extremely reducing conditions, and the occurrence of sinoite indicates that there was not sufficient oxygen available to bind all the excess silicon as SiO_2. The results of the quantitative microprobe analyses are shown in Figure 15. The data were taken by moving the section after each analysis in steps of 3 μm. The analyses of sinoite were performed with pure Si, SiO_2, $MgSiO_3$, and BN as standards for Si, N, and O. These analyses were among the first performed for the light elements nitrogen and oxygen.

In order to understand how to process a high-speed steel so that the steel will retain hardness and wear resistance at elevated temperatures, it is necessary to examine the as-cast microstructure of the high-speed steel. Such a study of microstructure was performed on a commercial M2 high-speed steel.[42] Low-magnification metallography shows three major constituents in as-cast M2. As the steel solidifies, the primary solidification element, called a dendrite, solidifies initially as ferrite, δ-bcc. As temperature decreases, the primary crystallization of δ-ferrite is followed by the formation of austenite, γ-fcc, around the original dendrite cores, or ferrite. At the lower temperatures the last remaining liquid solidifies as a eutectic containing many carbides (note Figure 6 and description). To examine more closely the phenomenon of δ-ferrite and γ-austenite formation, concentration profiles across a dendrite, from eutectic to dendrite to eutectic, were obtained for the major solute elements W, Mo, Cr, and V. The microprobe data are shown in Figure 16. A slight alloy enrichment at the dendrite core and a relatively uniform distribution of chromium in comparison with more pronounced segregation of tungsten, molybdenum, and vanadium is observed. The segregation gradients occur due to a balance between ferrite-forming and austenite-stabilizing elements.[42]

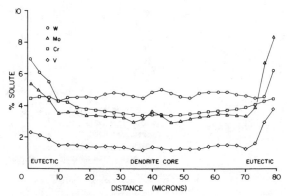

FIGURE 16.   Concentration profiles of tungsten, molybdenum, chromium, and vanadium across a primary dendrite axis in a unidirectionally solidified M2 steel. (From Barkalow, Kraft, and Goldstein.[42])

Many processes such as homogenization, phase growth, and oxidation in semiconductors, metals, and ceramics are controlled by mass transport in the solid, that is, diffusion. The value of the rate constant, called the diffusion coefficient D, of a given material can be obtained in many cases by a Matano analysis.[43] A concentration vs. distance profile is developed under certain specified boundary and initial conditions and the microprobe is used to measure such a profile. Figure 17 shows the diffusion profile developed when a Au vs. Au–6.1 wt % Ni diffusion couple was heat-treated for one week at 875°C. The concentration profile was taken by moving the sample under the electron beam and the concentration was determined by comparison with well-characterized Au–Ni alloys.[44] Both sides of the couple were face-centered cubic and the gradient obtained was processed to obtain diffusion coefficients using the Matano analysis. Diffusivity data in the Au–Ni system are of interest when Au–Ni bonds are made during semiconductor processing.

Electron probe analysis has some important advantages over other conventional methods (quantitative metallography, x-ray diffraction, etc.) used for phase diagram analysis. If the alloy phases are at equilibrium at the temperature of interest, the electron probe can measure the composition

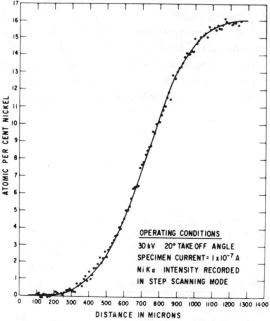

FIGURE 17.   Diffusion profile developed when a Au vs. Au–6.1 wt % Ni couple was heat-treated for one week at 875°C. (From Lifshin and Hanneman.[44])

of these phases directly. Tie lines can be obtained directly by measuring the composition of the two coexisting phases at equilibrium. Also, in three-phase regions the composition of the three coexisting phases can be measured directly and only one alloy is necessary to determine the phase field. Even if the various phases are not totally in equilibrium, phase equilibrium data can still be obtained by measuring the interface compositions of coexisting phases.[45] This procedure is suitable so long as equilibrium is maintained at the phase interfaces.

Figure 18 illustrates the type of data obtained with the probe for an Fe–Ni–P phase diagram study.[46] An alloy with a nominal composition of 92 wt % Fe, 6 wt % Ni, 2 wt % P was annealed at two different temperatures, 1000 and 875°C. As seen in Figure 18, the α and γ phases at 1000°C and the α, γ, and Ph [phosphide, $(FeNi)_3P$] phases at 875°C are well equilibrated. Compositions for the two-phase tie line at 1000°C and the three-phase field at 875°C were therefore quite accurately measured. Repeated interface measurements establish reproducible trends that allow measurement of the equilibrium compositions to be made. The Fe–Ni–P phase diagram can be used to understand the nucleation and growth of phosphide $(FeNi)_3P$ in FeNi alloys, as discussed previously (see Figure 13).

The final example of applications involving compositional analysis involves a study of the redistribution of nickel in the serpentinization of olivine, Mg_2SiO_4.[47] The mineral olivine, from ultramafic rocks, contains minor amounts of Ni, 0.2–0.3 wt %, in substitution for Mg. In this form it is not, at present, economically worthwhile to extract the Ni for commercial purposes. However, if Ni becomes redistributed or concentrated, then the situation could change. On alteration by hydration to serpentine, $Mg_3Si_2O_5(OH)_4$, the Ni is transferred into some other phase, either sulfide or oxide, but it rarely substitutes for Mg in the silicate.[47]

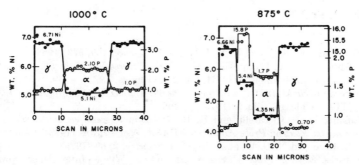

FIGURE 18. Electron microprobe data taken on a ternary alloy of nominal composition 92 wt % Fe-6 wt % Ni-2 wt % P heat-treated at 1000°C for 93 hr and 875°C for 168 hr. Ni $K\alpha$ and P $K\alpha$ radiations at 20 kV were measured. (From Doan and Goldstein.[46])

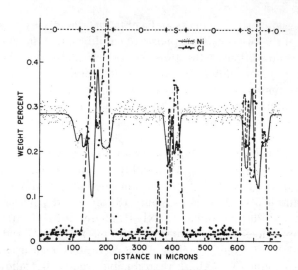

FIGURE 19.   Composition profile showing the distribution of Ni and Cl between
olivine (O) and serpentine (S) in partially serpentinized dunite.
(From Rucklidge.[47])

Figure 19 shows the distribution of Ni and Cl between olivine and serpentine
in partially serpentinized dunite. Apparently the Ni segregates to the
central area or feature of the serpentine. This central feature, when viewed
in three dimensions, is actually a surface where serpentinizing fluids enter
the crystal and by which Ni and Fe leave it. Nickel may be concentrated
in very fine grains of sulfide. It is probable, however, that in many cases Cl
is a controlling factor in transporting the metals.

REFERENCES

1. R. Fitzgerald, K. Keil, and K. F. J. Heinrich, *Science,* **159,** 528 (1968).
2. E. Lifshin and M. F. Ciccarelli, in *SEM/1973 Proceedings of the 6th Annual SEM
   Symposium* (O. Johari, ed.), IITRI, Chicago, Illinois (1973), p. 89.
3. J. M. Short in *SEM/1973 Proceedings of the 6th Annual SEM Symposium* (O. Johari,
   ed.), IITRI, Chicago, Illinois (1973), p. 106.
4. E. Lifshin, M. F. Ciccarelli, and R. B. Bolon, in *Proceedings of the 8th National
   Conference on Electron Probe Analysis, EPASA, New Orleans* (1973), Paper 29.
5. S. J. B. Reed and N. G. Ware, *X-Ray Spectrometry,* **2,** 69 (1973).
6. L. S. Birks, *Electron Probe Microanalysis,* 2nd ed., Wiley—Interscience, New York
   (1971), p. 41.
7. K. F. J. Heinrich and M. A. Giles, NBS Technical Note 406 (1967).
8. K. F. J. Heinrich, NBS Technical Note 278 (1967).

9. L. Curgenven and P. Duncumb, Tube Investments Research Labs Report No. 303 (1971).

10. R. H. Barkalow, R. W. Kraft, and J. I. Goldstein, *Met. Trans.*, **3**, 919 (1972).

11. H. Yakowitz and K. F. J. Heinrich, *J. Res. NBS, A. Phys. Chem.* **73A**, 113 (1969).

12. H. Hahn, A. P. Divecha, P. Lare, and B. Dennison, Melpar Report. 5192 (1966), unpublished.

13. H. Yakowitz, W. D. Jenkins, and H. Hahn, *J. Res. NBS, A. Phys. Chem.*, **72A**, No. 3 (1968).

14. O. R. Bergman, G. R. Cowan, and A. H. Holzman, *Trans. Met. Soc. AIME*, **236**, 646 (1966).

15. B. Z. Weiss, *Z. Metallkunde*, **62**, 159 (1971).

16. H. Yakowitz and K. F. J. Heinrich, *Mikrochim. Acta*, **5**, 182 (1968).

17. H. Yakowitz, in *Fifty Years of Progress in Metallographic Techniques*, ASTM-STP 430, Am. Soc. Testing Materials (1968), p. 383.

18. M. L. Picklesimer and G. Hallerman, Report ORNL-TM-1591 (1966).

19. R. Kiessling and N. Lange, *Non-Metallic Inclusions in Steel*, Special Report 90, The Iron and Steel Institute, London (1964).

20. E. C. W. Perryman, *Metal Industry, London*, **79**, 23, 71, 111, 131 (1951).

21. M. L. Picklesimer, Report ORNL-2296 (1957).

22. ASTM, *Methods of Metallographic Specimen Preparation*, ASTM-STP 285 (1960).

23. R. L. Anderson, *Revealing Microstructures in Metals*, Westinghouse Research Laboratories, Scientific Paper 425-COOO-P2 (1961).

24. D. E. Cadwell and P. W. Weiblen, *Economic Geology*, **60**, 1320 (1965).

25. G. L. Kehl, *Principles of Metallographic Laboratory Practice*, 3rd. ed., McGraw-Hill, New York (1949).

26. C. M. Taylor and A. S. Radtke, *Economic Geology*, **60**, 1306 (1965).

27. W. Tegart, *Electrolytic and Chemical Polishing of Metals in Research and Industry*, 2nd rev. ed., Pergamon Press, London and New York (1959).

28. A. T. Nelms, *Energy Loss and Range of Electrons and Positions*, National Bureau of Standards, Circular 577 (1956).

29. I. Adler, in *X-Ray and Electron Probe Analysis*, ASTM-STP 349, Am. Soc. Testing Materials (1963), p. 183.

30. J. V. Smith, "Production of X-Rays," notes of a course taught at California Institute of Technology, Pasadena, California (1965).

31. J. I. Goldstein, R. E. Hanneman, and R. E. Ogilvie, *Trans. Met. Soc. AIME*, **233**, 812 (1965).

32. J. I. Goldstein, F. J. Majeske, and H. Yakowitz, in *Advances in X-Ray Analysis*, Vol. 10, Plenum Press, New York (1967), p. 431.

33. R. E. Michaelis, H. Yakowitz, and G. A. Moore, *J. Res. NBS*, **68A**, 343 (1964).

34. K. F. J. Heinrich, R. L. Myklebust, S. D. Rasberry, and R. E. Michaelis, NBS Special Publication 260-28 (1971).

35. H. Yakowitz, R. E. Michaelis, and D. L. Vieth, in *Advances in X-Ray Analysis*, Vol. 12, Plenum Press, New York (1969), p. 418.

36. H. Yakowitz, C. E. Fiori, and R. E. Michaelis, NBS Special Publication 260-22 (1971).

37. H. Yakowitz, D. L. Vieth, K. F. J. Heinrich, and R. E. Michaelis, in *Advances in X-Ray Analysis*, Vol. 9, Plenum Press, New York (1966), p. 289.

38. H. Yakowitz, A. W. Ruff, Jr., and R. E. Michaelis, NBS Special Publication 260-43 (1972).

39. J. I. Goldstein and R. E. Ogilvie, *Geochim. Cosmochim. Acta*, **27**, 623 (1963).

40. J. I. Goldstein and H. J. Axon, *Naturwiss.*, **60**, 313 (1973).

41. K. Keil and C. A. Andersen, *Geochim. Cosmochim. Acta*, **29**, 621 (1965).

42. R. H. Barkalow, R. W. Kraft, and J. I. Goldstein, *Met. Trans.* **3**, 919 (1972).

43. C. Matano, *Japan. J. Phys.*, **8**, 109 (1932-3).

44. E. Lifshin and R. E. Hanneman, General Electric Research Laboratory Reports 65-RL-3944M (1965) and 66-C-250 (1966).

45. J. I. Goldstein and R. E. Ogilvie, in *X-Ray Optics and Microanalysis*, (R. Castaing, P. Deschamps, and J. Philibert, ed.), Hermann, Paris (1966), p. 594.

46. A. S. Doan, Jr. and J. I. Goldstein, *Met. Trans.*, **1**, 1759 (1970).

47. J. Rucklidge, in *Proceedings of the Sixth International Conference on X-Ray Optics and Microanalysis* (G. Shinoda, K. Kohra, and T. Ichinokawa, eds.) University of Tokyo Press (1972), p. 743.

48. H. Yakowitz and K. F. J. Heinrich, *Metallography*, **1**, 55 (1968).

XII

SPECIAL TECHNIQUES IN THE X-RAY ANALYSIS OF SAMPLES

J. I. Goldstein and J. W. Colby

The x-ray signals obtained from the SEM–EPMA are most often used either to identify the elements present in a sample or to measure the relative or actual amounts of these elements in localized areas of the sample. The methods for determining the presence of a given element by the wavelength- or energy-dispersive method have been described in previous chapters. Similarly, the methods of scanning x-ray analysis and quantitative analysis have been discussed in some detail (Chapters IX–XI).

Several specific types of quantitative x-ray data are, however, difficult to obtain in practice. Often more sophisticated methods of sample preparation and/or data analysis are needed. Among these techniques are light element analysis, the measurement of precision and sensitivity and of compositions at interfaces, soft x-ray analysis, and the analysis of thin films. This chapter will discuss each of these techniques in order, outline how they may be used, and provide examples of their application.

I. LIGHT ELEMENT ANALYSIS

Quantitative x-ray analysis of the long-wavelength $K\alpha$ lines of the light elements (Be, B, C, N, O, and F) as well as of long-wavelength $L\alpha$ lines

J. I. GOLDSTEIN—Metallurgy and Materials Science Department, Lehigh University, Bethlehem, Pennsylvania
J. W. COLBY—Bell Telephone Laboratories, Allentown, Pennsylvania

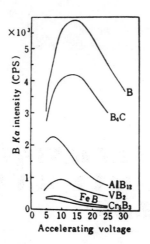

FIGURE 1.   Boron $K\alpha$ intensity in counts per second vs. operating voltage E_0
for several borides (from Shiraiwa *et al.*[1]).

(Ti, Cr, Mn, Fe, Co, Ni, Cu, and Zn) is difficult. The attenuation of the
primary radiation is large when these long-wavelength, low-energy x-rays
are measured. Furthermore, the correction models developed for quantitative analysis may not be applicable in the light element range. A large
absorption correction is usually necessary; unfortunately, the mass absorption coefficients for long-wavelength x-rays are not well known.

One can reduce the effect of absorption by choosing to analyze using
low operating voltages, E_0, and by using high x-ray takeoff angles, Ψ. The
higher the takeoff angle of the instrument, the shorter will be the path
length for absorption within the specimen. The penetration of the electron
beam is decreased when lower operating voltages are used, and x-rays are
produced closer to the surface (see Chapter III). Figure 1 shows the variation of boron $K\alpha$ intensity with voltage E_0 for several borides.[1] A maximum in the boron $K\alpha$ intensity occurs when E_0 is 10–15 kV, depending on
the sample. This maximum is caused by two opposing factors: (1) an increase in x-ray intensity due to increasing voltage, and (2) an increase in
absorption due to the fact that x-rays are produced deeper in the sample
as incident energy increases. One factor just offsets the other at the maximum intensity. Light element analysis for a given sample can be carried out
using an operating voltage equivalent to this maximum intensity. Even if
one uses these procedures, selecting the optimum E_0 and maximizing Ψ, the
effect of absorption within the sample is still significant. Other considerations for light element analysis include overlapping x-ray peaks, chemical
bonding shifts, surface contamination on the specimen, and availability of

standards. These problems will be considered in some detail in the following material.

Reliable quantitative results can be obtained by comparison with standards whose composition is close to that of the specimen. Figure 2 shows the calibration curves for the x-ray analysis of carbon $K\alpha$ using 10 kV operating voltage in standard alloys of Fe, Fe–10% Ni, and Fe–20% Ni each containing specific concentrations of carbon.[2] A lead stearate dodecanoate analyzing crystal with a d spacing of 50.15Å was used. The carbon intensity ratio is the carbon $K\alpha$ line intensity from a given alloy standard less its background, divided by the line intensity from a Cr_3C_2 carbon standard less its background. At a given carbon level, the addition of Ni to the steel standard decreases the C intensity ratio. The presence of Ni in iron probably lowers the C $K\alpha$ intensity because it has a greater mass absorption coefficient for C $K\alpha$ than does iron. Figure 3 shows a C gradient in a carburized gear steel containing about 3.5 wt % Ni[2] The gear steel was heat-treated at 925°C for 3 hr in a high carbon atmosphere. In this process, C enters the steel at the surface and diffuses by a mass transport mechanism into the body of the part. The carbon distribution was determined by stepping the sample under the beam and recording the C intensity as a function of distance. The calibration curves

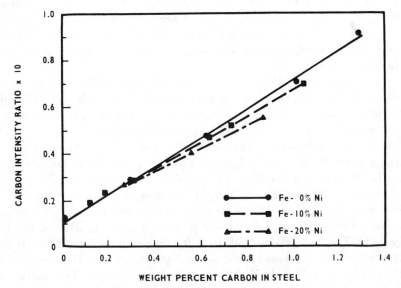

FIGURE 2.   Calibration curves for microprobe analysis of carbon in nickel steels (from Fisher and Farningham[2]).

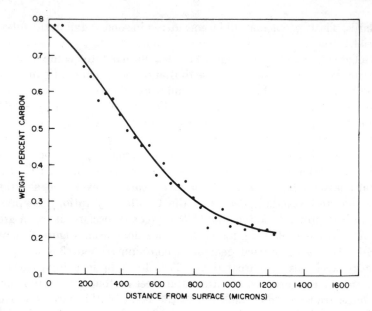

FIGURE 3.   Carbon distribution in a carburized nickel gear steel. Carbon is measured from the surface of the steel into the bulk (from Fisher and Farningham[2]).

in Figure 2 were used to determine the C concentration levels. Diffusion occurred over about 1200 μm, at which point the original C level of the gear steel (0.18 wt %) was reached. This carbon-enriched case increases the hardness of the steel, the desired property for the outer edge of a steel part to be used as a gear.

Complications in light element analysis may arise because of the presence of the L spectra from heavier metals. Duncumb and Melford[3] have shown that even if such overlapping occurs, a qualitative analysis can be obtained. In steels, for example, titanium carbonitride inclusions, TiNC, which are probably solid solutions of TiN and TiC, are found. Figure 4 shows a comparison of the Ti L spectra obtained from TiN, TiC, and pure Ti. The titanium carbonitride phase gave a more intense peak at the Ti Ll wavelength than that of pure Ti. This peak contains mainly Ti Ll whose peak is at 31.4 Å (18.3° θ) together with a small amount of nitrogen $K\alpha$ emission indistinguishable from it with a peak at 31.6 Å (18.5°θ) (Figure 4). The Ti $L\alpha$ line at 27.4 Å (16.0°θ) is heavily absorbed by nitrogen and is about one-third as intense as that from pure Ti. The titanium Ll emission, however, is only slightly absorbed by nitrogen. The analysis of this type of inclusion may appear to be impossible, but Duncumb and

Melford[3] analyzed the Ti content by using the $K\alpha$ radiation and analyzed the C content by use of the TiC standard (Figure 4). The results indicate about 80 wt % Ti and 4 wt % C. Meaningful analysis for nitrogen was not possible, for reasons already stated. Nitrogen composition was obtained by difference from 100%. An analysis procedure such as this may have to be used if overlapping lines occur and if such interferences cannot be eliminated by pulse height analysis (see Chapter VII).

For the light elements, the x-ray emission spectra consist mainly of a single band produced by the transition of a valence electron to a vacancy in the K shell. As pointed out by Fischer and Baun,[4] the valence electrons are the ones most affected by chemical combination and the emission band can and does reflect the often large effects of changes in chemical bonding between atoms. These changes are signified by wavelength shifts, by increases or decreases in the relative intensities of various lines or bands, and by alteration of shape. Such shifts may cause problems when quantitative light element analysis is desired.

Figure 5 shows the C K band from carbon deposited by the electron beam as well as the C K band of electrode-grade graphite and various carbides.[5] The wavelength shift for the carbides relative to graphite is significant and can easily be observed with the wavelength-dispersive spectrometer. This shift is important since, in order to accomplish a quantitative analysis, measurement of $K\alpha$ peak intensity must be made at the position of

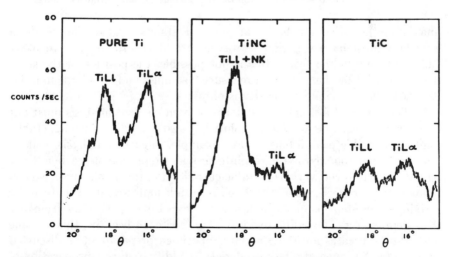

FIGURE 4.   Comparison of the Ti L spectra (intensity vs. diffraction angle θ) obtained from pure Ti (left), TiN (center), and TiC (right) at 10 kV operating potential (from Duncumb and Melford[3]).

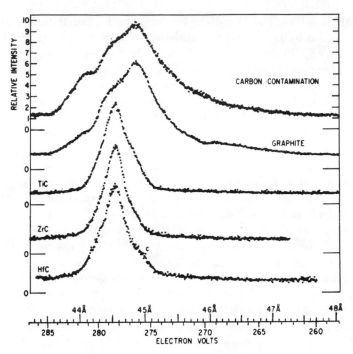

FIGURE 5.   C K spectra at 4 kV from carbon deposited by the electron beam, electrode grade graphite, and various carbides (from Holliday[5]).

maximum intensity for both sample and standard alike. Therefore a standard must have a negligible wavelength shift with respect to other standards and unknowns. If this is not possible, the position of the spectrometer must be changed when measuring samples and standards in order to assure that maximum intensity is obtained.

The choice of a primary standard for light element analysis must not only show a negligible wavelength shift but also must give a strong, stable, and reproducible peak intensity. For example, in carbon analysis, neither spectrographic nor pyrolytic graphite provides reproducible carbon x-ray standards. However, various metallic carbides do provide adequate carbon standards. For steels, cementite Fe_3C is an adequate standard and can be produced by step-cooling a high-purity Fe–~1 wt $\%$ C alloy from the high-temperature austenite region. The ferrite, α-bcc, in this sample can also be used as an iron standard. Specimen preparation, as discussed in Chapter XI, must also be considered. In addition, during preparation of the specimen surfaces, abrasives containing the light elements should either be avoided or, if this is impractical, the sample should be carefully cleaned to ensure total removal of these materials. After the final polish, polishing material may be removed by ultrasonic cleaning. One may not wish to etch a sample since etching can leave a residual contamination layer. Ideally, specimens should be placed in the instrument immediately after preparation. If this is inconvenient, storage in a vacuum dessicator is usually satisfactory.

A sample subjected to electron bombardment in a diffusion pumped vacuum gradually becomes covered with a "contamination" layer due to polymerization, under the action of the beam, of organic matter adsorbed on the surface.[6] The organic molecules come from the oil vapors of the vacuum pumps and the outgassing of any organic material present in the instrument. The effect is not very troublesome unless the deposited layer absorbs the emitted x-rays to a great extent. For light element radiation, particularly that of Be, B, and C, the absorption of the long-wavelength x-radiation can be severe. The problem, in the case of carbon analysis, is increased because the "contamination" layer contains carbon in large measure. This circumstance leads to the observation of an increasing carbon $K\alpha$ count as a function of beam impingement time.

Two methods have been used to avoid the contamination layer. Castaing and Deschamps[7] as early as 1954 showed that directing a low-pressure jet of gas onto the specimen at the region bombarded by the electron beam suppresses contamination. When air is introduced into the vicinity of the sample, the hot carbon deposit is oxidized and the high-energy electron beam acts to produce an ion bombardment-sputtering condition. Air jets have been installed on various SEM–EPMA instruments and can be built for a nominal cost. The design of such an air jet has been described in detail by Duerr and Ogilvie.[8] Not only is the contamination rate reduced nearly to zero, but previously adsorbed surface layers are often removed. Figure 6 shows a plot of C intensity versus time for an iron sample bombarded by the electron beam. The C intensity increases with time but after the air jet is turned on, the contamination layer is removed and the C intensity is reduced to background levels. The line through the points in Figure 2 for the Fe–Ni–C standards does not pass through the origin. The carbon intensity ratio (background already subtracted) does not go to zero as expected. Probably, a very thin carbon film present on the standard is the cause. Apparently the growth of the carbon film due to the breakdown of hydrocarbons by the electron beam is just balanced by its removal rate by the air jet.[1] This carbon film can attain different thicknesses, depending on the quality of the vacuum in a particular system. Room air dried by passing through a desiccant, appears to work very well. Inert gases also are effective.[9,10] For normal operation of the jet using room air, the operator allows enough air into the operating vacuum of the EPMA to raise the vacuum to about 3×10^{-4} Torr. Although the vacuum level in the EPMA

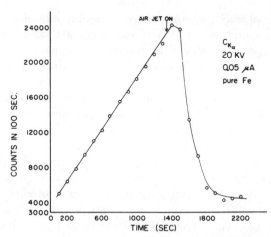

FIGURE 6.   Operation of air jet to eliminate C contamination on a pure Fe sample at 20 kV. Carbon intensity is plotted versus time.

is degraded by this process, the accompanying decrease in filament life is not severe. Another means of reducing contamination is to provide a surface within the SEM–EPMA that is cold relative to the surface of the specimen. Organic molecules will then tend to collect on the colder surface rather than on the specimen. The cold surface or cold finger must, however, be placed very close to the specimen. Cold fingers have been installed on various instruments and have effectively reduced the contamination rate to nearly zero. In one case,[11] both an air jet and a liquid N_2 cold finger have been used. Light element quantitative analysis probably ought not to be attempted without some kind of decontamination device or use of a high-vacuum system to obtain the data. The decontamination methods discussed also may be useful in reducing C contamination during scanning in a SEM.

It is difficult to obtain accurate light element analyses in multi-component alloys. Multicomponent standards are not easily produced and the major input parameters for the calculation schemes, e.g., mass absorption coefficients of the light elements in heavy element matrices, are not well known. As an example of the problem, Table I lists the carbon $K\alpha$ mass absorption coefficients used for a study of carbide compositions in tool steels,[12] those used for a study of binary carbides,[1] and the values from the tables of Henke and Ebisu[13]. The discrepancies are significant, particularly for Mo and V. Calculation of composition, using long-wavelength x-rays, is rarely better than $\pm10\%$ of the amount present in complex alloys.

TABLE I. Mass Absorption Coefficients for C $K\alpha$

Absorber	μ/ρ for C $K\alpha$ (g/cm²)		
	Barkalow *et al.*[12]	Shiraiwa *et al.*[1]	Henke and Ebisu[13]
Fe	15,000	14,300	13,300
W	18,000	—	18,750
Mo	19,000 *a*	—	32,420
Cr	10,000	11,000	10,590
V	15,000	9,300	8,840
C	2,300	2,270	2,373

a Estimated.

II. PRECISION AND SENSITIVITY IN X-RAY ANALYSIS

One of the most important objectives in quantitative x-ray analysis is to obtain not only the analysis of a sample, but also the sensitivity or precision of that measurement. By precision of an analysis, we mean the scatter among the test results that occurs without any prior assumptions concerning the tested population. This scatter exists as a result of the nature of the x-ray measurement process. Conceivably, a result could be obtained in which the errors associated with the calculation of composition (the accuracy of the analysis, Chapters IX and X) are smaller than the precision of the measurement itself. Therefore the number of measurements and the measurement times for a particular analysis should be chosen so as to ensure a precision better than the analytical error expected from the computational schemes. The following sections will discuss methods used for obtaining precision and sensitivity in quantitative analysis. The objective of this section will be to point out practical analysis strategy, i.e., methods in which the analyst can ascertain, before analysis, the level of precision and sensitivity that can be expected from a given analysis method.

A. Statistical Basis for Calculating Precision and Sensitivity

X-ray production is statistical in nature; the number of x-rays produced from a given sample and interacting with radiation detectors is completely random in time but has a fixed mean value. The distribution or histogram of the number of determinations of x-ray counts from one point on a sample vs. the number of x-ray counts for a fixed time interval can be closely approximated by a continuous normal (Gaussian) distribution. Individual x-ray count results from each sampling lie upon a unique Gaussian curve for which the standard deviation is the square root of the mean ($\sigma_c = \bar{N}^{1/2}$). Figure 7 shows such a Gaussian curve for x-ray emission spectrography and the standard deviation $\sigma_c = \bar{N}^{1/2}$ obtained under ideal conditions. Here $\bar{N}$ is considered to be the most probable value of N, the total number of counts obtained in a given time t. Inasmuch as σ_c results from fluctuations that cannot be eliminated as long as quanta are counted, this standard deviation σ_c is the *irreducible* minimum for x-ray emission spectrography. Not only is it a minimum, but fortunately it is a predictable minimum. The variation in percent of total counts can be given as $(\sigma_c/\bar{N})100$. For example, to obtain a number with a minimum of a 1% deviation in N, at least 10,000 counts must be accumulated.

As Liebhafsky *et al.*[14] have pointed out, the actual standard deviation of the experiment S_c is given by

$$S_c = \left[\sum_{i=1}^{n} (N_i - \bar{N}_i)^2 / (n-1) \right]^{1/2} \qquad (1)$$

where N_i is the number of x-ray counts for each determination i and

$$\bar{N}_i = \sum_{i=1}^{n} N_i \Big/ n \qquad (2)$$

where n is the number of determinations of i. The standard deviation S_c equals σ_c *only* when operating conditions have been optimized. In most EPMA–SEM instruments, drift of electronic components and of specimen position (mechanical stage shifts) create operating conditions that are not necessarily ideal. The high voltage-filament supply, the lens supplies, and other associated electronic equipment may drift with time. After a speci-

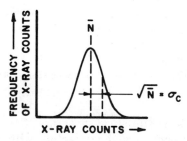

FIGURE 7.   Gaussian curve for x-ray emission spectrography and the standard deviation ($\sigma_c = \bar{N}^{1/2}$) obtained under ideal conditions (from Liebhafsky *et al.*[14]).

men is repositioned under the electron beam, a change in measured x-ray intensity may occur (1) if the effective "depth of focus" for the crystal x-ray spectrometers is less than the "depth of focus" of the light optical system in the EPMA or (2) if the takeoff angle Ψ of the specimen varies, as it easily may when an energy-dispersive detector in the SEM is used for analysis. In practice, for typical counting times of 10–100 sec/point, the actual standard deviation S_c is often about twice σ_c. If longer counting times are used, S_c/σ_c increases due to instrument drift. Only when counting times are short and the instrument is electronically stable does S_c approach σ_c. Besides the sample signal, sources of variation may also occur if data from reference standards and/or background standards are required.[15] These, as well as faulty specimen preparation, may also affect the precision of an analysis. Therefore, both instrumental factors and signal variations must be considered when the precision of an analysis is determined.

B. Sample Homogeneity

An analyst is often asked if a sample and/or a phase is homogeneous. In order to answer this question, the x-ray data obtained in a SEM–EPMA must be obtained so that it can be treated statistically. One can either establish criteria for homogeneity and apply them or one can measure the range of composition variation of a sample, for a certain confidence level, and report that number. Either method allows a more quantitative statement to be made than just a simple "yes" or "no" to questions concerning homogeneity. The following material discusses the methods for obtaining x-ray data, several homogeneity criteria that can be used, the calculation of the range and level of homogeneity, and finally a method for selecting operating conditions necessary to determine the level of homogeneity.

1. METHODS FOR OBTAINING X-RAY DATA

The usual procedure for investigating the homogeneity of a given sample or phase is first to determine by any suitable means whether inclusions and secondary phases are present. After this is done, a preliminary check on homogeneity of the matrix can be made by obtaining x-ray output as a function of position as the specimen is mechanically driven under the beam (mechanical line scans). These results will point out any gross inhomogeneities ($>10\%$ of the amount present) at the 1–100-μm level. The possibility of gross inhomogeneities on the 1 mm to 1 cm level can be investigated by conventional means, such as x-ray fluorescence methods. To check for inhomogeneities of less than 10% of the amount present, a static probe is used and x-ray quanta are accumulated at each point. The procedure normally used is to accumulate data at many points, usually between 10 and 200 spread across the sample.

Several different methods have been devised for obtaining such data. The objective in all these cases is to obtain data from enough points to be representative and to obtain a minimum number of counts per point for each element of interest. This must be done within a reasonable time interval to avoid stability problems due to instrument drift. Systems for automatic displacement of the electron beam[16,17] or the specimen stage by some sort of automation equipment, such as a matrix generator or programmed stepping motors, are available. A matrix of a single string of points along a line or a square or rectangular area can therefore be developed. The data must be handled efficiently (computer, multichannel analyzer, or tape) and considerable computation is usually needed to transform the mass of data into a format that permits meaningful interpretation.

Data are usually collected with crystal spectrometers. If the electron beam is displaced far enough from the electron optical axis of the SEM–EPMA, the emitted x-rays will no longer be on the focusing circle of the x-ray spectrometer, and a loss of x-ray intensity will occur. Typically, electron beam displacements greater than 50 μm will cause defocusing effects. An interesting feature of the solid state detector as compared to the crystal spectrometer system is that the counting rates are not altered by displacement of the beam on the specimen. Therefore, the use of a solid state detector for the analysis of points in a raster as large as 1 mm² simplifies both instrument operation and interpretation of data by eliminating a major source of error.[16]

2. Criteria for Homogeneity

A simplified criterion that has been used to establish the homogeneity of a phase or a sample is that all the data points n must fall within the $\overline{N} \pm 3\overline{N}^{1/2}$ limits.[18-20] If this criterion is satisfied, one assumes then that the sample is homogeneous. The variation

$$(\pm 3\overline{N}^{1/2}/N)100 \ (\%) \tag{3}$$

for the element of interest in the sample represents the level of homogeneity in percent that is measured for the sample, remembering that there must be an irreducible minimum level due to the fact that x-ray production is statistical in nature. If 100,000 counts are accumulated at each point in a sample and all these points fall within the limits $\overline{N} \pm 3\overline{N}^{1/2}$, the sample is homogeneous and the level of homogeneity is, according to equation (3), $\pm 0.95\%$. A level of homogeneity of $\leq \pm 1\%$ is often desired. If the concentration in the sample C is 10 wt %, the range of homogeneity, that is, the minimum variation of concentration that can be validly measured, is ± 0.1 wt %.

If determinations are made of the actual standard deviation S_c, more meaningful criteria for homogeneity can be applied. Experience on many different types of homogeneous samples, obtained by performing repeated analyses in small areas, refocusing at each point, and counting for 10–100 sec per point, demonstrates that S_c is usually about twice the value of σ_c. Presumably if an uncharacterized sample is investigated by the methods for obtaining x-ray data as outlined previously, the sample might be assumed homogeneous if $S_c \leq 2\sigma_c = 2\bar{N}^{1/2}$. A criterion for homogeneity based on this observation has been developed[21] as follows: if $S_c/2\sigma_c \leq 1$, the crystal is homogeneous; if $1 \leq S_c/2\sigma_c \leq 2$, the homogeneity is uncertain; and if $S_c/2\sigma_c > 2$, the crystal is inhomogeneous. The level and range of homogeneity can be calculated by statistical techniques as discussed in the next section.

3. Range and Level of Homogeneity

A more exacting determination of the range (wt %) and level (%) of homogeneity involves the use of (a) the standard deviation S_c of the measured values and (b) the degree of statistical confidence in the determination of $\bar{N}$. The standard deviation includes effects arising from the variability of the experiment, e.g., instrument drift, x-ray focusing errors, and x-ray production. The degree of confidence used in the measurement states that we wish to avoid a risk α of rejecting a good result a large percentage (say 95 or 99%) of the time. The degree of confidence is given as $1 - \alpha$ and is usually chosen as 0.95 or 0.99, that is, 95 or 99%. The use of a degree of confidence means that we can define a range of homogeneity in wt % for which we expect, on the average, only α (5% or 1%) of the repeated random points to lie outside the range.

The range of homogeneity in wt % for a degree of confidence $1 - \alpha$ is

$$W_{1-\alpha} = \pm C \left(\frac{t_{1-n}^{1-\alpha}}{n^{1/2}}\right) \frac{CV}{100} = \pm C \left(\frac{t_{1-n}^{1-\alpha}}{n^{1/2}}\right) \frac{S_c}{\bar{N}} \qquad (4)$$

where C is the true weight fraction of the element of interest, n is the number of measurements, $\bar{N}$ is the average number of counts accumulated at each measurement, CV is the percent coefficient of variation, and $t_{n-1}^{1-\alpha}$ is the student t value for a $1 - \alpha$ confidence level and for $n - 1$ degrees of freedom.[22] Student t values for t_{n-1}^{95} and t_{n-1}^{99} for various degrees of freedom $(n - 1)$ are given in Table II.[23] It is clear from Table II that at least four

TABLE II. Values of Student t Distribution for 95 and
99% Degrees of Confidence[23]

n	$n-1$	t_{n-1}^{95}	t_{n-1}^{99}
2	1	12.71	63.66
3	2	4.304	9.92
4	3	3.182	5.841
8	7	2.365	3.499
12	11	2.201	3.106
16	15	2.131	2.947
30	29	2.042	2.750
∞	∞	1.960	2.576

measurements, $n = 4$, should be made to establish the range of homogeneity.
If less than four measurements are made, the value of $W_{1-\alpha}$ will be meaningless.

The level of homogeneity, or homogeneity level, for a given confidence
level $1 - \alpha$ is given by

$$\pm W_{1-\alpha}/C = \pm (t_{n-1}^{1-\alpha})S_c(100)/(n^{1/2}\,\overline{N})\ (\%) \tag{5}$$

It is more difficult to measure an equivalent level of homogeneity as the concentration present in the sample decreases. Although $W_{1-\alpha}$ is directly
proportional to C, the value of $S_c/\overline{N}$ will increase as C and the number of
x-ray counts per point will decrease. To obtain the same number of x-ray
counts per point, the time of the analysis must be increased.

An example of the application of the statistical equations for the
range of homogeneity [equation (4)] and the level of homogeneity [equation
(5)] can be obtained from a study of NBS Standard Reference Material 479,
an austenitic iron–chromium–nickel alloy.[21] Using a 40-sec counting
interval on the specimen, 900,000 Fe counts, 340,000 Cr counts, and 270,000
Ni counts were obtained at each point. In no case was the coefficient of
variation $(S_c/\overline{N})100$ greater than 1.5% for any element present. Therefore,
if 1.5% is used as a conservative value for the coefficient of variation, and,
if n is set equal to 16 in equation (4), the confidence interval for $1 - \alpha =
0.99$ (99%) is

$$W_{99} = \pm C(2.947/\sqrt{16})(0.015) = \pm 0.011C \text{ wt } \%$$

or the level of homogeneity is $\pm 1.1\%$ of C, the weight fraction of the
element of interest. The time to collect data for the 16 points was not
prohibitive, being about 30 min.

4. Operating Procedures—Predicting the Level of Homogeneity

Using the statistical analysis presented in the previous section, it is now possible to develop a method to select operating procedures for measuring a desired level of homogeneity. An expected level of homogeneity for a sample can be calculated if the value of S_c can be determined or approximated before the analysis is begun. Assuming $S_c \simeq 2\sigma_c$ to be the case for most analyses, the range of homogeneity $W_{1-\alpha}$ can be calculated as follows:

$$W_{1-\alpha} = \pm \frac{Ct_{n-1}^{1-\alpha}}{n^{1/2}} \frac{S_c}{\bar{N}} \simeq \pm \frac{Ct_{n-1}^{1-\alpha}2\sigma_c}{n^{1/2}\bar{N}} \simeq \pm \frac{2Ct_{n-1}^{95}}{n^{1/2}\bar{N}^{1/2}} \text{ wt } \% \qquad (6)$$

The predicted level of homogeneity is therefore

$$\pm W_{1-\alpha}/C \simeq \pm (2t_{n-1}^{95}/n^{1/2}\bar{N}^{1/2})(100) \% \qquad (7)$$

Graphs can thus be constructed for the expected level of homogeneity versus accumulated counts per point, the number of points analyzed n, and the confidence level $1 - \alpha$. One such graph for a confidence level of 95% is

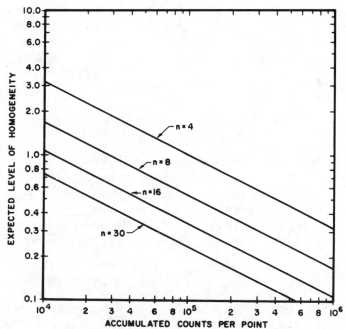

FIGURE 8.   Expected level of homogeneity versus accumulated counts per point for a confidence level of 95%.

given in Figure 8, which shows that to measure a 1% expected level of homogeneity, about 100,000 counts must be accumulated per point if $n = 4$, 28,000 counts per point for $n = 8$, and 11,500 counts per point for $n = 16$. It appears that as the number of point analyses n increases, less total x-ray counts, as given by the product $n\overline{N}$, must be accumulated to obtain the same level of homogeneity. It is therefore advantageous to analyze more points per sample because (a) more areas are analyzed and (b) the total time for analysis can be decreased. In fact if one is willing to invest enough analysis time to accumulate 10^6 total counts ($n\overline{N}$), the expected degree of homogeneity can be improved (decreased) as more points are measured; for example, 0.65% for $n = 4$, 0.47% for $n = 8$, 0.43% for $n = 16$, etc. For homogeneity measurements, 16–30 analysis points would appear to be sufficient for most analytical purposes.

C. Analytical Sensitivity

Analytical sensitivity concerns the ability to distinguish, for a given element, between two compositions C and C' that are nearly equal. X-ray counts for both compositions $\overline{N}$ and $\overline{N}'$ therefore have a similar statistical variation. If one determines two compositions C and C' by n repetitions of each measurement taken for the same fixed time interval, then these two values are significantly different at a certain degree of confidence $1 - \alpha$ if

$$\overline{N} - \overline{N}' \geq \sqrt{2}\, t_{n-1}^{1-\alpha}\, S_c/n^{1/2} \tag{8}$$

and

$$\Delta C = C - C' \geq \frac{\sqrt{2}\, C t_{n-1}^{1-\alpha}\, S_c}{n^{1/2}(\overline{N} - \overline{N}_{\mathrm{B}})} \tag{9}$$

in which C is the composition of one element in the sample, $\overline{N}$ and $\overline{N}_{\mathrm{B}}$ are the average number of x-ray counts of the element of interest for the sample and the continuum background on the sample, respectively, $t_{n-1}^{1-\alpha}$ is the "student factor" dependent on the confidence level $1 - \alpha$ (Table II), and n is the number of repetitions. Ziebold[15] has shown that the analytical sensitivity for a 95% degree of confidence can be approximated by

$$\Delta C = C - C' \geq \frac{2.33}{n^{1/2}} \frac{C\sigma_c}{(\overline{N} - \overline{N}_{\mathrm{B}})} \tag{10}$$

The above equation represents an estimate of the maximum sensitivity that can be achieved when signals from both compositions have their own errors but instrumental errors are disregarded. Since the actual standard deviation S_c is usually about two times larger than σ_c, ΔC is in practice approximately twice that given in equation (10).

If $\overline{N}$ is much larger than $\overline{N}_B$, equation (10) can be rewritten as

$$\Delta C = C - C' \geq 2.33 C/(n\overline{N})^{1/2} \tag{11}$$

and the analytical sensitivity in percent that can be achieved is given as

$$\Delta C/C \; (\%) = 2.33(100)/(n\overline{N})^{1/2} \tag{12}$$

For an analytical sensitivity of 1%, $\geq 54{,}290$ accumulated counts, $n\overline{N}$ from equation (12), must be obtained from the sample. If the concentration C is 25 wt %, then $\Delta C = 0.25$ wt %, and if the concentration C is 5 wt %, then $\Delta C = 0.05$ wt %. Although the analytical sensitivity improves with decreasing concentrations, it should be pointed out that the x-ray intensity decreases directly with the reduced concentration. Therefore longer counting times become necessary in order to retain a 1% sensitivity level.

Equation (12) is particularly useful for predicting necessary procedures to obtain the sensitivity desired in a given analysis. If a concentration gradient is to be monitored over a given distance in a sample, it is important to predict how many data points should be taken and how many x-ray counts should be obtained at each point. For example, if a gradient from 5 to 4 wt % occurs over a 25-μm region, and 25 1-μm steps are taken across the gradient, the change in concentration per step is 0.04 wt %. Therefore ΔC, the analytical sensitivity at a 95% degree of confidence, must be ≤ 0.04 wt %. Using equation (11), since $\overline{N}$ is much larger than $\overline{N}_B$, $n\overline{N}$ must be at least 85,000 accumulated counts per step. If only ten 2.5-μm steps are used across the gradient, the change in concentration per step is 0.1 wt % and now $n\overline{N}$ need only be $\geq 13{,}600$ accumulated counts per step. By measuring 10 as opposed to 25 steps, the analysis time is cut down much more than the obvious factor of 2.5 since the number of required accumulated counts per step, due to sensitivity requirements, also decreases.

D. Trace Element Analysis

As the elemental composition C approaches the order of 0.1 wt % in EPMA analysis, $\overline{N}$ is no longer much larger than $\overline{N}_B$. This composition range, below 0.1 wt % (1000 ppm), is often referred to as the trace element analysis range. For the light elements, the trace element range begins at about the 1 wt % level (10,000 ppm). The analysis requirement in trace element analysis is to detect significant differences between the sample and the continuum background generated from the sample.

To develop a useful procedure for trace detection, we need a criterion that will guarantee that a given element is present in a sample. This criterion can be called the "minimum amount guaranteed observable" or the "detectability limit" DL. This so-called detectability limit is governed by the minimum value of the difference $\overline{N} - \overline{N}_B$ that can be measured with statistical significance. Liebhafsky et al.[24] have discussed this problem in some detail. They use the operating rule that one may safely ignore the occurrence of errors greater than three standard errors $(3S_c)$ in "guaranteeing" an analytical result. In a Gaussian distribution, only 0.135% of the values will exceed the mean by more than three standard deviations. To guarantee that an element is present, the value of $\overline{N}$ must then be greater than the value of the background $\overline{N}_B$ by $3\overline{N}_B^{1/2}$, assuming operating conditions are optimized. However, since the values of $\overline{N}$ and $\overline{N}_B$ are comparable, each individual count is subject to the same statistical fluctuations. To determine the effect of these fluctuations on the difference $\overline{N} - \overline{N}_B$, one must combine the errors for both $\overline{N}$ and $\overline{N}_B$. The total standard deviation S_c for the two measurements $\overline{N}$ and $\overline{N}_B$ can be obtained from the standard error of a difference,[24] that is,

$$S_c = (\overline{N} + \overline{N}_B)^{1/2} \tag{13}$$

where each term under the square root is the square of the standard deviation for each measurement assuming operating conditions are optimized. Since $\overline{N}$ and $\overline{N}_B$ have very similar values in trace analysis,

$$S_c \simeq (2\overline{N}_B)^{1/2} \tag{14}$$

Since the DL is equal to $3S_c$, then

$$\text{DL} = 3(2\overline{N}_B)^{1/2} \tag{15}$$

By analogy with equation (8), we can also define the detectability limit DL as $(\overline{N} - \overline{N}_B)_{DL}$ for trace analysis as

$$(\overline{N} - \overline{N}_B)_{DL} \geq \sqrt{2}\, t_{n-1}^{1-\alpha} S_c / n^{1/2} \tag{16}$$

where S_c is essentially the same for both the sample and background measurement. In this case, we can define the detectability limit at any confidence level $1 - \alpha$ (Table II) the analyst chooses. The 95 or 99% confidence level is usually chosen in practice. In most cases, equations (15) and (16) yield quite similar values of DL. If we assume for trace analysis that the x-ray calibration curve of intensity vs. composition is expressed as a linear function, then C, the unknown composition, can be related to $\overline{N}$ by the equation

$$C = [(\overline{N} - \overline{N}_B)/(\overline{N}_S - \overline{N}_{SB})]C_S \tag{17}$$

where $\overline{N}_S$ and $\overline{N}_{SB}$ are the mean counts for the standard and the continuum background for the standard, respectively, and C_S is the concentration in wt % of the element of interest in the standard. The detectability limit

C_{DL}, that is, the minimum concentration that can be measured, can be calculated by combining equations (13) and (14) to yield

$$C_{\mathrm{DL}} = \frac{C_S}{\overline{N}_S - \overline{N}_{SB}} \frac{\sqrt{2} \; t_{n-1}^{1-\alpha} S_c}{\dot{n}^{1/2}} \qquad (18)$$

The relative error or precision in a trace element analysis is equal to C/C_{DL} and approaches $\pm 100\%$ as C approaches C_{DL}.

The background intensity $\overline{N}_B$ must be obtained accurately so that trace analysis can be accomplished [note equation (17)]. It is usually best to measure the continuum background intensity directly on the sample of interest. Other background standards may have different alloying elements or a different composition, which will create changes in absorption with respect to the actual sample. Also, such background standards may have different amounts of residual contamination on the surface, which is particularly harmful in carrying out light element trace analysis. The background intensity is obtained after a careful wavelength scan is made of the major peak to establish precisely the intensity of the continuum on either side of the peak. Spectrometer scans must be made to establish that these background wavelengths are free of interference from other peaks in all samples to be analyzed. It is extremely difficult to measure backgrounds using the energy-dispersive spectrometer with the accuracy needed to perform trace element analysis. (Measurements of continuum background with the EDS are discussed in Chapter VII.) Therefore, trace element measurements are almost always made in the EPMA–SEM with wavelength-dispersive spectrometers (WDS). The following example illustrates the difficulty in determining background values and hence in obtaining accurate trace element analyses.

In a study to determine the distribution of Ni and Co between a metal phase (fcc γ-Fe–Ni) and the mineral olivine, $(MgFe)_2SiO_4$, in lunar basalt rocks,[25] trace element analysis was necessary. For the analysis of Ni and Co in olivine, the operating conditions were selected as 15 kV and 0.2 μA sample current. By obtaining four determinations of Ni peak and background readings taken on each side of the Ni $K\alpha$ peak and by using 100-sec counting times, a detectability limit of 36 ppm was deduced for Ni and 25 ppm, for Co. Olivine crystals were traversed by moving the sample in steps under the electron beam in order to analyze metal inclusions and the areas of olivine adjacent to them. At each point Fe and Mg counts were collected, followed by minor element Ni and Co peak and background readings.

In Figure 9, the values for Ni peak intensity and background intensity on each side of the peak as a function of distance in a lunar olivine from the Apollo 12 sample 12004,8 are displayed. As observed, the count rate on the Ni $K\alpha$ line is greater than the mean background count rate. Note, how-

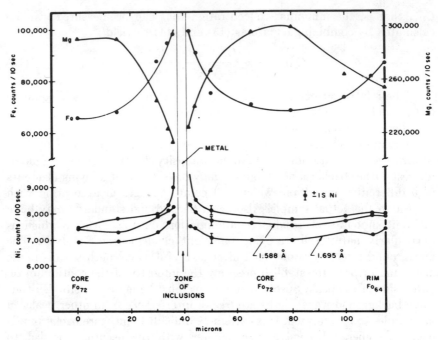

FIGURE 9.   Peak and background counts for Ni as a function of position within an olivine $(MgFe)_2SiO_4$ in lunar rock 12004,8. The Fe and Mg contents of the olivine crystal are also plotted. (From Hewins and Goldstein.[25])

ever, that the continuum background varies greatly with wavelength. This variation points out the necessity for fully characterizing the background–continuum radiation (see Chapter VII). The Ni concentration values in this complex crystal, described below, are as high as 300 ppm. As the Fe content and the average atomic number of the olivine increase, the background intensity of Ni also increases (Figure 9). Because of this increase in Ni background with increasing Fe–Mg content it was necessary to measure Ni background at every point. After careful background measurements were made, the calculated Ni content decreased from 312 ± 36 ppm in the high-Mg core to 207 ± 36 ppm in the low-Mg rim of the olvine crystal. Cobalt background readings had to be collected at each point; the spectral position for these readings was very close to the Co peak so as to avoid interference from the Fe $K\beta$ peak whose intensity varied with position in the crystals. When dealing with the trace element range, many subtle effects can occur, and these must be investigated before meaningful measurements can be made.

Ziebold[15] has shown the trace element sensitivity or the detectability limit to be

$$C_{DL} \geq 3.29a/(n\tau P \cdot P/B)^{1/2} \qquad (19)$$

where τ is the time of each measurement taken, n is the number of repetitions of each measurement, P is the pure element counting rate, P/B is the peak/background ratio of the pure element, i.e., the ratio of the counting rate of the pure element to the background counting rate of the pure element, and a relates composition and intensity of the element of interest through the Ziebold–Ogilvie[26] empirical relation (see Chapter IX).

To illustrate the use of this relation, the following values were used for calculating the detectability limit for Ge in iron meteorites.[27] The operating conditions were as follows:

operating voltage, 35 kV $\qquad \tau = 100$ sec

specimen current, 0.2 μA $\qquad n = 16$

$P = 150,000$ counts/sec $\qquad a = 1$

$P/B = 200$

With these numbers, (19) gives $C_{DL} \geq 15$ ppm; the experimental detectability limit obtained after calculating S_C and solving equation (18) was 20 ppm.[27] Counting times of the order of 30 min were found to be necessary to achieve this detectability limit. In measuring the carbon content of steels, detectability limits of the order of 300 ppm are more typical if one uses counting times of the order of 30 min, and the instrument is set up so as to operate at 10 kV with a 0.05-μA specimen current. The chief problem facing the analyst with respect to light element detectability limits is the huge amount of absorption occurring within the sample.

Several points should be considered when attempts are made to carry out x-ray measurements, with the wavelength dispersive spectrometer (WDS) in the trace element range, ≤ 0.1 wt %. The detectability limit C_{DL} must be minimized by careful selection of operating parameters. Equation (19) is useful for predicting operating conditions suitable for trace analysis. One needs to employ long counting times, high peak intensities, and a high peak-to-background ratio. Because of instrumental drift and sample contamination, a practical limit on the counting time is 15–30 min. Peak intensities can be raised by increasing the beam current; however, beam currents above 0.2 μA may heat the sample locally; electron-beam size increases as well. Unfortunately, the peak-to-background ratio cannot be reduced below a certain limit since continuum x-ray radiation is always produced. Pulse height analysis provides no additional advantage unless interfering lines are present. The peak-to-background ratio is often quite high, >100, for pure elements. With the specimen currents typically used, the peak intensity is also quite high, $>10^4$–10^5 counts/sec from the pure elements.

A criterion which is often used to compare wavelength and energy dispersive detectors is the product $P \cdot P/B$ in the Ziebold relation, equation (19). This product P^2/B is similar for both WDS and EDS systems. (See Chapter VII for a discussion of the two spectrometer systems and the P^2/B criterion.) One might be tempted to argue that the detectability limit, DL, is similar for both spectrometers and that trace element analysis can be accomplished in the SEM at much lower beam currents. However, as pointed out earlier, it is difficult to measure the background continuum intensity $\overline{N}_B$ accurately with the EDS (Chapter VII). Therefore detectability limits of 1000 ppm (0.1 wt %) are reasonable practical estimates for the energy-dispersive spectrometer.

III. X-RAY ANALYSIS AT INTERFACES

One of the major applications of quantitative EPMA analysis is the determination of compositions at or near phase interfaces. The results of these measurements can indicate the last temperature of equilibration of a given system, the composition of various phases in an equilibrium phase diagram, or the discontinuous change of the composition of a component near an interface. As discussed by Reed and Long,[28] there are three major effects which must be considered when measurements at or near interfaces must be made.

1. Spatial resolution—x-ray excitation volume. If there are variations in the concentration of the analyzed element within the volume excited by the electrons, the apparent concentration will be an average of the concentration within that volume. Therefore accurate measurements at phase boundaries where part of the beam is in both phases are difficult to obtain.

2. X-ray absorption. The absorption correction in quantitative analysis (Chapter IX) can only be calculated unambiguously for homogeneous specimens. An uncertainty may be introduced when the region through which the generated x-rays pass is of a different composition from the analyzed area.

3. X-ray fluorescence. When measurements are made near interfaces, not only may part of the electron beam excite both continuum and characteristic x-radiation in the major phase, but the x-radiation produced could possibly cause secondary fluorescence of elements in the second phase

across the boundary. The measured x-ray intensity may thus be modified by such a fluorescence effect.

In order to perform x-ray analysis at interfaces, each of these three major effects must be taken into proper account. The following sections will describe each of these effects in detail and suggest various methods designed to minimize the contribution of each to the resultant analysis.

A. Spatial Resolution

Since the volume in which primary x-rays are produced is much larger than the size of the focused electron beam, direct measurement of the phase boundary composition is impossible within one x-ray source size diameter of a two-phase interface. If concentration gradients near phase boundary interfaces are steep, meaningul extrapolations to these interfaces are also impossible to make. Therefore, the size of the x-ray source must be minimized. The x-ray source sizes for various x-ray lines in matrices of different atomic numbers have been discussed in Chapter III in some detail. Figure 19 in that chapter shows the various x-ray ranges in μm that can be expected as a function of incident energy E_0, average specimen atomic number, and the characteristic x-ray line. X-ray source size can be decreased below 1 μm by lowering E_0 and by choosing an appropriate x-ray line for measurement. For samples of geological interest, such as silicates, the incident energy must be less than 10 keV in order to obtain a source size smaller than 1 μm for Al, Mg, Si, etc., K radiation. Although the K lines of elements such as Fe, Ni, Cu, etc., are just barely excited by 10-keV electrons (see Table II, Chapter III), it is possible to use the L lines of these elements for analysis. However, the L lines of Fe, Ni, Cu are of long wavelength. Hence, accurate analyses are difficult to obtain; all the precautions usually considered for light element analysis (Section I of this chapter) must be employed. For metallurgical samples, an x-ray source size of <1 μm can be obtained for many elements of interest by employing an operating voltage of 20 kV or less. An x-ray source size of $<1/2$ μm can be obtained if the L lines are used for elements such as Ti, Cr, Mn, Fe, Ni, Co, and Cu; if the K lines are used for elements such as Na, Mg, Al, Si, P, and S and the operating voltage is set at 10 kV, source sizes of <1 μm result.

When steep concentration profiles are encountered, such as at a phase boundary, the finite volume from which x-rays are excited in a sample causes the true concentration profile to appear "smeared." A schematic representation of this process of apparent smearing in an x-ray intensity vs. distance plot is shown in Figure 10 as given by Rapperport.[29] The

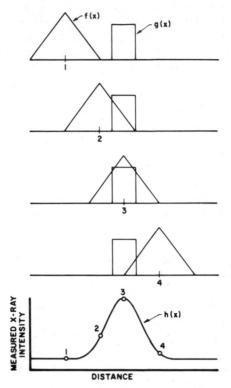

FIGURE 10.   Convolution technique. A concentration step $g(x)$ measured by a
                   triangular-shaped x-ray distribution $f(x)$ will result in the observed
                   profile $h(x)$ as the beam passes over the concentration step (after
                   Rapperport[29]).

x-ray excitation volume (probe function) is described by a triangular-shaped
function $f(x)$ in this example, while the true concentration profile is in the
form of a step function $g(x)$. The curve $h(x)$ represents the profile that
would be measured with intermediate-numbered points corresponding to
several positions of the concentration step during the translation of the
beam having a probe function $f(x)$.

Mathematically, the relationship between the observed x-ray intensity
$h(x)$ and the "probe function" $f(x)$ as the beam is passed over a true concentration function $g(x)$ is expressed by the integral[29,30]

$$h(x) = \int_{-\infty}^{\infty} g(x - x_0)f(x)\ dx \qquad (20)$$

The observed profile $h(x)$ is therefore the mathematical convolution of the
probe function and true concentration profile function. In reality, the
lateral distribution of electrons in the material can be considered to be

Gaussian in two dimensions about the beam axis and radially symmetric about the axis.[31] If x-ray absorption and fluorescence effects are small, the lateral x-ray distribution can be assumed to be of the same form.

One way in which the probe function can be experimentally determined is by passing the electron beam across a known concentration step and plotting the observed profile. By assuming a probe function which is Gaussian,[31] the resulting probe trace should be in the form of a mathematical error function. Figure 11 shows measured values of nickel and phosphorus taken in $1/4$-μm steps across an interface between ferrite (α phase) and a stoichiometric phosphide $(FeNi)_3P$.[32] The compositions of the two phases are constant as a function of distance and are well known from previous measurements. The Ni and P $K\alpha$ radiations were monitored and the instrument operating voltage was 20 kV with a sample current of 0.05 μA. The interface was positioned parallel to the spectrometer used to monitor phosphorus so that the phosphorus x-rays reached the detector after passing through only the phase in which they were generated. This permitted the probe function to be represented as a two-dimensional function. An error function curve is indicated in Figure 11 for both the

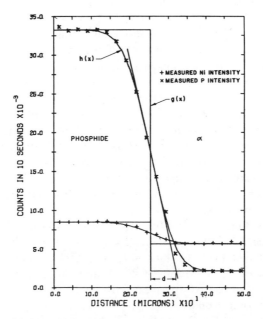

FIGURE 11.   Measured Ni and P concentrations across an equilibrated α/phosphide interface. The data are used to determine the parameter d in the probe function $f(x)$. (From Norkiewicz.[32])

Ni $K\alpha$ and P $K\alpha$ radiation. The Gaussian probe function can be described by[30,33]

$$f(x) = \frac{1}{d} \exp \left[-\frac{\pi}{4} \left(\frac{x}{d} \right)^2 \right] \tag{21}$$

The probe function can be thought of as the spatial intensity distribution of the emitted characteristic x-ray line being investigated, suitably normalized so as to be concentration independent. Since the probe function will be dependent on the material being analyzed, after the x-ray line and beam voltage are chosen, it must be determined on materials similar in composition to that of the samples to be analyzed.

The parameter d in the probe function can be determined by drawing a tangent to the curve at the midpoint and measuring the intercept on the distance axis (Figure 11). Once the probe function $f(x)$ is known, it should be possible to determine $g(x)$ after the measurement of $h(x)$ is made. This technique, called deconvolution, is in fact very difficult to carry out in practice. The best method available is to use the known probe function and a predicted true concentration profile $g(x)$ to calculate the expected probe result $h(x)$. A comparison of $h(x)$ with experimental data then permits an evaluation of the predicted true concentration profile $g(x)$.

This convolution technique can be performed mathematically by representing the functions $f(x)$ and $g(x)$ by their ordinates at small intervals (the same interval for both functions) and multiplying these ordinates together as one would the coefficients of polynomials. This operation for high-order polynomials can be carried out using a standard computer polynomial multiplication subroutine.[30] The probe function, however, must first be normalized such that the sum of its ordinates—at the interval decided upon—is unity. Figure 12 shows the probe function $f(x)$ normalized for 0.2-μm intervals, determined from the two-phase interface. Figure 11 shows the result of a convolution $h(x)$ performed by numerical methods on a digital computer for both nickel and phsophorus with $d = 0.65$ μm across the α/phosphide interface. This value of $d = 0.65$ μm is reasonable when compared to the x-ray source sizes described previously in this section.

The convolution technique was used in one case to determine the peak nickel concentrations in the γ-taenite phase at the α–γ interface in the Tucson meteorite.[34] This meteorite was cooled rapidly, and large Ni gradients developed due to the lack of time for diffusion of Ni in the taenite phase. With operating conditions of 20 kV and 0.01 μA, an average d value of 0.45 μm was measured from an interface. By computer simulation, the probe function [equation (21)] was generated across a predicted true concentration profile such as that illustrated in Figure 13. The resulting convoluted profile $h(x)$ is also shown in Figure 13. The agreement between

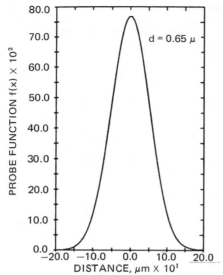

FIGURE 12.   Probe function $f(x)$ normalized for 0.2-μm intervals (from Norkie-wicz[32]).

the convoluted and the predicted profile is good, and therefore the predicted profile can be assumed to be correct. In the analysis shown in Figure 13, a peak nickel concentration of 21.7 wt % was measured although a true nickel value of 24.6 wt % is present at the interface. Since the actual measured data points across the γ–α interface correspond closely to the calculated convoluted profile, it is reasonable to assume that the true composition of taenite in equilibrium with kamacite is $\sim$25 wt % nickel. Using the α and γ compositions at the two-phase interface and the ternary Fe–Ni–P phase diagram, it was concluded that the last temperature of equilibration between phases was about 500°C. In this example, the inter-face equilibrium composition could be accurately determined. The convolu-tion technique therefore results in an apparent increase in the spatial resolution of the electron microprobe. Interface compositions as well as steep compositional gradients can be more precisely measured with the EPMA if the convolution technique is properly applied. However, this method can only be directly applied if the effect of fluorescence is unim-portant. The effect of fluorescence is discussed later in this chapter.

B. X-Ray Absorption

A large analytical error may be introduced when the region through which the generated x-rays pass is of different composition from the

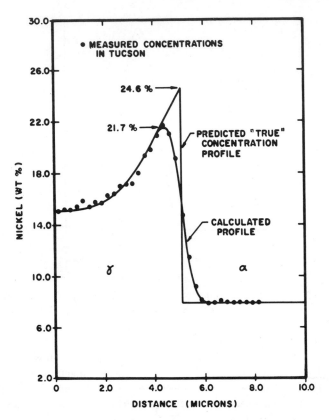

FIGURE 13.   Deconvolution technique applied to the Ni concentration gradient in the taenite (γ) phase of the Tucson meteorite (from Miyake and Goldstein[34]).

analyzed area. Such errors can be minimized (1) by having a high takeoff angle Ψ in order to make the absorption path in the sample smaller and (2) by choosing the orientation of the specimen with respect to the spectrometer so that the x-rays generated in a given phase leave parallel to the phase interface in material of similar composition. Since the absorption correction increases with increasing mass absorption coefficient, it is particularly important to ascertain that those characteristic x-rays from elements that are highly absorbed are only absorbed in material of similar composition. For a specimen consisting of a one-dimensional concentration gradient, as in a diffusion couple, or having a concentration gradient normal to the interface of two phases, the sample should be oriented so that the excited x-rays collected by the x-ray spectrometer lie in a plane of constant composition, i.e., a plane that is normal to the direction of the concentration gradient. This

configuration is illustrated in Figure 14.[35] When simultaneous recording
of elements is required, two spectrometers 180° apart can satisfy the
correct specimen orientation condition for a given takeoff angle. In the
case of an α/phosphide interface (Figure 11), P $K\alpha$ and Ni $K\alpha$ were
measured simultaneously. The P $K\alpha$ is a less energetic x-ray and is more
strongly absorbed in the sample. Therefore the specimen was oriented such
that the α/phosphide interface was parallel to the x-ray path to the
detector for P $K\alpha$. In the EPMA instrument employed, the x-ray spectrometers were not 180° apart and the correct orientation could only be obtained
for one element. Since light-element long-wavelength x-rays are more highly
absorbed, particular attention must be paid to these x-ray lines.

C. X-Ray Fluorescence

Possible excitation of characteristic x-rays by the continuum and the
characteristic lines from elements a considerable distance away from the
electron beam impact point can also occur. The effect is most serious when
the analyzed element is present in large concentration in the adjacent phase,

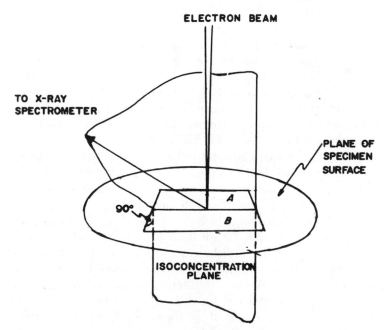

FIGURE 14.   Specimen orientation to minimize x-ray absorption effect (from
Koffman[35]).

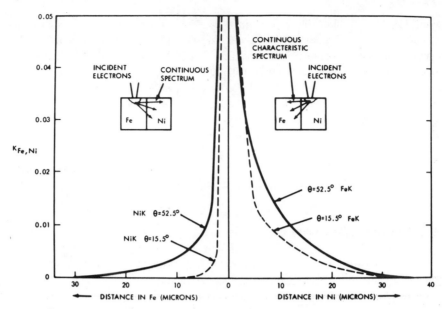

FIGURE 15.   Effect of x-ray fluorescence, Ni $K\alpha$, and Fe $K\alpha$ intensity from an undiffused Fe–Ni couple (from Goldstein and Ogilvie[36]).

as illustrated in Figure 15. In this figure, an undiffused couple of pure Fe vs. pure Ni was polished and microprobe traces were taken using two takeoff angles, $\theta(\Psi) = 15.5°$ and $52.5°$, for Fe K and Ni K radiation across the interface.[36] The pure iron is on the left-hand side and pure nickel is on the right-hand side of the Fe–Ni interface. The intensity ratios k_{Fe} and k_{Ni} with respect to pure Fe and Ni are plotted vs. distance. The Ni $K\alpha$ radiation produces secondary fluorescence of Fe $K\alpha$. Ideally when the data are taken across the interface, a sharp gradient influenced only by the excitation volume should be observed. In this example, when the electron beam is situated in pure Ni (Figure 15), both the continuum radiation and the Ni K characteristic radiation travel across the interface and produce, by secondary fluorescence, Fe $K\alpha$ radiation, which is registered by the x-ray spectrometer. The effect extends up to 30 μm from the interface into the Ni, but is very minor ($k_{Ni} = 0.01$) at distances greater than 10 μm into the Ni. When the electron beam is situated in the Fe (Figure 15), the continuum radiation travels across the Fe–Ni interface and produces some Ni $K\alpha$ secondary fluorescence radiation, which is also registered by the spectrometer. A similar fluorescence effect is shown in Figure 9 describing Ni measured in olivine, $(FeMg)_2SiO_4$. Since the Fe content of the olivine is about 25 wt % and the metal contains less than 30 wt % Ni, the continuum fluorescence effect is much less than is observed in the Ni vs. Fe case

(Figure 15). The effect of fluorescence for Ni in the α/phosphide and α/γ cases, shown in Figures 11 and 13, is very small, since the discontinuity in Fe content at the interfaces is small.

The effect of secondary fluorescence is difficult to calculate, although a few attempts have been made.[28,37] If one is to attempt measurements of compositions near interfaces, calculations of possible fluorescence effects should be made. It is not difficult to prepare undiffused couples; therefore in some cases experimental measurements may be justified.

IV. SOFT X-RAY EMISSION SPECTRA

Soft x-ray spectroscopy is a logical and extremely useful ancillary technique of the electron microprobe which allows one to determine the chemical and structural role of elements in complex materials, such as corrosion layers, thin films, glasses, and complex mineral phases. As an example, Figure 16 shows the aluminum $K\beta$ emission band from aluminum metal, α-Al_2O_3, and γ-Al_2O_3.[38] The differences between the spectra of the metal and the two polymorphs are clearly shown. The spectrometer was scanned across the peak and the spectra recorded on a strip chart or x–y recorder. By looking at the shapes and/or peak positions of the emission spectra, one can determine the exact phases that are present.

Another example of the use of soft x-ray spectra includes the cor-

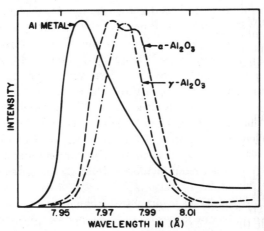

FIGURE 16.   Aluminum $K\beta$ x-ray spectra from aluminum, α-alumina, and γ-aluminum showing significant differences in peak position and band shape (from White[38]).

relations between peak positions and mean Si–O and Al–O bond distances in aluminosilicates.[39] This correlation is so consistent that it may be useful in predicting similar bond distances in unknowns. In certain samples, the peak position of the $K\beta$ satellite peak on the low-energy (long-wavelength) side of the $K\beta$ emission band is characteristic of the ligand (oxide, nitride, or carbide), and the peak intensity is indicative of the relative amounts of the ligands present.[40] In addition, the ferrous to ferric ratio for minerals has been determined by correlation with the iron $L\beta$ to $L\alpha$ intensity ratios.[41]

These variations in peak positions and shapes of soft x-ray emission spectra are a natural consequence of the emission process itself. In the electron microprobe, the focused beam of electrons striking the sample has sufficient energy to eject electrons from the inner shells of the atoms. An atom with a K electron removed from the K shell is usually a singly charged positive ion but differs from an ion having a valence electron missing. If the electron is completely ejected from the atom, then the ion so formed is in a highly energetic state. In attempting to return to the ground state, an electron drops from one of the outer shells to fill the vacancy, with the resulting emission of an x-ray photon. The energy of the photon is a function of the difference in energy between the K and L states ($K\alpha$ emission) or between the K and M states ($K\beta$ emission). Transitions from the M shell to the K shell ($K\beta$ radiation) result in higher energies than transitions from the L shell to the K shell ($K\alpha$ radiation). Thus the $K\beta$ radiation will have a shorter wavelength than the $K\alpha$ radiation (see Figure 14, Chapter III).

In elemental crystalline silicon, for example, these photon energies or wavelengths are fixed and well known. However, when silicon atoms are chemically bound to other atoms such as oxygen, valence electrons in the M shell may in fact be closer in energy to the K shell than those in the L shell, depending on the nature of the bond. Consequently, an M electron in dropping from the M shell to fill K-shell vacancies may in some cases have more energy than the L to K transition, thus resulting in shorter or longer emitted wavelengths than would occur in the unbound atom. This is the basis for the chemical or wavelength shift. The amount of observed shift, then, is a function of the nature of the chemical bond, or the valence state. The L shell in silicon is not involved in chemical bonding, but a measurable shift in the $K\alpha$ emission band is observed due to the screening of the inner shells by the M shell.

Although these effects occur most frequently in x-ray spectra from light elements, they are also observed in heavier elements when x-ray spectra associated with the bonding are examined. Since these effects are

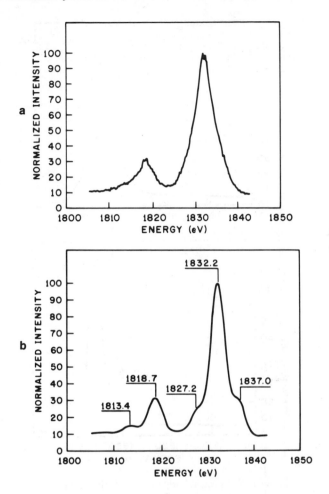

FIGURE 17.   (a) Recorded silicon $K\beta$ x-ray spectrum from silicon dioxide (from Colby[42]). (b) Filtered, resolution-enhanced silicon dioxide spectrum, with peaks labeled.

usually quite small, they require high-precision, wavelength-dispersive spectrometers for detection. The spectrometers should be of the fully focusing type (Chapter VII), carefully aligned, and should always be scanned in the same direction when recording peaks (usually from long to short wavelength). The electronics should be stabilized (no drifts), and the beam should be defocused or the beam current kept as small as practical to reduce the probability of sample damage. In the electron microprobe the

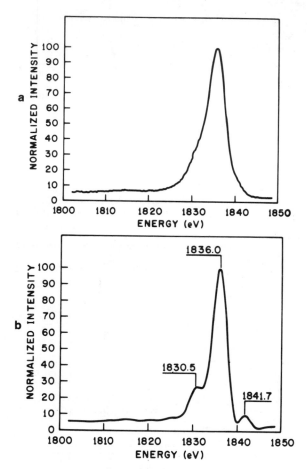

FIGURE 18.   Silicon $K\beta$ spectra from elemental silicon. (a) Raw data; (b) filtered and resolution-enhanced (from Colby[42]).

sample should be "optically" focused as accurately as possible, using an optical microscope with as short a working distance as possible. (See Chapter XI.)

The data obtained from the examples given earlier do not require any special experimental arrangements, and serve to illustrate wavelength shift in general. However, for other applications it may be necessary to resolve fine structure in the spectra. The required resolution may be beyond the conventional recording method. In such cases, smoothing and deconvolution of the emission spectra[42] caused by instrumental broadening is required.

The recorded data consist of the emission spectra as smeared by the recording system, a background component, and a noise component. Al-

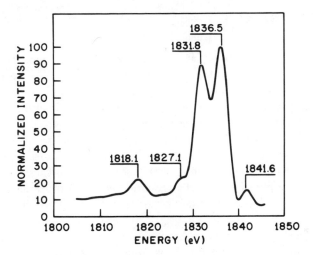

FIGURE 19.   Silicon $K\beta$ spectrum from "silicon monoxide" (from Colby[42]).

though signal averaging techniques (in which the spectrometer is repetitively scanned) may be employed to reduce the noise, they cannot eliminate noise entirely. A typical silicon $K\beta$ spectrum from silicon dioxide is shown in Figure 17a. The presence of noise is obvious upon inspection.

Figure 17b shows the same spectrum after filtering out the noise and applying deconvolution to remove the system response function. The removal of the broadening has clearly enhanced the resolution and facilitates comparison of spectra. The details of the deconvolution process are given in the appendix. For comparison purposes, Figure 18 shows the emission spectrum from elemental silicon and Figure 19 the emission spectrum from silicon monoxide. It should readily be apparent, by comparing Figures 17–19, that silicon monoxide is a mixture of silicon and silicon dioxide. The high precision and accuracy of the technique leave no ambiguities. Another example in which it is necessary to precisely locate peaks is the analysis of steam-grown silicon dioxide. This oxide is formed when silicon is exposed to steam at $\sim$1200°C and a layer of silicon dioxide is formed on the surface. As the film is formed, it is essentially amorphous, but after a relatively short time (about two weeks) it begins to crystallize. As a consequence of this crystallization (which begins at the interface), the low-energy $K\beta$ satellite shifts $\sim$1 eV away from the main silicon $K\beta$ band, which remains fixed. Such fine differences can only be seen through the use of digital techniques as described above. The peak positions of the main $K\beta$ emission bands and the $K\beta$ satellites are summarized in Table III. The precision and accuracy of the measurements is typically ±0.1 eV.

V. THIN FILMS

For many applications, it is necessary to analyze extremely thin films. In some instances, the accelerating potential may be sufficiently lowered, or the mass thickness of the film may be sufficiently large, that the electron beam may be entirely stopped within the film. Under these conditions, conventional ZAF techniques (Chapter IX) can be employed. Frequently, however, films must be analyzed that do not meet these boundary conditions, and thus require special techniques. One approach developed by Colby[43,44] has been used by a number of investigators, and has been reasonably successful. It was modified by Warner and Coleman[45] for biological specimens, and more recently has been improved by Oda and Nakajima.[46] The original technique proposed by Colby[43,44] required a prior knowledge of the mass thickness. However, the extension of Oda and Nakajima[46] calculates the mass thickness as well as the composition of the substrate.

The following development assumes familiarity with the material contained in Chapters III and IX. Assuming initially that the substrate upon which the film of interest is deposited does not contain any of the elements to be analyzed in the film, we may proceed as follows. A portion of the electrons impinging on the surface are backscattered (electron 3 in Figure 20), while those that penetrate the films may take either of two courses. The number n of ionizations produced by electrons of type 2 in Figure 20 is

$$n = (C_A N_0 / A) \int_{E_L}^{E_0} (Q/S^*) \, dE \qquad (22)$$

where C_A is the weight percent of element A, N_0 is Avogadro's number, A is the atomic weight, E_0 is the accelerating potential, Q is the ionization

TABLE III. Peak Position of $K\beta$ Emission Bands and Low-Energy Satellites

Material	Main band		Satellite	
	eV	Å	eV	Å
Silicon	1835.99	6.7628	—	—
Silicon nitride	1834.48	6.7584	1823.35	6.7996
Fused silica	1832.13	6.7670	1818.37	6.8182
Steam silica ^{a}	1831.92	6.7678	1817.91	6.8200
Steam silica ^{b}	1831.91	6.7679	1818.72	6.8169
Aluminum	1557.13	7.9621	—	—
Alumina	1553.78	7.9793	1538.43	8.0589

^{a} Less than two weeks old.

^{b} At least two weeks old.

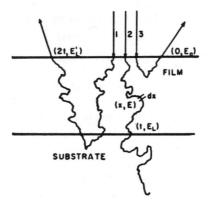

FIGURE 20.   Electron trajectories in thin film on thick substrate (from Colby[43]).

cross section (see Chapter IX), S is the electron stopping power (see Chapter IX), and E_L is the mean energy of electrons at the film–substrate interface. The asterisk denotes values for the thin-film sample.

Both Q and S are functions of electron energy as discussed in Chapter IX. The number of ionizations produced by electrons of type 1 in Figure 20 is

$$n = (N_0C_A/A) \int_{E_L}^{E_0} (Q/S^*)\ dE + (N_0C_A/A) \int_{E_{L'}}^{E_L} (Q/S^*)\ dE \qquad (23)$$

where E_L' is the energy of the electrons as they leave the film at the upper surface. The number of electrons backscattered from the substrate is given by the backscattered electron yield of the substrate, and the number of ionizations lost because of backscattering electrons can be taken into account by introducing the backscatter loss factor R (Chapter IX). Consequently, the total number of ionizations produced in the film is

$$n = (N_0C_A/A) \left[R^* \int_{E_L}^{E_0} (Q/S^*)\ dE + \eta_s \int_{E_{L'}}^{E_L} (Q/S^*)\ dE \right] \qquad (24)$$

where η_s is the backscattered electron fraction from the substrate and the asterisk represents the film. If the intensity from the film is compared to the intensity from a pure bulk standard, using the intensity ratio k_A, then, in the absence of other effects,

$$k_A = C_A \frac{R^* \int_{E_L}^{E_0} (Q/S^*)\ dE + \eta_s \int_{E_{L'}}^{E} (Q/S^*)\ dE}{R_A \int_{E_c}^{E_0} (Q/S_A)\ dE} \qquad (25)$$

X-ray range equations of the type given by Castaing[6] and discussed in Chapter III that consider the excitation potential E_c are of the form

$$\rho R(x) = \text{const} \times (A/Z)(E_0{}^n - E_c{}^n) \tag{26}$$

in which ρ is the density (g/cm^3) and $R(x)$ is the projected x-ray range. Equation (26) also gives the depth of the electrons when their mean energy has been reduced from E_0 to E_c. It is assumed, then, that when the electrons have penetrated a mass thickness $\rho R(x)$, their mean energy has been reduced from E_0 to E_L, where E_L is the mean energy at the film–substrate interface, and that

$$\rho t = \text{const} \times (A/Z)(E_0{}^n - E_L{}^n) \tag{27}$$

where t is the film thickness. The mean electron range follows a 3/2-power voltage dependence, as shown by Cosslett and Thomas,[47,48] and the constant was found[43,44] to be 330 when t is given in angstroms. Consequently, we may write

$$E_L = [E_0^{3/2} - (\rho t Z/330A)]^{2/3} \tag{28}$$

$$E_L{}' = [E_0^{3/2} - (\rho t Z/165A)]^{2/3} \tag{29}$$

The constant may change with sample inclination to the electron beam. The value of $f(\chi)$ is calculated by employing the electron absorption coefficient of Duncumb and Shields,[49] which takes into account the fact that the x-rays are not generated so deeply in the film and therefore are less absorbed. For instance, using this model for a 100 Å aluminum film on a silicon substrate with 20 keV incident energy $f(\chi)$ (see Chapter IX) for the film is 0.991, while for bulk aluminum under the same conditions, $f(\chi)$ is 0.836.

The proposed model was tested for pure aluminum films evaporated onto silicon substrates, and the agreement between theory and experiment was found to be satisfactory. The results obtained through the use of the thin-film model are very sensitive to thickness and density, and the composition so obtained is only as accurate as the values of thickness and density used in equations (28) and (29). In fact, this can be used to advantage. If the films are sufficiently thick to permit analyses in the usual fashion ($\sim$2500 Å) and the thickness is sufficiently well known, then by making analyses at two accelerating potentials, both composition and density can be obtained. Table IV lists results obtained with this "thin-film" model for conditions of an accelerating potential of 20 keV and a depth of penetration of 1–2 μm. Good agreement exists between calculated and theoretical mass concentrations. Table V illustrates the use of the model to determine the densities of alumina films deposited at different temperatures. The density of the film and index of refraction both appear to be proportional

***TABLE IV*.** Results of Microprobe Analysis Using Thin-Film Model

Film	Thickness, Å	Element analyzed	Intensity ratio	Mass concentration	
				Calc.	Theor.
Andalusite (Al_2O_3–SiO_2)	1800	Al	0.039	34.5	31.9–32.4
Aluminum nitride	2800	Al	0.124	66.9	65.8
Al_2O_3	540	Al	0.021	52.9	52.9

to the temperature of deposition. Thus the technique can be employed to determine either composition or mass thickness.

However, this technique suffers from an inability to accurately analyze thin films whose mass densities are not known. No information is provided on the substrate composition. To extend the technique to account for these cases, Oda and Nakajima[46] assume that the energy in equation (22) can be replaced by an average value $\bar{E} = (E_0 + E_c)/2$. (See Chapter IX for similar assumption in the atomic number correction.) With this assumption the integration can be eliminated from equations (22)–(25) and one obtains

$$k_A = C_A(R^*/R_A)(\bar{S}_A/\bar{S}^*) \tag{30}$$

where $\bar{S}_A$ and $\bar{S}^*$ [see Chapter IX, equation (22)] represent the ionization-penetration correction in the solid standard and thin film, respectively, obtained through the use of a mean energy $\bar{E}$. However, Oda and Nakajima maintain the integration with respect to absorption losses, instead of using the approximation introduced by Colby,[43] so that

$$I_A = C_A(R^*/\bar{S}^*)\int_0^{\rho t} \phi(\rho z) \exp[(-\mu/\rho)\rho t \cos \Psi] \, d(\rho z) \tag{31}$$

where I_A is the measured intensity from element A in the film, $\phi(\rho z)$ is the distribution in depth of x-rays, μ/ρ is the mass absorption coefficient of A

***TABLE V*.** Density of Alumina Films

Temperature, °C	Density	Index of refraction
550	2.44	1.59
650	2.25	1.61
750	2.62	1.65
850	3.04	1.73
900	3.05	1.73
1000	3.24	1.75

in the film, ρt is the mass thickness of the film, Ψ is the x-ray emergence angle, and, z is the variable depth of integration.

A similar relation can be developed for the standard, as is done in equation (25), giving the k ratio in terms of a straightforward integral equation for each element. A similar set of equations can be derived for the substrate, but the integration limits then must be ρt to infinity. One also can use the fact that the sums of the concentrations in the film and in the substrate are both unity (100%). Consequently one has $N + n + 1$ unknowns and $N + n + 2$ equations, where n is the number of elements in the film and N is the number of elements in the substrate. The redundant set of equations can be solved simultaneously in the least squares sense (form the normal equations) using conventional matrix techniques.

Table VI lists the Oda–Nakajima[46] analysis of two films of silver–copper on substrates of 25 wt % Ni–75 wt % Fe. In film A, the composition was determined by x-ray diffraction to be 90 wt % Ag–10 wt % Cu and for film B, the composition was 50 wt % Cu–50 wt % Ag. The composition and thickness of the film determined by electron probe analysis agree well with results obtained by conventional techniques.

The analytical techniques described above appear to be adequate from a practical standpoint; they yield reasonably accurate results. Monte Carlo methods for thin-film studies were described in Chapters VIII and IX. Here results obtained by the Monte Carlo-based method of Kyser and Murata[50,51] will be discussed.

The sequence of events in this Monte Carlo simulation is illustrated in Figure 21 for a single electron trajectory. An electron with energy E_0 impinges at the origin at normal incidence to the surface of a semiinfinite solid. The first scattering event is assumed to occur at the origin. The scattering angle θ_0 and step length Λ_0 are calculated with Monte Carlo technique as described in Chapter VIII. Since Rutherford scattering is axially symmetric about the incident direction of the electron being scattered, a uniformly generated random number is calculated in order to assign a value to the azimuthal angle φ_0. The spatial position of the next scattering event is then determined, and the electron energy at this point

TABLE VI. Silver–Copper Films on Iron–Nickel Substrates

Sample		Ag	Cu	Fe	Ni	Thickness, Å Multiple beam	Electron probe
A	k	0.562	0.061	0.261	0.086	4800	4700
	C	0.908	0.092	0.753	0.247	—	—
B	k	0.023	0.018	0.768	0.244	400	420
	C	0.510	0.490	0.743	0.257	—	—

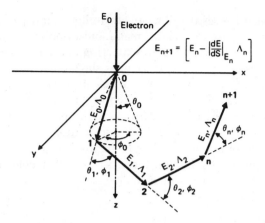

FIGURE 21.   Geometry of initial steps in Monte Carlo simulation in thick target (from Kyser and Murata[51]).

(point 1) is then calculated by decrementing the energy with respect to its value at point 0. At point 1, the sequence is repeated to calculate a new Λ_1, θ_1, and φ_1, and the energy is again decremented. This process is continuously repeated until the energy has decreased to a value close to the mean ionization potential, at which point the process is terminated. The calculation is then repeated for additional electrons until a statistically valid result is obtained.

When the target is composed of a thin film with mass thickness ρt on top of a thick substrate, the sequence of events is as described in Figure 22.

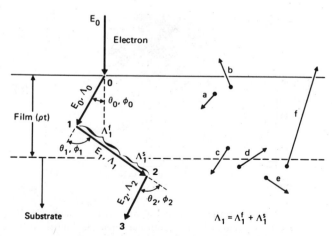

FIGURE 22.   Geometry of initial steps in Monte Carlo simulation in thin film on a thick substrate (from Kyser and Murata[51]).

As the electron traverses the film–substrate boundary, parameters such as the atomic number, atomic weight, mean ionization potential, etc. must be altered to describe the appropriate scattering and energy loss. In addition the various fluxes of electron scattering, as shown by paths a–f in Figure 22, must be taken into account. When an electron crosses a boundary, the scattering and energy loss parameters appropriate to the initial point are used to calculate θ_i, Λ_i, and φ_i. However, the parameters appropriate to the terminal point are used in the next calculation.

Figure 23 shows 100 simulated electron trajectories for a 20-keV electron beam incidently normally on (a) a 1000-Å gold film on a thick aluminum substrate and (b) on a bulk gold target. Figure 24 shows the

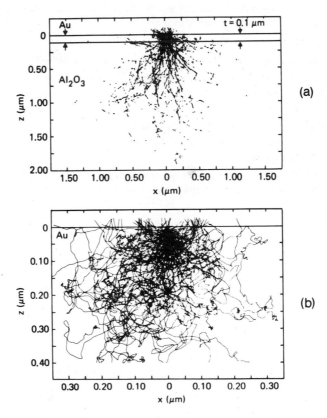

FIGURE 23.   Simulated trajectories of 100 normally incident 20-keV electrons for (a) 1000-Å gold film on alumina substrate, and (b) thick gold target. Scale is expanded by 5 × in (b). (From Kyser and Murata.[51])

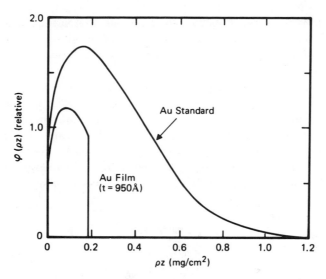

FIGURE 24.   Simulated depth distribution of gold $M\alpha$ x-ray generation produced in a gold film on an alumina substrate and in a thick gold target by 20-keV electrons at normal incidence (from Kyser and Murata[51]).

ionization distribution $\phi(\rho z)$ with depth ρz (see Chapter IX) for these same conditions. Note that the distribution of gold $M\alpha$ radiation terminates abruptly at the film–substrate boundary, as it should, and that the shapes of the two curves are different, due to the decreased backscattering from the alumina substrate.

In order to convert the experimental data into values of elemental weight fraction C_i and total film mass thickness ρt, the Monte Carlo computer program for simulation of electron scattering and energy loss is utilized to generate calibration curves of k_i vs. C_i, with ρt as a parameter.[50,51] The Monte Carlo program is not arranged to iterate the matrix effects and converge to a unique solution in the same manner as the ZAF model for thick targets. Hence Monte Carlo data must be used to generate numerical points on a calibration curve for a particular set of experimental conditions. The results of such calculations are shown in Figure 25 for the case of Mn_xBi_y films on a SiO_2 substrate with $E_0 = 20$ keV and $\Psi = 52.5°$. Calculations of $k(Mn\ K\alpha)$ and $k(Bi\ M\alpha)$ have been made for particular combinations of C_{Mn}, $C_{Bi} = 1 - C_{Mn}$, and ρt. A smooth line has then been drawn through those points with common values of ρt. Note that the curves are nonlinear with respect to C_i, as a consequence of the difference in scattering and energy loss properties of Mn and Bi atoms. The analysis of the data then reduces to graphical iteration and interpolation of $k(Mn\ K\alpha)$

and $k(\text{Bi } M\alpha)$ experimentally measured for each film within the calibration curves of Figure 25 in order to arrive at a unique fit for both C_i and ρt. This is possible because there are two unknowns (ρt and C_{Mn} or C_{Bi}) and two knowns (k_{Mn}, k_{Bi}).

Graphical convergence is easily accomplished by replotting the calibration curves of Figure 25 as shown in Figure 26. The parameters have been interchanged, and calibration curves of $k(\text{Mn } K\alpha)$ and $k(\text{Bi } M\alpha)$ have been plotted versus ρt for particular values of C_i. The experimental intensity ratios for each film are then compared with Figure 26 and the intercept points provide two separate values for ρt. The object is to find a particular value of C_{Mn} for which these two values of ρt coincide. Then a

TABLE VII

| Sample | Monte Carlo | | Nuclear backscatter | |
	C_{Mn}, %	ρt, $\mu\text{g}/\text{cm}^2$	C_{Mn}, %	ρt, $\mu\text{g}/\text{cm}^2$
X115-1	0.0	54	0.0	48.6
X115-3	25.5	53	26.6	48.2
X115-5	51.0	38	53.5	33.6
X115-7	80.0	29	80.1	26.2
X115-9	100.0	42	100.0	40.1

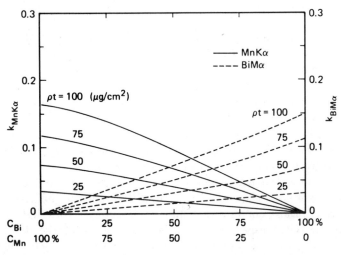

FIGURE 25.   Theoretical calibration curves for Mn and Bi x-ray fluorescence from thin films on SiO_2, 20 kV, $\Psi = 52.5°$. (From Kyser and Murata.[51])

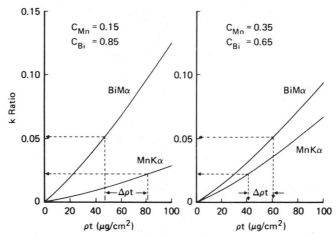

FIGURE 26.   Theoretical calibration curves for Mn and Bi x-ray fluorescence obtained from Figure 25 for two particular values of composition. 20 kV, $\Psi = 52.5°$, substrate SiO_2. (From Kyser and Murata.[51])

unique solution is obtained. Results obtained in this manner for several $Mn_x Bi_t$ films are shown in Table VII. Also shown in Table VII are analytical results obtained via nuclear backscattering energy analysis.[52]

The Monte Carlo technique also yields composition and mass thickness results from thin films. The Monte Carlo technique is more elegant than the Colby thin-film method, but is also much more costly and time-consuming, and always requires a large computer facility in order to test enough trajectories to obtain meaningful results. For the Kyser–Murata method described above, binary films of unknown thickness on known substrates can be analyzed.

APPENDIX. DECONVOLUTION TECHNIQUE

The background is first removed from a spectrum by fitting a least squares first or second degree polynomial to the minima on both sides of the emission spectrum. The backgrounds are then calculated at each point and subtracted from the data, leaving only a signal composed of the superposition of the smeared emission spectra and the noise. Mathematically this can be described as

$$y(E) = s(E) + n(E) \tag{32}$$

where $y(E)$ represents the recorded spectrum (background subtracted), $s(E)$ is the smeared emission spectrum, and $n(E)$ is the noise. The smeared

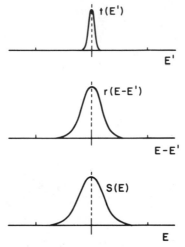

FIGURE 27.   A schematic of the convolution process in which emission spectrum $t(E')$ is convolved with the instrument response function $r(E - E')$ to yield the recorded spectrum $s(E)$.

emission spectrum is, in fact, a convolution of the true emission spectrum $t(E')$ with the recording system response function $r(E - E')$, hence[42]

$$s(E) = \int_{-\infty}^{+\infty} r(E - E') \cdot t(E') \, dE' \tag{33}$$

Equation (33) can be thought of as representing a transformation which takes a function $t(E')$ in E' space and maps it into E space. The result is $s(E)$. In practical cases, $s(E)$ is defined to exist over some finite interval (a, b) and to be positive, real, and continuous in that interval. These constraints imply that $s(E)$ goes to zero at $E = a$ and $E = b$. Now r and t can be assumed to be functions defined in the same way as $s(E)$; then equation (33) becomes

$$s(E) = \int_{a}^{b} r(E - E') t(E') \, dE' \tag{34}$$

Figure 27 illustrates, schematically, a convolution as described by equation (34).

In all but very special practical cases, the system response function is not known in closed form, but must be determined empirically. Then the problem becomes one of solving equation (34) numerically, in which case one must describe $s(E)$ and $r(E - E')$ with a finite number of points and solve for a finite number of points, which will describe $t(E')$. If one takes N equally spaced points over the interval (a, b) in which $s(E)$ is defined and replaces the integral by a sum, equation (34) can be reexpressed as

$$S_n = K \sum_{m=1}^{N} R_{nm} T_m \tag{35}$$

Equation (35) is a matrix equation, which can be used to describe a system of N simultaneous linear equations in N unknowns. The T_m represent N unknowns of an N vector; the R_{nm} are elements of the $M \times N$ coefficient matrix and the S_n are the elements of a measured M vector. Equation (35) and Figure 28 provide a numerical and graphical illustration, respectively, of the convolution process.

To solve for T, then, one must invert the R_{nm} matrix and premultiply both sides of equation (35). As the order of R gets large, it becomes more singular, and the number of computations becomes extremely large. Hence in practice other techniques must be employed. The technique most commonly used is to transform the data digitally, using a discrete fast Fourier transform (FFT). The FFT used is due to Bergland,[53] and is quite rapid and conservative of computer core storage. The Fourier trans-

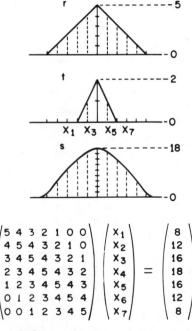

FIGURE 28. A numerical example of convolution using discrete points and the matrix technique.

formation considerably reduces the number of computations required[54] and eliminates much of the truncation error resulting from the singularity of equation (35) as M becomes large.

In the above representation, the spectra are in the energy domain, which might be analogous to the time domain in ordinary signal processing language (see Chapter IV). If the data are then transformed into the frequency domain, using a discrete fast Fourier transform,[53] conventional signal processing techniques can be employed. Thus

$$Y(\omega) = S(\omega) + N(\omega) \qquad (36)$$

where $Y(\omega)$ is now the spectrum as transformed into the frequency domain, $S(\omega)$ is the useful signal, and $N(\omega)$ is the noise. In actual practice, the useful signal $S(\omega)$ is entirely a low-frequency component and the noise is a high-frequency component; hence the two are easily separated. One need only apply a low-pass filter to the transformed data to strip the noise from the spectrum.

Next the convolution theorem[55] is applied to the spectrum. The Fourier transform of the convolution of two functions is the product of the Fourier transforms of the individual functions. Thus

$$S(\omega) = T(\omega)R(\omega) \qquad (37)$$

where $T(\omega)$ is the Fourier transform of the true emission spectrum and $R(\omega)$ is the Fourier transform of the system response function. To get the true spectrum deconvoluted from the recorded spectrum (subsequent to noise removal and background subtraction), it is necessary to divide the Fourier transform of the recorded spectrum by the Fourier transform of the system response function and perform an inverse transform.

Figure 17a shows the digitally recorded x-ray spectrum from SiO_2.[42,56] The spectrum is instrumentally distorted and has noise superimposed on it. The Fourier transform of the spectrum is shown in Figure 29. It may be noted that all of the useful information of the spectrum is essentially contained in the first 25 channels; the noise is contained in the remaining channels. To remove the noise, we can multiply the transformed data (from Figure 29) by an "ideal" filter, such as that shown in Figure 30. This filter has a value of unity up to some cutoff frequency, so as not to distort the signal-containing channels; the filter value is zero in the region of the noise. The spectrum must be retransformed after filtering. The Fourier transform of an ideal filter is shown in Figure 31; note that considerable oscillation occurs. The transform of a rectangular function (impulse response) (Figure 31) is of the form $(\sin x)/x$,[42,54] which continuously oscillates, and so does not make a good practical filter. A filter that has a better impulse response is shown in Figure 32 and its transform

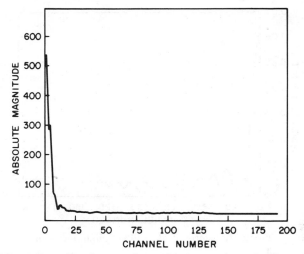

FIGURE 29. Fourier transform of spectrum shown in Figure 17a.

in Figure 33. This filter is of the form of an exponential rolloff from unity after some cutoff frequency.[41,55]

Multiplying point by point the transformed spectrum (Figure 29) by the filter function (Figure 32), the filtered transformed spectrum shown in Figure 34 is obtained. Upon retransformation, the spectrum shown in Figure 35 is obtained, which is seen to be completely free of noise. Before retransformation, however, it is desirable to remove the instrument response function (deconvolution). The system response function can be

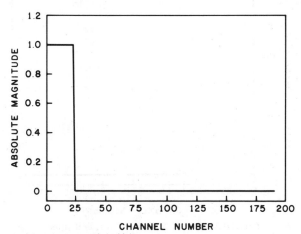

FIGURE 30. Ideal filter with sharp cutoff.

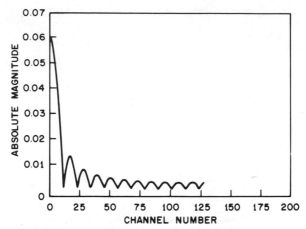

FIGURE 31.   Fourier transform of ideal filter (impulse response) shown in Figure 30.

approximated by a Gaussian,[42] whose transform is shown in Figure 36. The filtered and transformed data (Figure 34) are divided (point by point) by the Fourier transform of the system response function (Figure 36). Since these transforms are in fact complex (real and imaginary components), the division must also be complex. The data are then retransformed and the peaks labeled (described below) resulting in the resolution-enhanced spectrum shown in Figure 17b.

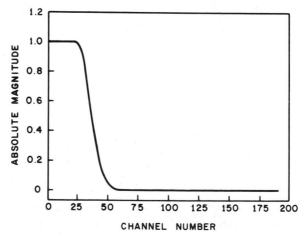

FIGURE 32.   Exponential rolloff filter.

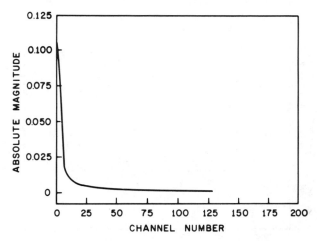

FIGURE 33. Fourier transform of exponential rolloff filter.

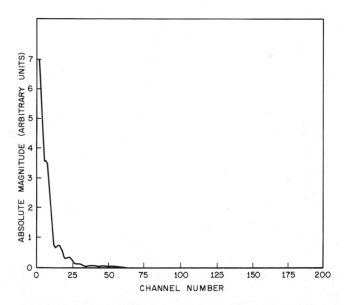

FIGURE 34. Filtered, transformed spectrum of silicon dioxide.

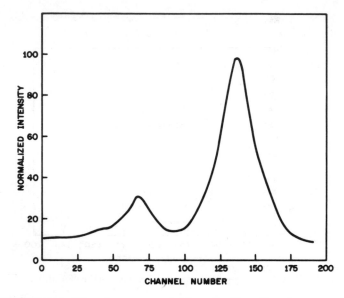

FIGURE 35. Filtered spectrum of silicon dioxide, with noise removed.

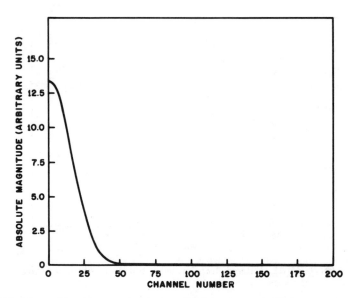

FIGURE 36.   Fourier transform (amplitude spectrum) of instrument response function.

Morrey[57] recently described a technique for determining spectral peak positions from either emission spectra or absorption spectra. The technique is completely general and permits accurate determination of peak positions that appear only as shoulders on stronger peaks. According to Morrey,[57] at peak positions, the following conditions exist:

$$dy = 0, \qquad d^2y < 0, \qquad d^3y = 0, \qquad d^4y > 0 \qquad (38)$$

It has been verified that adjacent peaks will have much less effect on the third and fourth derivatives than on the first and second derivatives. Thus, the use of the third and fourth derivatives to determine peak positions should produce more accurate results. Morrey[57] has shown that these criteria are applicable to functions having Lorentzian, Gaussian, or student T_3 distributions, and has determined their limitations. He has verified the utility of the technique by analyzing a synthetic spectrum in which there was sufficient overlap to obscure the peaks. The method successfully determined nine of ten peaks. The single peak not detected was, as predicted, outside the limitations of the method (too close to a much stronger band).

The data are numerically differentiated, and the arrays of derivatives scanned. When the conditions contained in equation (38) are met, the position of a peak is confirmed, and the peak is annotated.

REFERENCES

1. T. Shiraiwa, N. Fujino, and J. Murayama, in *Proceedings of the Sixth International Conference on X-Ray Optics and Microanalysis* (G. Shinoda, K. Kohra, and T. Ichinokawa, eds.), University of Tokyo Press (1972), p. 213.
2. G. L. Fisher and G. D. Farningham, Quantitative Carbon Analysis of Nickel Steels with the Electron Probe Microanalyzer, ASM Materials Engineering Congress, Cleveland, Ohio, October (1972).
3. P. Duncumb and D. A. Melford, in *X-Ray Optics and Microanalysis, Fourth International Conference on X-Ray Optics and Microanalysis* (R. Castaing, P. Deschamps, and J. Philibert, eds.), Hermann, Paris (1966), p. 240.
4. D. W. Fisher and W. L. Baun, *Norelco Reporter*, **14**, 92 (1967).
5. J. E. Holliday, *Norelco Reporter*, **14**, 84 (1967).
6. R. Castaing, in *Advances in Electronics and Electron Physics*, (L. Marton, ed.), p. 317, Academic Press, New York (1960).
7. R. Castaing and J. Deschamps, *Compt. Rend.*, **238**, 1506 (1954).
8. J. S. Duerr and R. E. Ogilvie, *Anal. Chem.*, **44**, 2361 (1972).
9. G. W. Bruno and S. H. Moll, Second National Microprobe Conference, Boston, Massachusetts (1967), Paper 57.
10. R. Theisen, *Quantitative Electron Microprobe Analysis*, Springer, Berlin (1965).
11. V. E. Kohlhaas and F. Scheiding, *Arch. Eisenhüttenwessen*, **40**, 1 (1969).

12. R. H. Barkalow, R. W. Kraft, and J. I. Goldstein, *Met. Trans.*, **3**, 919 (1972).
13. B. L. Henke and E. S. Ebisu, in *Advances in X-Ray Analysis*, Vol. 17, Plenum Press, New York (1974), p. 150.
14. H. A. Liebhafsky, H. G. Pfeiffer, and P. D. Zemany, *Anal. Chem.* **27**, 1257 (1955).
15. T. O. Ziebold, *Anal. Chem.*, **39**, 858 (1967).
16. H. Yakowitz, C. E. Fiori, and R. E. Michaelis, NBS Special Publication 260-22 (1971).
17. F. Kunz, E. Eichen, and A. Varshneya, in *Proceedings of the Sixth National Conference on Electron Probe Analysis EPASA, Pittsburgh* (1971), Paper 20.
18. R. E. Michaelis, H. Yakowitz, and G. A. Moore, *J. Res. NBS*, **A68**, 343 (1964).
19. H. Yakowitz, D. L. Vieth, K. F. J. Heinrich, and R. E. Michaelis, NBS Special Publication 260-10 (1965).
20. J. I. Goldstein, F. J. Majeske, and H. Yakowitz, in *Applications of X-Ray Analysis*, Vol. 10 (J. B. Newkirk and G. R. Mallett, eds.), Plenum Press, New York (1967), p. 431.
21. P. R. Buseck and J. I. Goldstein, *Geol. Soc. Am. Bull.*, **80**, 2141 (1969).
22. H. Yakowitz, A. W. Ruff, and R. E. Michaelis, NBS Special Publication 260-43 (1972).
23. E. L. Bauer, *A Statistical Manual for Chemists*, 2nd ed., Academic Press, New York (1971), p. 189.
24. H. A. Liebhafsky, H. G. Pfeiffer, and P. D. Zemany, in *X-Ray Microscopy and X-Ray Microanalysis* (A. Engström, V. Cosslett, and H. Pattee, eds.), Elsevier, Amsterdam (1960), p. 321.
25. R. H. Hewins and J. I. Goldstein, "Metal-Olivine Associations and Ni-Co Contents in Apollo 12 Mare Basalts," *Earth and Planet. Sci. Lett.* **24**, 59 (1974).
26. T. O. Ziebold and R. E. Ogilvie, *Anal. Chem.*, **36**, 322 (1964).
27. J. I. Goldstein, *J. Geophys. Res.*, **72**, 4689 (1967).
28. S. J. B. Reed and J. V. P. Long, in *X-Ray Optics and X-Ray Microanalysis*, Academic Press, New York (1963), p. 317.
29. E. J. Rapperport, in *Advances in Electronics and Electron Physics*, Supplement 6, Academic Press, New York (1969), p. 117.
30. J. B. Gilmour, "The Role of Manganese in the Formation of Proeutectoid Ferrite," Ph.D. Thesis, McMaster University (1970).
31. P. K. Gupta, *J. Phys. D. Appl. Phys.*, **3**, 1919 (1970).
32. A. S. Norkiewicz, "Dissolution of Phosphides in the Ternary Fe–Ni–P System," M.S. Thesis, Lehigh University (1972).
33. J. B. Gilmour, G. R. Purdy, and J. S. Kirkaldy, *Met. Trans.*, **3**, 3213 (1972).
34. G. T. Miyake and J. I. Goldstein, *Geochim. et Cosmochim. Acta*, **38**, 1201 (1974).
35. D. M. Koffman, *Norelco Reporter*, **11**, 59 (1964).
36. J. I. Goldstein and R. E. Ogilvie, in *X-Ray Optics and Microanalysis, IVth International Congress on X-Ray Optics and Microanalysis* (R. Castaing, P. Deschamps, and J. Philibert, eds.), Hermann, Paris (1966), p. 594.
37. M. J. Henoc, F. Maurice, and A. Zemskoff, in *Fifth International Congress on X-Ray Optics and Microanalysis* (G. Möllenstedt and K. H. Gaukler, eds.), Springer Verlag, Berlin (1969), p. 187.
38. E. W. White, in "Tutorial Session," *7th National Conference on Electron Probe Analysis* (1972).
39. E. W. White, in *Microprobe Analysis* (C. A. Andersen, ed.), Wiley New York (1973), p. 349.

40. J. W. Colby, D. R. Wonsiddler, and A. Androshuck, in *Proceedings of the 4th National Conference on Electron Probe Analysis*, EPASA (1969), p. 26.
41. A. L. Albee and A. A. Chodos, *Am. Mineral.*, **55**, 491 (1970).
42. J. W. Colby, in *Proceedings of the 6th International Conference on X-Ray Optics and Microanalysis* (G. Shinoda, K. Kohra, and T. Ichinokawa, eds.), University of Tokyo Press (1972), p. 247.
43. J. W. Colby, in *Advances in X-Ray Analysis*, Vol. 11, Plenum Press, New York (1968), p. 287.
44. J. W. Colby, in *Thin Film Dielectrics* (F. Vratny, ed.), The Electr. Chem. Soc., New York (1969).
45. R. R. Warner and J. R. Coleman, *Micron*, **4**, 61 (1973).
46. Y. Oda and K. Nakajima, *J. Jap. Inst. Met.*, **37**, 673 (1973).
47. V. E. Cosslett and R. N. Thomas, in *The Electron Microprobe* (T. D. McKinley, K. F. J. Heinrich, and D. B. Wittry, eds.), Wiley, New York (1966), p. 248.
48. V. E. Cosslett and R. N. Thomas, *Brit. J. Appl. Phys.*, **15**, 1283 (1964).
49. P. Duncumb and P. K. Shields, in *The Electron Microprobe* (T. D. McKinley, K. F. J. Heinrich, and D. B. Wittry, eds.), Wiley, New York (1966), p. 284.
50. D. F. Kyser and K. Murata, in *Proceedings of the 8th National Conference on Electron Probe Analysis*, EPASA (1973), p. 28.
51. D. F. Kyser and K. Murata, *IBM J. Res. Dev.*, **18**, 352 (1974).
52. M. A. Nicolet, J. Mayer and I. Mitchell, *Science*, **177**, 844 (1972).
53. G. D. Bergland, *IEEE Trans., Audio and Electroacoustics*, **AU-17**, 138 (1969).
54. G. D. Bergland, *IEEE Spectrum*, **6**, 41 (1969).
55. R. Bracewell, in *The Fourier Transform and Its Applications*, McGraw-Hill, New York (1965), p. 108.
56. J. W. Colby, in "Tutorial Session," *7th National Conference on Electron Probe Analysis* (1972).
57. J. R. Morrey, *Anal. Chem.*, **40**, 905 (1968).

XIII

BIOLOGICAL APPLICATIONS: SAMPLE PREPARATION AND QUANTITATION

James R. Coleman

I. SAMPLE PREPARATION

The most common source of uncertainty in the use of electron beam analytical techniques with biological materials is the preparation technique: the steps that intervene between the normal *in vivo* condition of the sample and the state in which it is analyzed. Generally, these steps involve removing material from an aqueous, atmospheric pressure environment to one that is dry and at relatively high vacuum, and may also include sectioning, sawing, or fracturing in order to expose the portions of interest. During this treatment, material must not be lost, redistributed, or gained, and the original structural relationships of the material must be conserved. These are quite restrictive requirements, and are usually satisfied only in part. Indeed, establishing the degree to which they are met is often a major problem in itself. There may be no universally satisfactory preparation technique, although some of the freezing techniques approach this. Thus it is usually the responsibility of each analyst to establish criteria for the problem of interest. In general, the properties of the sample determine which preparative techniques can be employed. Thus, if one wishes to analyze tissue fluids, crystalline inclusions, or cell organelles, it is

JAMES R. COLEMAN—Department of Radiation Biology and Biophysics, University of Rochester Medical Center, Rochester, New York.

likely that different methods will be useful. A brief survey of some of the methods that are available follows, along with some suggestions about their advantages as well as the hazards that are likely to be associated with them. A general overview was given by Coleman and Terepka.[1]

A. Liquid Samples

Ingram and Hogben[2] and Lechene and colleagues[3-6] have employed electron probe analysis to determine the composition of very small volumes, e.g., hundred of picoliters, of fluids. Drops of the aqueous solution to be analyzed are placed on the surface of a polished beryllium block, which is ruled into squares to facilitate identification of the samples. The block is submerged in a bath of paraffin oil and a micromanipulator is used to position a micropipette over a square on the block, using a microscope to view the process. The aqueous solutions do not mix with the hydrophobic paraffin oil, and remain as separate drops on the block. When all the samples have been positioned, the block is washed with xylene or chloroform to remove the paraffin oil. For whatever reason, this washing process does not wash away any material from the droplets. The droplets are then frozen rapidly and lyophilized. This procedure dries the nonvolatile material in the sample without producing large crystals that would complicate the x-ray counting geometry and make quantitation capricious. This technique has been shown to be quite reproducible and accurate. Undoubtedly it will be more widely employed in the future to analyze biological fluids that have been in contact with a few cells, or fluids that have been extracted from a single or at most a few cells. It has a further advantage in that the electron probe analysis can be readily automated since the droplets are deposited in a regular, defined pattern.

B. Thick Samples

Some thick samples in which the composition of the already exposed surface is of interest can be examined with a minimum of preparation, e.g., shell, tooth, microfossils. Usually these require only cutting or cleaving to a size convenient for insertion into the instrument or onto the specimen stage. If one is interested in only the mineral portions of such a sample, organic matter can be removed by treatment with a weak solution (2–5%) of sodium hypochlorite or by low-temperature oxygen plasma ashing. The samples can be mounted in a thermoplastic, or fastened to a substrate with a conductive adhesive. Such samples are usually sufficiently thick that they require a conductive coating. This coating may be vacuum-evaporated aluminum, carbon, or Aquadag, an aqueous suspension of colloidal graphite.

Such specimens may be adequate for qualitative analysis, but more often than not the surfaces are so irregular that the x-ray counting geometry is uncertain. The latter is a greater problem when employing several wavelength-dispersive spectrometers rather than one energy-dispersive spectrometer. With wavelength-dispersive spectrometers positioned at various angles around the sample, the x-rays going to each spectrometer may travel different absorption path lengths in a rough sample. Thus, elements present in low concentrations but located in depressions may emit x-rays that will be almost completely absorbed by the surrounding elevations, while the x-rays from elements present in higher concentration will still be detected. With energy-dispersive spectrometers, all x-rays travel the same absorption path to the single detector so that differential absorption is not as great a hazard. However, one is still liable to conclude that an element is not present if it occurs in low concentration in a depression so that x-rays are absorbed by an elevated region intervening between the volume analyzed and the detector.

First, examination of the secondary electron image will provide a good indication of whether the surface is rough or not. Then the extent of absorption due to surface roughness can be assessed by rotating the sample while carefully keeping the beam in the same position on the sample and determing whether the signal varies (more than two or three standard deviations of the peak counting rate) as the sample rotates. If the specimen stage is not capable of precise rotation, so that the rotation causes the specimen to change in position with regard to the beam location or focus, this test is unreliable. If wavelength-dispersive spectrometers are available, one can set two spectrometers to detect the same wavelength and see whether the ratios of peak intensity on sample and standard (k ratio) on both spectrometers are comparable. If not, one must be concerned about differential absorption due to surface roughness.

When using an energy-dispersive spectrometer, tilting the specimen may reduce the effects of surface roughness, although this complicates x-ray counting geometry. More frequently, polishing must be employed. For this, the sample can be embedded in a thermoplastic and polished according to any of the available protocols. Of course, the abrasives employed and any vehicles for the abrasives must not contain the elements one wishes to analyze. Furthermore, the abrasive particles must be thoroughly removed from the sample before analysis. Usually several washes with organic solvents or aqueous solutions are used for this purpose, but one must be careful that these solvents do not permit a differential extraction removing one element to a greater extent than another.[7]

C. Particles

Particles can be mixed in a solution of 1–2% collodion in amyl acetate and spread in a thin film over a substrate. When the solvent evaporates, the particles will be attached to the substrate by the resulting collodion film. This treatment presupposes that there is nothing of interest in the particle that will be extracted by the treatment with amyl acetate, a relatively good solvent for many organic materials, and that the batch of collodion and amyl acetate contain no elements that are likely to interfere with the analysis. When pressed for time, two rapid methods are available: sprinkle particles over a thin layer of conductive paint as it is drying; or attach cellophane tape that has adhesive on both faces to a substrate and place the particles on the other adhesive surface. The latter preparation has a high vapor pressure and requires a longer pumping time before reaching operating pressures. These preparations can then be coated with a conductive layer and examined.

For relatively small particles (e.g., up to 6 or 7 μm), semiquantitative and even quantitative data can be obtained from such preparations. For larger particles it is better practice to have the particles in a more homogeneous matrix. This can be accomplished by attaching the particles to the substrate first, and then polymerizing an epoxy matrix in a layer at least as thick as the particle height on the substrate. This preparation is then polished in the conventional fashion until the polished faces of the samples can be observed at the surface.

Particles can occur in which organic matter is an undesirable constituent. Low-temperature oxygen plasma ashing can be used to remove organic matter, leaving behind only the inorganic material of interest. It should be noted that the microincineration procedure will also oxidize any organic matter used as an adhesive in making the preparation, and may also oxidize the surface of the substrate, reducing its conductivity. Thus an additional conductive coating after microincineration may be necessary and careful handling of the sample will be required to avoid losing the particle from the substrate.

In some cases the particle may not occur in a free form, but is found included in tissue, e.g., asbestos in lung. If so, it may be desirable to remove the particle from its surroundings for analysis. Langer et al.[8,9] have utilized micromanipulation to remove individual particles from sections of lung tissue for analysis and have employed both low-temperature ashing as well as hydrolysis with 40% KOH at 100°C to remove the organic matter of the lung from particles prior to analysis. They showed, at the same time, that asbestos particles of known composition were not altered by these treatments, providing evidence that the composition of asbestos fibers isolated by this method would reflect the composition of the fibers as they occurred *in vivo* and would not be an artifact of the preparative method.

D. Single Cells and Cell Organelles

Although cells and particles may be approximately the same size, there are several important differences between them that must be taken into account in preparing them for analysis. First, cells and their organelles are fairly elastic and therefore more easily deformed than most particles. Second, the different chemical phases of cells and organelles are maintained by lipoprotein membrane boundaries which are relatively fragile, and which depend on "hydrophobic bonds" and therefore the presence of water for their integrity. These characteristics place severe restrictions on any preparation procedure.

Simple cells with no internal membrane-bound phases, such as bacteria and nonnucleated erythrocytes, can be air-dried on a silicon disk prior to analysis.[10] We have also employed the same procedure with nucleated cells grown in culture and with cells isolated from various tissues. Two major influences affect such preparations. First, one must dry the cell without contamination from the suspending medium, which frequently has an elemental composition quite different from the cell. Many of the ionic species present in the medium will precipitate as salts as water evaporates, and the cell membrane often serves as a nucleus for crystallization to occur, so that an air-dried cell may be surrounded by a matrix of dried medium. Sodium chloride crystals form readily and can almost cover the surface of a cell. The formation of this surrounding layer of dried medium can be minimized by washing the cells with isotonic sucrose[11] or with solutions of ammonium acetate, or Tris-HCl buffer. Ammonium acetate has the advantage of volatilizing in the vacuum of the electron beam column, and Tris does not readily crystallize under these circumstances. Air-drying does not seem to cause disruption of the plasma membrane. If this did occur, one would expect that potassium (about 150 mM inside) would leak out and would be detected in the region around the cell. However, this has not been found for erythrocytes, baby hamster kidney cells, isolated intestinal epithelial cells, or cells from embryonic calvaria. In addition, mitochondria isolated from liver cells have been prepared by air-drying.

In the above cases, the purpose of the analysis was to determine the composition of the total cell volume irrespective of the number or types of phase boundaries that the cell might contain. Thus if intracellular membrane systems were disrupted during preparation, it made no difference in the interpretation of results. However, the determination of the composition of such phases may be the purpose of the experiments. If this is so, then methods to preserve their integrity and demonstrations that their integrity has been maintained are necessary. It is this last requirement that is often the most difficult to satisfy, usually because electron probe analysis is the only feasible method to establish the composition of these phases, and yet one must use the preparative techniques in question to establish the composition of the phases. It is known that with nucleated cells present in blood smears, the distribution of the macromolecules that are responsible for the staining properties of these cells is maintained, and certain intracellular organelles, e.g., nuclei, mitochondria, and granules, can be recognized in the light microscope. It is not known whether smaller molecules, which can diffuse more rapidly, retain their original distribution. Some larger cells, *Tetrahymena pyriformis* and *Amoeba proteus*, do not respond well to simple air-drying.[13–15] These cells tend to fragment when air-dried, producing quite obvious cell destruction. Such cells react much better to rapid heat-drying. When spread in a drop on the surface of a silicon disk and then passed through the flame of a propane torch, these cells dry in a flattened but intact state, their integrity being ascertained by the lack of any potassium leakage. Furthermore, the nucleus remains intact, judged from reflected light microscopy and sample current images, as do calcium-rich lipid droplets throughout the cytoplasm. It is thought that the heat-drying is successful because of the relatively good heat conduction of the silicon disk and the property of the cells that causes them to spread on the surface of the silicon. It may also be due to the fact that the plasma membrane is relatively permeable to water. Whatever the reasons, the same procedure has not been successful when thin glass coverslips are substituted for the silicon disks.

Several methods have been employed to preserve fine surface structure of cells. It has been shown that simple air-drying, and presumably heat-drying as well, distorts the finer structural features of single cells. The use of critical point drying has generally preserved these features quite well.[16–18] In this procedure, the cells are fixed, typically with 3% glutaraldehyde in a pH 7.4 buffer (0.05 M cacodylate) in a saline solution (Puck's Saline G) and then post-fixed in 1% osmium tetroxide in 0.2 M cacodylate buffer, pH 7.2, and dehydrated through several changes of acetone. The cells are dried in a critical point apparatus and then coated with a 200 Å thick layer of vacuum-evaporated gold.[19]

It appears unlikely that the fixation and dehydration treatments can preserve the normal intracellular distribution of soluble diffusible elements and in the few cases investigated, similar fixation routines have caused severe redistribution artifacts.[15,20,21] However, the distribution of elements incorporated into macromolecules or tightly bound to macromolecules (e.g., P in DNA; Fe in hemoglobin) will probably be preserved. In spite of this drawback, which precludes the use of the technique for studies involving diffusible elements, it does offer superior preservation of surface structure and is useful for elucidation of morphology with the SEM. The major mechanism involved in the preservation of surface structure is the elimination of liquid–gas interfaces during the drying; this mechanism has been recently reviewed.[18] As a liquid interface moves through drying tissue, relatively large surface tension forces can be exerted on delicate cell structures, causing severe distortions. Drying with the critical point technique avoids this interface in the following way. At any given temperature, and at equilibrium in a closed vessel, the space above a fluid will be saturated by molecules of the fluid, according to the partial pressure of the liquid. If the temperature is increased, more molecules will be driven from the liquid phase into the vapor phase, resulting in an increase in the density of the vapor phase and a decrease in density of the liquid phase. As the temperature continues to rise, at the so-called "critical point" the density of the two phases will be equal, and the phase boundary between liquid and vapor will disappear. At temperatures and pressures above the critical point only one phase, a vapor phase, will exist.

Critical point drying is carried out in the following manner. A sample immersed in a fluid such as acetone or amyl acetate is placed in the chamber of the pressure bomb (see Figure 1). This fluid is replaced by a liquified gas (such as carbon dioxide or one of the Freons), referred to as the "transi-

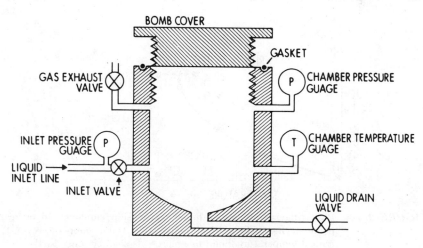

FIGURE 1.   Diagram of typical critical point drying apparatus (adapted from Lewis and Nemanic[18]).

tional fluid," until the chamber is almost or entirely full. The cover is closed, and the exhaust and drain valves are closed, the chamber being at room temperature. The bomb is then heated, elevating both pressure and temperature to the critical point. At this point the chamber contains only vapor. The vapor then is released through an exhaust valve, leaving behind a specimen that has been dried without experiencing the distortion resulting from passing through a vapor–liquid interface.

If the chamber has been only partly filled prior to elevating the temperature of the bomb, then a phase boundary will exist and as the temperature rises, this boundary will recede at first, then become ill-defined, and finally disappear. Thus one must be careful that there is sufficient fluid covering the sample so that the transient receding interface does not contact and distort the sample. On the other hand, if the bomb chamber is completely filled, the temperature and pressure will rise in the pattern seen in Figure 2. The pressure and temperature will not pass through the critical point; instead their relationship will be described by a line similar to 1234 in Figure 2. In this case one exceeds the critical pressure first, then keeps this pressure constant by venting the exhaust valve. The temperature continues to be elevated until the temperature of the chamber exceeds the critical temperature. At this point the chamber is full of vapor, without an interface having been formed, and the vapor can be vented through the exhaust valve.

Water is not employed as a transitional fluid because of its high critical temperature, 374°C. At this temperature water will not only distort tissue, but it may cause a critical point bomb made of aluminum

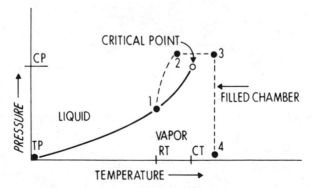

FIGURE 2.   Phase diagram of vapor-liquid relationship employed in critical point drying. CP, critical point; RT, room temperature; CT, critical temperature. Solid line represents chamber only partly filled at beginning, dashed line represents relationship in chamber completely full at beginning. (Adapted from Lewis and Nemanic.[17])

to corrode and disintegrate. Consequently, water must be removed from the tissue before critical point drying. This dehydration can be accomplished by immersing the sample in a graded series of acetone or ethanol concentrations until the sample is in the pure solvent. Several workers recommend the use of an intermediate fluid between the dehydration fluid and the transition fluid. If the transition fluid is to be Freon 13, then Freon TF (also known as Freon 113 and Genetron) is the intermediate fluid of choice, while amyl acetate is employed when the transitional fluid is carbon dioxide.

For convenience and economy, dehydration and substitution with intermediate fluids are not carried out in the critical point bomb, but in smaller separate vessels. The sample is transferred to the bomb for exposure to the transitional fluid. However, caution is necessary at this point to prevent the dehydration of intermediate fluids, which are quite volatile, from evaporating and drying the tissue below the critical point and thus distorting it. Cooling the bomb permits one to place either intermediate or transitional fluids in it without excessively rapid evaporation. Then the bomb can be sealed and filled with transitional fluid. Carbon dioxide and Freon have been employed because of their low critical temperatures (about 31 and 29°C, respectively) and ready availability. After filling and sealing, the temperature is elevated, either by a built-in heater or hot plate, or by contact with, or immersion in, hot water. The critical pressure for carbon dioxide is about 1080 psi, and is about 560 psi for Freon 13. During heating, if the bomb has been filled with carbon dioxide at room temperature, the pressure will rise to 1400–1600 psi and somewhat higher if the bomb was precooled while being filled. If Freon 13 is the transitional fluid, the pressure will rise to above 1000 psi, and if the bomb has been precooled, the pressure rise will be even greater. Most bombs are designed to withstand 2500 psi so it is not absolutely necessary to regulate the pressure by venting the exhaust valve while drying with carbon dioxide. However, there seems little to be gained from letting the pressure rise above 1400 psi with carbon dioxide, and it is conceivable that excess pressure may have deleterious effects. For the same reason it is recommended that, when drying with Freon, the pressure be regulated so that it does not rise above 850 psi.

The same reasoning governs the choice of drying temperatures. The low critical temperatures of carbon dioxide and Freon permit relatively low drying temperatures, with 45–50° sufficient to provide a safety margin of 15–20°, which is enough to prevent recondensation if the transitional fluid is pure. Drying at higher temperatures risks heat damage to the sample. When the drying temperature has been reached, the bomb can be vented through the exhaust valve. The rate of release should be slow, about 100 psi/min, to prevent recondensation at high pressures and steep pressure gradients at low pressures. If fluid forms in the exhaust line, it can evaporate and cause adiabatic cooling; thus it may be also necessary to heat the exhaust valve to the drying temperature.

While glutaraldehyde fixation preserves many macromolecules, most lipids are not well preserved by this fixative. Thus it is likely that there are substantial lipid losses during the subsequent exposure to osmium, dehydrating, intermediate, and transitional fluids. Lipids and elements soluble in these agents are probably extracted from the tissue. The crosslinking properties of glutaraldehyde are probably responsible for stabilizing cell structures so that they do not respond to the increased temperature and pressure encountered in the preparative process. This seems a likely conclusion since microtubules, which are responsible for the maintenance of many surface features, are depolymerized by pressure and yet these surface features are visualized in material that has been dried by the critical point method. It is also true that microtubules are more stable at higher temperature since their structure depends on the presence of hydrophobic bonds, but to what extent heat is responsible for retaining the structure of microtubules is not known.

At present there is no obvious answer to the dilemma of preserving the distribution of soluble, diffusible elements by methods that distort surface fine structure of interest to SEM studies, and methods that preserve this surface structure for SEM but permit the loss and redistribution of the elements of interest in electron probe studies. A further difficulty arises from the fact that methods of preparation for SEM analysis preserve the surface roughness that one tries to avoid in x-ray analysis.

A preparative method that promises a reasonable compromise between these conflicting requirements is the examination of frozen or frozen-dried specimens. Several workers have adapted the specimen stages of available instruments so that the stage can be kept cold enough to permit examination of frozen material with the water still in place. Boyde and Echlin[22] have viewed frozen specimens in the SEM, using beam currents of $10-100 \times 10^{-12}$ A at accelerating voltages of 1–15 kV without a conductive coating, and found no evidence that the specimens melted. They did find that the frozen sample acted as a condenser and was rapidly contaminated by oil and water vapors in the SEM. Adding other cold surfaces, e.g., liquid nitrogen-cooled baffles, between the vacuum pump and the sample space could be used to reduce this. Echlin and Moreton[23] have described apparatus that can be used to freeze tissue rapidly, evaporate a conductive coat on the tissue, and transfer the tissue to the freezing stage of a SEM, without letting the temperature of the specimen rise above $-130°$. These specimens have been used for both SEM and x-ray analysis.

A major problem with such techniques appears to be the difficulty associated with determining whether the tissue remains frozen during exposure to the electron beam. Thermocouples are too large to be useful, but observation of a frozen sample by light microscopy often can provide useful information. For example, our laboratory wished to analyze frozen red blood cells. The cells had been spread on a silicon disk substrate, in a layer only one cell thick, and rapidly frozen in liquid nitrogen. The substrate was fastened to a large aluminum block which had been cooled to liquid nitrogen temperature and contained a thermocouple. The block and substrate were transferred to the electron probe microanalyzer and the temperature of the preparation was monitored through the thermocouple. With light microscopic observation it could be seen that the cells were frozen. When the electron beam (operating voltage 22 kV, sample current 3×10^{-8} A) bombarded the cells, the cells and the surrounding matrix were immediately dried. The thermocouple registered no increase in temperature. Apparently the drying process occurred so rapidly that it could not be observed in the electron image. By viewing the cells prior to and during their exposure to the electron beam it was demonstrated that the cells had actually dried before being imaged. Unfortunately, most SEM instruments do not permit simultaneous light and electron imaging of a sample.

The drying was most probably a result of localized heating by the electron beam (see Section G). It is worth noting that the operating conditions for this experiment were more drastic than those used by Boyde and Echlin[22] in their examination of frozen red blood cells.

In the case of frozen-dried specimens, cells are rapidly frozen in a cryogenic medium, e.g., propane or Freon 13 cooled by liquid nitrogen, or by liquid nitrogen alone. The first two media are preferred even though their boiling points are higher than that of liquid nitrogen.[24] When cells or tissue samples are placed ("quenched") into liquid nitrogen alone, the nitrogen boils, forming a gas envelope around the sample. Nitrogen gas is a relatively poor heat conductor and consequently the sample is surrounded by a bubble of insulating gas and tends to freeze rather slowly. The heat conduction rates with propane and Freon are much higher (see Table I). This means that even though propane and Freon 13 are employed at temperatures higher than that of liquid nitrogen, samples quenched in these media freeze more rapidly.

Once frozen, the samples are transferred rapidly to a freeze-drying apparatus, which consists of a chamber connected to a vacuum pump and a cold trap (see Figure 3). Under the reduced pressure produced by the vacuum pump, water sublimes from the frozen specimen and is deposited on the walls of the cold trap. The cold trap is usually maintained at $-50°$ or below, but the lower the temperature of the trap, the greater will be the tendency of water molecules trapped on the walls to remain on the walls and not reenter the gas phase of the system. Thus liquid nitrogen cold traps are commonly employed. The frozen sample remains frozen due to evaporation. The temperature of the sample must be higher than that of the cold trap because an equilibrium will be established between the water associated with the sample and the cold trap, with the colder body condensing more water in proportion to the temperature differential between the two. Finally, water will stop evaporating from the sample when the partial pressure of the water in the sample equals the partial pressure of water in the rest of the vacuum system. When this point is reached the pressure in the system will be at a minimum and continued pumping will produce no further pressure drop.

Although evaporation keeps the sample frozen in a freeze-drying apparatus, some workers feel that the temperature of the drying specimen must not exceed the vitreous transition temperature of water, about $-130°$. Ingram *et al.*[25] find that it is necessary to maintain the tissue at -60 to $-85°C$ during the drying process to prevent translocation artifacts due to ice crystal formation. Our experience has been that thermocouples placed in frozen samples that are being freeze-dried in the apparatus shown in Figure 3 remain this cold or colder for most of the drying period, presumably due to heat loss from evaporation. The disadvantage of maintaining a drying sample at such a low temperature is that the time to dry even a relatively small sample may be extended from hours to weeks, due to the greatly decreased tendency of water molecules to leave the sample at these temperatures. Similarly, when the cold trap (the cold trap has dimensions of 12.5 cm diameter and 15.5 cm depth with a surface area of about 10^3 cm²) is filled with liquid nitrogen, a final pressure of 0.4–0.5 mm Hg is usually sufficient to signify drying. These conditions must be determined for each piece of apparatus. When the dried sample is removed

TABLE I. Cryomedia and Their Rate of Heat Transfer at $-79°C$ [a]

Medium	Rate, °C/sec
Propane	5860
Isopentane	2415
Propane:isopentane 2:1	4330
Freon 13	1200
Freon 14	473
Genetron 23	5410

[a] Adapted from Rebhun.[24]

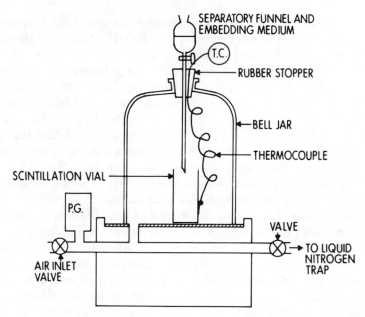

FIGURE 3.   Diagram of simple freeze-drying apparatus. P.G.—pressure gauge; T.C.—thermocouple gauge.

from the apparatus, it should be protected from atmospheric humidity by storage in a desiccator, since many such samples show a tendency to absorb water from the atmosphere. If the sample is to be embedded for sectioning, it may be immersed in the fluid embedding material before being removed from the chamber. A 1:1 mixture of propylene oxide and epoxy monomer may be used to infiltrate the tissue overnight. The sample is then transferred to fresh Araldite which has been polymerized at 60°C for one to three days (Table II).

Freezing methods are employed to fix the water of the sample in a vitreous state, thus immobilizing large and small molecules in a solid matrix. Hence no extrinsic agents are added to alter the structure or composition of the sample, and nothing is removed from the sample, unless it is freeze-dried, in which case only water and similarly volatile materials will be lost. Freeze-drying appears to offer the same advantage as critical point drying, that is, the lack of a receding liquid–vapor interface and its associated distorting forces due to surface tension. In freeze-dried material, there is a receding solid–vapor interface, which does not seem to disrupt fine details of structure. However, as the supporting matrix of frozen water is vaporized, materials suspended in the matrix may be redistributed. This process may be analogous to that of salt dissolved in water being deposited on the walls and floor of the container and not remaining suspended in space when the water is evaporated. While the rationale behind the method is sound, there are difficulties encountered when one tries to freeze biological samples rapidly enough to obtain water in a vitreous, as opposed to a crystalline, state. In order to accomplish water vitrification, the temperature must be lowered rapidly throughout the sample and be retained below the temperature at which the vitreous–crystal transition of water occurs. Studies on yeast cells and red blood cells have indicated that the rate of temperature drop must be 2000°C/sec or greater to form vitreous (instead of crystalline) ice. Obviously, cells must be frozen in media having rather large heat transfer capacities. Whatever medium is employed, large samples may not be able to attain such rates of heat transfer throughout the sample from center to periphery. The consequence of not obtaining vitreous ice is the formation of ice crystals. As these crystals form and grow, they disrupt surrounding structures and redistribute even large molecules. Zingsheim[26] has shown that the growth of such ice crystals can concentrate polystyrene particles as large as 2600 Å in diameter at the junctions where adjacent crystals meet. The deleterious effects of these ice crystals necessitate maintaining the sample temperature below the point at which vitreous ice will transform to crystalline ice and so produce disruption.

It is also possible to examine cells and particles in a hydrated state. Morgan[27] has described an SEM chamber which incorporates an electron

---

TABLE II. Procedure for Glutal Fixation, Dehydration, and Epoxy Embedment

4% glutaraldehyde in 0.1 M cacodylate buffer (pH 7.2)	60 min
0.1 M cacodylate buffer (pH 7.2)	5 min
1% osmium tetroxide in 0.1 M cacodylate buffer (pH 7.2)	30 min
50% EtOH	5 min
70% EtOH	5 min
95% EtOH	4 × 15 min
Propylene oxide (P.O.)	4 × 15 min
P.O.:Spurr (1:1)	1–4 hr
Spurr	overnight
Embed in fresh Spurr Cure 8 hr in a 70°C oven	8 hr

Standard Spurr low-velocity embedding medium: Prepared by
  gravimetrically adding the components singly into a container
  as follows:
Vinylcyclohexene dioxide (VCD)	10.0 g
Diglycidyl ether of polypropyleneglycol (D.E.R. 736)	6.0 g
Nonenyl succinic anhydride (NSA)	26.0 g
Dimethylaminoethanol	0.4 g
Mix thoroughly. May be stored for several months in a deep freeze	

detector as the base. The walls of the chamber are designed to hold a 200 mesh support for a thin, transparent film of Formvar which forms the top. A water reservoir causes the chamber atmosphere to be saturated at any given pressure. In addition, the chamber is differentially pumped from outside the column so that its pressure is greater than that of the column but less than atmospheric pressure. Cells are placed on the Formvar top (on the inside) and examined. The electron beam penetrates the Formvar to irradiate the cells with little loss of energy or gain in beam size. X-rays generated by the beam exit through the Formvar top to the spectrometers, while the scattered electrons from the beam continue through the cells to the electron detector at the base, providing a configuration for scanning transmission imaging.

E. Sections

Many samples too large to be examined directly have interior portions that are of interest. These samples must be sectioned; depending on the purpose of the analysis, they can be conveniently considered in two categories: thick and thin sections. Sections are considered thick or thin in terms of ρt, the product of density and thickness, and the accelerating voltage employed. The range of densities encountered in biological studies is from about 0.8 to about 3.0, as can be seen in Table III. As can be seen from Figure 4, a section with a density of 1 g/cm³ and a thickness of 15 μm will transmit some 20-keV electron beam and be "thin." The same section would completely absorb a 10-keV beam and be "thick." Consequently, the terms thick and thin can only be used when the density of the specimen, the accelerating voltage employed, and the actual thickness are specified.[28]

1. THICK SECTIONS

Thick sections can consist of only one prepared surface, for example, a small piece of fractured bone or shell. Unless the fractured surface is polished, the surface may be so rough as to prevent accurate x-ray analysis, and may even exceed the depth of focus of the electron beam in the SEM. Thus, fracture surfaces are usually polished prior to analyses. Sawn surfaces of hard samples may also have to be polished before analysis.[29] Thick sections can also be produced by microtomy, but this technique is more commonly associated with thin sections and will be discussed under that heading. Replicas of freeze-fractured and/or freeze-etched material can also be examined with the SEM, but replicas are not useful for x-ray

TABLE III. Densities of Representative Biological
Materials: Anhydrous and Embedding Materials [a]

Material	Density
Gelatin	1.27
Silk	1.56
Amino acid crystals	1.1–1.8
Fat	0.9
Nucleic acids	1.6–1.8
Starch	1.5
Cellulose	1.3–1.6
Sucrose	1.6
Bone	1.7–2.0
Shell (calcium carbonate)	2.8
Butyl polymethacrylate	1.0
Methyl polymethacrylate	1.2
Epoxies	1.2–1.3
Polyester	1.3–1.4
Carbowax	1.2
Paraffin	0.9–1.0
Collodion	1.7

[a] Adapted from Wachtel *et al.*[30]

analysis. Both freeze-fracturing and freeze-etching have many steps in common, freeze-etching being a variation of the more general freeze-fracturing techniques (see Figure 5). In both techniques, the sample to be analyzed is rapidly frozen in a small container (often the sample is first suspended in a "cryoprotectant" such as 2% glycerol, which hinders the formation of disruptive ice crystals during freezing). Once frozen, the sample is transferred to a cold stage in a vacuum evaporator and is evacuated to about 10^{-4} Torr. The sample is then fractured or cleaved by any of a variety of methods. A microtome knife or a scalpel blade can be used. The sample container itself can be hinged so that the sample can be fractured near the center in order to preserve both complementary fracture faces. For a freeze-etched preparation, after fracturing, and prior to replication, the temperature of the sample can be raised slightly so as to evaporate a thin layer of water. Thus, portions of the sample that would be covered by the frozen water can be exposed. Then, for both freeze-fractured and freeze-etched preparations, a replica of the frozen surface is made. Usually a fine coating of platinum–carbon is evaporated onto the sample from a single angle, producing a strong shadowing effect. Next, a heavy backing coating of carbon is evaporated from directly above or with a rotating sample, producing a replica which is still attached to the sample. The sample is removed from the vacuum and digested,

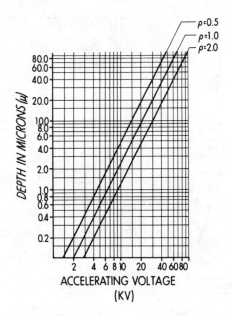

FIGURE 4.   Depth of electron beam penetration as a function of accelerating voltage and density (calculated from Andersen and Hasler[57]).

usually by immersing the whole preparation in 5% sodium hypochorite. The replica is picked up on a copper grid or aluminum stub for analysis. These preparations can be analyzed with TEM as well as SEM and are very useful for evaluating surfaces not readily visible within cells and tissues. Because of the strong shadowing effect, these preparations, when examined with TEM, appear similar to those viewed with the SEM.

2. Thin Sections

Figure 4 shows that most thin sections will be thinner than 15 μm. To produce such sections microtomy must be employed. The major difficulty encountered with microtomy is that most animal and plant tissue is too soft to resist the severe distorting forces encountered when cutting thin slices. For this reason, the tissue must be strengthened by being infiltrated with a supporting medium. The tissue may have its water content removed and replaced by another fluid which can be hardened and thus form a supporting medium; or the tissue may be frozen, in which case the watery cytoplasm and extracellular fluid become the supporting medium.

The biological materials most often thin-sectioned are of variable density (see Table III) and have low elasticity and plasticity. Dry tissue

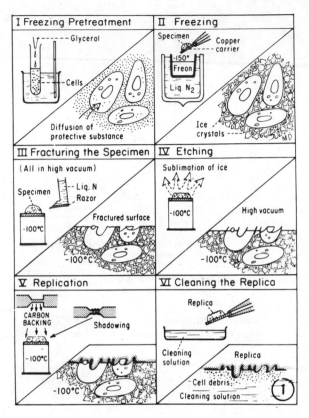

FIGURE 5.   Diagrammatic representation of freeze-fracture technique (adapted from Koehler[60]).

would deform permanently under the pressure of the knife edge advancing through the tissue. Further, the deformation would be highly variable, being slight in hard, stiff regions and relatively great in soft, plastic regions. The use of an embedding material circumvents these difficulties. The embedding material fills all the spaces of the tissue with hard material. This prevents hard regions of the sample from being pushed into softer regions. Further, the elasticty of the embedding material is such that any deformation suffered by the section during cutting can be reversed. A comprehensive discussion of the mechanical properties of various embedding materials is given by Wachtel et al.[30] and is beyond the scope of this chapter. Suffice it to say that epoxy resins such as Epon, Araldite, and Spurr's low-viscosity mixture[31] have proven quite useful and reliable for preparing sections as thin as 50–60 nm and as thick as several micrometers.

Freezing is employed in three general types of sectioning procedures: freeze-drying, freeze-substitution, and frozen sectioning. Freeze-drying, as discussed previously, usually involves freezing the tissue in a cryogenic medium, placing the frozen tissue in a vacuum to evaporate the water in the tissue, and finally infiltrating the dry tissue with a fluid anhydrous embedding medium, which is subsequently hardened. A representative freeze-drying procedure is given in Table IV.

Freeze-substitution removes the water from the frozen tissue by replacing it with a water-miscible solvent followed by infiltration and embedding with a material miscible with this solvent. Frequently, acetone is used as the freeze-substitution solvent. The tissue is frozen by rapid quenching in Freon 22 or Genetron 23 cooled by liquid nitrogen. Once frozen, the specimens are transferred to liquid nitrogen in scintillation vials. The vials are closed with rubber stoppers lubricated with a thin coat of petroleum jelly. The stoppers are pierced by two hypodermic needles, one for venting and the other to permit the addition of fluid. A needle smaller than No. 20 makes fluid addition difficult. The stoppered vial is then transferred to a cold bath, either CO_2–ethanol, CO_2–acetone, or an ice chest containing solid CO_2. The vaporizing nitrogen is replaced with acetone precooled in a hypodermic syringe. (Acetone that contains osmium or other fixative may be added at this point.) After two weeks, or more, fresh, cold acetone is added to the vial to flush out the old acetone, and the vial is then brought to room temperature. Epoxy monomer can then be added and mixed with the acetone. The acetone is permitted to evaporate overnight and the next day the sample is transferred to fresh epoxy monomer plus accelerator and polymerized at 60°C for one to three days.[24]

Frozen sectioning involves freezing the tissue rapidly and keeping it frozen to provide a supporting matrix for microtomy.[32,33] The section itself is then usually freeze-dried or examined while in the frozen state. The

TABLE IV. Procedure for Freeze-Drying Followed by Embedding

1. Cut specimen into small pieces, <1 mm
2. Rapidly quench pieces by submerging in Genetron 23 or Freon 22 cooled by liquid nitrogen
3. Transfer in cryogen to chamber of freeze-drying apparatus, and seal chamber
4. Evacuate chamber slowly to prevent cryogen from vigorous boiling
5. Continue pumping until vacuum reaches plateau level
6. Tissue sample may be exposed to osmium vapor for several hours
7. Add Spurr's low-viscosity medium and propylene oxide (1:1) under vacuum
8. After 1 hr, remove sample from vacuum and permit propylene oxide to evaporate overnight
9. Transfer to fresh Spurr, polymerize at 70°C for 8 hr

temperature of the sample must be maintained below $-60°C$ in order to prevent the growth of ice crystals; conventional cryostats do not meet this requirement. There are several commercially available attachments for ultramicrotomes that can accomplish this task. These attachments are usually cooled by liquid nitrogen or nitrogen gas vaporized from liquid nitrogen reservoirs. Temperatures of $-140°C$ can usually be maintained in these instruments. Sections are microtomed using dry glass or diamond knives and transferred to substrates with a brush made from a single hair attached to a glass rod or with fine tungsten needles. Christensen[32] recommends fastening sections to an EM grid by "rolling" the rounded end of a chilled copper rod over the grid. Portions of the section that make contact with the grid bars will be attached by being crushed onto the grid. This procedure fails with opaque substrates such as silicon disks and it has been difficult to devise a satisfactory, reliable method of attaching a majority of 1–4-μm thick sections to such substrates. The only recourse at present is to cut more sections than are needed, and to press the corners and edges of the sections to the silicon substrate using fine tungsten needles. Of 10–15 sections so treated, one or two will usually be flat and in sufficiently good contact with the substrate so that no conductive coating is necessary.

The sections, still frozen and on the silicon substrate, are then transferred to a massive aluminum block maintained at the temperature of liquid nitrogen. The block is then placed in the vacuum chamber of the freeze-drying apparatus seen in Figure 3. The chamber is quickly evacuated and the pressure achieves a plateau at 5 mm Hg within 30 min.

Tissue can also be fixed prior to embedding and sectioning. The purpose of fixation is to prevent postmortem changes in structure of the tissue during any remaining manipulations. There appear to be two processes of importance in fixation: inactivation of catabolic enzymes and crosslinking of macromolecules to each other. The first process prevents the degradation of the tissue by the proteases, nucleases, lipases, phosphatases, etc. that are within a cell but are normally kept under control until vital functions fail. The second process induces a structural matrix of neighboring macromolecules by forming chemical links between them. The most commonly used chemical fixation is glutaraldehyde, and a typical fixation protocol employing this fixative is given in Table II. Osmium tetroxide is frequently added after the glutaraldehyde to assure complete fixation. Furthermore, OsO_4 provides a material of high atomic number that will produce and enhance contrast in the TEM as well as in backscattered electron and sample current images.

A comparison of the results obtained with three of the methods described is given Table V. These results will not be discussed in greater detail, but it is worth noting here that the conventional preparative technique has caused a loss of most of the potassium as well as a substantial portion of the sodium that is found in the cryosection. The freeze-dried preparation has lost most of the potassium and about 40% of the sodium found in the cryosection. The calcium and phosphorus contents are more difficult to assess and probably reflect an apparent increase in these materials due to a loss of lipid and other material during preparation.

F. Microincineration

A useful ancillary technique for electron probe analysis is microincineration.[34] A section or particle is "ashed" to oxidize and volatilize organic materials, leaving only inorganic residues of nonvolatile materials. Thus, one can take a relatively thick sample, ash it, and produce a thin inorganic skeleton. This procedure effectively increases the concentration of the inorganic material and also "thins" the sample, which in turn may help increase resolution or penetration by the beam. A typical example of the use of this technique is the preparation of sections too thick to permit useful TEM at conventional (<100 kV) accelerating voltages but providing sufficient material to be detectable with electron probe analysis. The distribution of elements can be mapped in the section by x-ray techniques. Then the section can be ashed and examined with TEM to determine what subcellular structures are present at sites where elements are concentrated. Schraer and Hohman[35] have used this technique to investigate calcium transport by the shell gland of chicken. Microincineration has been reviewed in detail by Thomas.[34]

TABLE V. Analysis of Mineralized Portions of Thin (1 μm) Sections of 3-Day-Old Chick Calvarium

	Glutal[a] (N = 8)	Freeze-dried[a] (N = 12)	Cryosection[a] (N = 12)	Bulk[b]
Wt % Ca	21.22 ± 0.37	23.70 ± 1.40	18.93 ± 1.50	25.0
P	12.39 ± 0.25	11.94 ± 0.66	10.60 ± 0.87	12.4
K	0.03 ± 0.01[c]	0.02 ± 0.01[c]	0.14 ± 0.01	0.06
Na	0.11 ± 0.01[c,d]	0.27 ± 0.03[c]	0.45 ± 0.09	0.55
At % Ca/P	1.33 ± 0.01[c,d]	1.52 ± 0.01[c]	1.45 ± 0.01	1.56

[a] Electron probe analysis, x-ray correction according to BASIC procedure.
[b] The bulk analysis was of 16 calvaria pooled and with tissue (endosteum and periosteum) still attached and is thus not directly comparable to microanalysis of just mineralized portions.
[c] Significantly different ($p < 0.05$) than cryosection.
[d] Significantly different ($p < 0.05$) than freeze-dried.

TABLE VI. Amount of Various Elements Recovered from Whole Blood Ashed by Either Low-Temperature Oxygen Plasma Microincineration (LTM) or Thermal Microincineration (TM) [a]

Element	LTM	TM
Antimony	99	35
Arsenic	100	0
Cobalt	102	67
Copper	101	87
Chromium	100	85
Gold	70	0
Iron	101	52
Manganese	99	85
Silver	72	45
Zinc	99	69

[a] After Thomas.[34]

The technique of microincineration has been available for some time—it was first employed in 1833 according to Thomas[34]—but suffered from a serious drawback that prevented its widespread use. Until recently, microincineration has been carried out with heat; usually a sample was heated to 500°C in an oven for a period of minutes to hours to produce complete ashing. Heat induced thermal agitation in the samples, resulting in disruption of the fine structure of the ash and accompanied by severe translocation artifacts from crystallization or eruptive boiling. This drawback has been eliminated through the use of low-temperature incineration by an oxygen plasma. Briefly, a typical low-temperature microincineration device consists of a chamber surrounded by a radiofrequency coil and into which the specimen is placed. The chamber is sealed and connected to a vacuum pump and to a supply of oxygen. The oxygen is supplied to the chamber at about 50 cm³/min. and the chamber is maintained at a pressure of about 1 mm Hg by the vacuum pump. Atomic oxygen as well as other highly reactive species are created in the plasma that fills the chamber. As the organic molecules oxidize, the volatile products are swept out through the vacuum pump. A further advantage stems from the fact that materials that would have been volatilized by heat are retained. This advantage is illustrated in Table VI.

G. Coating

The purpose of coating is to reduce heat damage by thermal effects of the electron beam and to eliminate the buildup of surface charges, which can interfere with imaging and analysis.

The extent of electron beam heating can be estimated from the formula provided by Reed[36] and attributed to Castaing

$$\Delta T = 4.8 E_0 i / k d_p \tag{1}$$

where ΔT is the temperature rise in °C that occurs in the specimen at the point of impact of the beam, E_0 is operating potential in kV, i is the beam current in μA, k is the thermal conductivity in W/cm °C, and d_p is the probe diameter in μm. Metals frequently have a thermal conductivity $k = 1$, producing a temperature rise of only 10°C or so, but plastics have a value of $k \simeq 2 \times 10^{-3}$, which produces temperature rises of about 500–1000°C. Undoubtedly, this thermal effect is at least partially responsible for the mass loss in specimens exposed to electron beams. By way of example, the presence of a 20 nm (200 Å) thick aluminum coating can reduce ΔT from 1430 to 515°C[37] on certain nonconductive samples. The gross effects of a lack of electrical conductivity in a sample are readily recognized by great fluctuations in count rates from the same site, the appearance of "flashes" or "streaks" in an electron image on a cathode ray tube, and even on a fluorescent sample when observed with a light microscope.

It is worthwhile to remember that any conductive coating applied to a sample will act to absorb x-rays. The extent of this absorption can be calculated from the equation

$$I = I_0 \exp[-(\mu/\rho)\rho x] \tag{2}$$

where I_0 and I are the x-ray intensities before and after passing through the absorber of thickness x, density ρ, and mass absorption coefficient μ/ρ. Thin specimens mounted so that they make intimate contact with a conducting substrate frequently require no further coating. In our laboratory sections of epoxy-embedded tissue up to 2 μm in thickness mounted on silicon disks or aluminum-coated quartz slides have been successfully analyzed. If, however, the sample is thick or nonconducting, a surface coating may be necessary. The thickness of the coating can be monitored accurately through the use of quartz crystal oscillators, although once a satisfactory coating thickness has been found, simpler methods of estimating thickness may be sufficient. Among these are measuring the resistance of the coating on a slide placed near the sample, or observing the interference colors on a polished metallic surface. In order to ensure even coating, even of relatively smooth surfaces, it is recommended that the sample be rotated during coating, and several manufacturers offer devices for this purpose.

Pfefferkorn[38] has recently reviewed techniques for coating samples. He points out that any coating must be: (a) continuous, and not in aggregate form; (b) uniform in thickness and density, and, especially for SEM studies; (c) free of structural features of its own; and (d) thicker than 100 Å so that the SEM image comes entirely from the metal layer. Echlin and Hyde[39] have tabulated various coatings and their properties. A generally useful method of coating is to use a first thin coat of carbon— about 5–10 nm (50–100 Å) should be evaporated—followed by a 10 nm (100 Å) layer of gold, and a final layer, about 5–10 nm (50–100 Å) thick, of carbon. The first carbon layer seems to promote adhesion of the gold, and the last carbon layer reduces contamination. A single coating, about 20 nm (200 Å) thick, of platinum–carbon has also been found useful.[17,38]

Whenever evaporation by heat is employed it is recommended that the specimen be at about 10^{-4} Torr and about 20 cm from the source. While the metal to be evaporated is being heated, prior to evaporation, a baffle or shield is placed in front of the sample to protect it from heat. This is removed before evaporation. A second baffle close to the source remains in place during evaporation. The purpose of this shield is to ensure that the specimen is covered by atoms that have been scattered from the walls of the evaporation chamber and by the remaining atmosphere within the chamber. This will prevent a "shadowing" phenomenon and help to produce a continuous coating.

II. ANALYSIS

Three types of investigations are most common in electron probe analysis: qualitative, in which one is concerned with whether an element is present in a sample or at a certain site within a sample; semiquantitative, in which one wishes to know whether one site in a sample contains more, less, or the same amount of a certain element than another site or sample; and quantitative, in which one wishes to know the amount of an element present in either weight percent or atom percent. The principles behind analysis are the same whether one employs energy-dispersive or wavelength-dispersive spectrometers but the procedure one employs with each can differ somewhat. These will be mentioned where appropriate.

A. Qualitative Analysis

Qualitative analysis is the simplest to perform; the situation usually encountered is a sample in which one is asked to decide whether one or several elements are present. For most cases it is wise to request or perform a bulk analysis of the material first. This serves to provide a rough estimate of whether there is likely to be enough material present to exceed the minimum detectability limits of about 0.01%. If, however, the element of interest is likely to be concentrated in one portion of the sample, the concentration calculated from bulk analysis is not particularly useful. Bulk analysis assumes that the element is distributed homogeneously throughout the volume of the sample. If the element is evenly distributed through the sample, then a series of spot analyses should be performed. For wavelength-dispersive spectrometers, a selected site is positioned under the beam and the x-ray intensity of the peak of interest is measured. The criterion for establishing whether an element is present or absent is often taken to be that the intensity of the peak of interest must exceed the background intensity by at least three standard deviation units, as was discussed in Chapter XII.

The background intensity is measured by changing the spectrometer so that two portions of the spectrum, one at the high-wavelength side and one at the low-wavelength side, can be counted, and the average of these two intensities is used. These positions are chosen somewhat arbitrarily, usually by examining the spectrum in the region of the peak, and then moving anywhere from 0.01–0.5 Å away from the peak to regions of the spectrum free of specific peaks.

With energy-dispersive spectrometers one focuses the beam on a standard and records the entire x-ray spectrum. Then one moves the sample under the beam and compares the spectrum from the sample with that from the standard. If coincident peaks occur in both spectra, one can conclude that the element of interest is present in the sample, with the proviso, as before, that the peak intensity must exceed the background intensity by at least three standard deviations. The background intensity can usually be estimated from the display system of the energy-dispersive unit. Where overlapping peaks occur it may be necessary to look for the presence of several peaks from the element of interest in order to establish its presence. Spectrum stripping techniques available with energy-dispersive units must be used if such other peaks cannot be located. The factors influencing background measurement in energy-dispersive spectrometers have been discussed in detail in Chapter VII.

When there is uncertainty concerning the presence of a localized concentration of the element of interest one can move the sample under the beam in a regular pattern and monitor the output of the x-ray detector most conveniently by listening to the audio signal from a ratemeter. Where the concentrations are relatively low, localization may not be detected in this way. It is a better procedure to take x-ray images over periods as long as the stability of the instrument will allow (e.g., 30–45 min). If these images indicate the possible presence of localized concentrations, such areas may be tested with a static beam. With wavelength-dispersive spectrometers, the x-ray images should be recorded at high magnifications in order to minimize movement of the beam away from the Rowland circle, which would decrease sensitivity.

One may wish to determine, in a semiquantitative way, whether there is more of an element in one sample as compared to another, or at one site in the sample compared to another site in the sample. If the two samples or sites to be analyzed are similar in thickness, density, and average atomic number, then the analysis is straightforward. One first establishes the presence of the peak by the techniques outlined above, and then compares the relative x-ray intensity in each. It is generally useful to compare the peak-to-background ratios in each sample or site, rather than the x-ray intensities themselves. Then it can be concluded that the sample or site with the higher P/B ratio has the greater concentration of the element of interest.

If the two samples or sites are of different thickness (so that one transmits more of the beam than the other) or of different densities, then the x-rays may emit from volumes of significantly different sizes, and direct comparison may not be valid. Similarly, if one sample or site contains a higher proportion of high-atomic-number elements than the others, then the number of x-rays emitted by the element of interest may differ even if the same amount is present in both samples or sites. If these differences between samples do occur, then it may be necessary to perform a complete quantitative analysis.

B. Quantitative Analysis

1. Models

The method requiring the least computation involves making standards with properties similar to the sample and finding the standard that gives x-ray intensities identical (within limits acceptable to the operator) to the sample. The sample and standard with the same x-ray intensities will have the same concentrations. A specific example of this method has been described by Ingram and colleagues[25,41] in which the purpose was to measure the amounts of sodium and potassium in erythrocytes. Suspensions of the erythrocytes were freeze-dried and embedded in plastic. Solutions of gelatin were made which contained various amounts of sodium and potassium and these were freeze-dried and embedded in plastic just as the erythrocytes. Both erythocytes and gelatin standards were sectioned to the same thickness and the intensities were compared. This technique has the advantage of not requiring extensive computation and having very little dependence on theory or modeling. The disadvantages are that the accuracy depends on how much similarity there is between sample and standard. If the sample is complex, containing many elements, e.g., Na, K, P, Mg, Cl, Ca, C, N, O, Fe, and if the sample is inhomogeneous, having phases of different densities and compositions, then the number of standards required may be so large as to eliminate the advantages gained from lack of computation and independence of theory. Similar methods have been employed by others with reasonably good results.[42,43]

2. METHODS OF HALL

Hall,[44] Marshall and Hall,[45] and Hall[46] have reported a technique that is useful for the analysis of thin sections mounted on substrates of low-atomic-number material, and which are thin enough so that the Brehmsstrahlung radiation from them is very low. In practice, this substrate is usually a film of Formvar, vacuum-evaporated carbon, or nylon less than 2000 Å thick. The section must also be thin, in terms of the mass thickness ρt. In practice, 5–7 μm seems to be the upper limit of useful thicknesses.

The principle of the quantitation procedure used is based on the relationship between mass thickness and continuum radiation that was originally proposed by Kramers.[47] This relationship states that there is a proportionality between mass thickness ρt and intensity of continuum radiation I_λ:

$$I_\lambda \propto \rho t \tag{3}$$

This relationship is very important because it means that measuring I_λ will provide a monitor for any changes in ρ or t from place to place in a section, and neither ρ nor t will have to be measured directly.

The other relationship of importance is the familiar one that the intensity of characteristic radiation I_p is proportional to the concentration C (in weight percent) of the element of interest present:

$$I_p \propto C \tag{4}$$

Combining these two relationships permits one to estimate concentration, already corrected for variations in mass thickness, and to calculate the relative concentration from site to site:

$$I_p/I_\lambda \propto C/\rho t \tag{5}$$

In order to use this method, one must only measure the characteristic radiation using either a wavelength- or energy-dispersive spectrometer and then measure a representative portion of the continuous spectrum, free from characteristic peaks, using an energy-dispersive spectrometer.

In order to calculate the absolute concentration, a similar relationship is used. Unfortunately, the theoretical basis for this relationship has been called into question by Philibert and Tixier[28,48] and Lifshin et al.,[49] who have shown that there is a variable and not a constant relationship between I_λ and ρt. However, if one employs a standard with a matrix similar in mean atomic number and mean density to that of the sample, then the following modification of the method can be used.

The assumption is made that the relationship between I_λ and ρt will be the same if sample and standard are similar in average atomic number (see Figure 23, Chapter VII), and this assumption, though not theoretically supported, seems justified in terms of results. Thus for a limited set of conditions

$$I_\lambda = F\rho t \tag{6}$$

where F is an arbitrary constant of proportionality. One also assumes, and this too is supported in practice, that characteristic x-rays will be generated in and detected from sample and standard in the same way if sample and standard are sufficiently similar; thus

$$I_p = F'C \tag{7}$$

where F' is a constant of proportionality for a limited set of conditions.

Hence, one can state

$$\frac{I_p^*/I_\lambda^*}{I_p/I_\lambda} = \frac{F'C^*/F(\rho t)^*}{F'C/F(\rho t)} \tag{8}$$

where I_p^* and I_λ^* denote, respectively, characteristic and background intensity in the sample and I_p and I_λ denote the same for the standard. Here C^* and $(\rho t)^*$ denote, respectively, concentration of the element of interest and the mass thickness of the sample; C and (ρt) are the same for the standard. In this manner the relationship becomes

$$\frac{I_p^*/I_\lambda^*}{I_p/I_\lambda} = \frac{C^*/\rho t}{C/(\rho t)^*} \tag{9}$$

This is referred to as the "relative method" since it requires a comparison to a thin standard of similar composition. An "absolute method" has also been devised that varies slightly, depending on the standard employed.

In order to use this method, one first measures the intensities at the peaks of interest. The spectrometers are then reset to determine the mean background intensities to be subtracted from the peak intensities so as to provide the true intensity of the characteristic x-rays. The continuum intensity is measured with an energy-dispersive spectrometer while the beam is on the area of interest. Then the beam is moved to a region of the specimen substrate and the continuum intensity is measured again. This intensity is subtracted from the continuum intensity measured from the specimen to give a true value for the specimen continuum. This, of course, requires that the specimen substrate and the continuum intensity is measured again. This intensity is subtracted from the continuum intensity measured from the specimen to give a true value for the specimen continuum. This, of course, requires that the specimen substrate be uniform throughout the preparation. These measurements result in the true ratio of characteristic to continuum radiation. It is known, then, that the intensity I_p of characteristic radiation is proportional to the number of atoms n of the element of interest; Using the disputed Kramers' relationship, [47] one assumes that the intensity I_λ of continuum radiation is proportional to the mass thickness, that is, the sum of the product of the number of atoms of each element and the square of their atomic numbers, or:

$$I_p/I_\lambda = n/\sum_r n_r Z_r^2 \tag{10}$$

where n is the number of atoms of the element of interest, n_r is the number of atoms of each of r elements in the sample, and Z is the atomic number of each of the r elements in the sample.

A comparison of sample and standard gives

$$\frac{(I_p/I_\lambda)^*}{I_p/I_\lambda} = \frac{(n/\sum_r n_r Z_r^2)^*}{n/\sum_r n_r Z_r^2} \tag{11}$$

where the asterisk denotes values pertaining to the sample.

In order to introduce weight fractions, the following identity is used:

$$C = nA/\sum_r n_r A_r \tag{12}$$

where C is the weight fraction of the element of interest. Substitution of equation (12) for the expression $(n/\sum_r n_r Z_r^2)^*$ in equation (11) produces

$$\frac{(I_p/I_\lambda)^*}{I_p/I_\lambda} \left(\frac{A \sum_r n_r Z_r^2}{\sum_r n_r A_r} \right)^* = \frac{C}{n/\sum_r n_r Z_r^2} \tag{13}$$

which can be rewritten as

$$C = A^* \frac{(I_p/I_\lambda)^*}{I_p/I_\lambda} \frac{n}{\sum_r n_r Z_r^2} \sum_r \frac{n_r Z_r^2}{n_r A_r} \tag{14}$$

A value for $\sum_r (n_r Z_r^2 / n_r A_r)$ can be calculated by assuming a common composition for biological tissues, and Hall[46] gives a value of 3.28 for this expression. The value of $n / \sum_r n_r Z_r^2$ can be calculated from the stoichiometry of the standard, and A is known. Thus, to employ this technique, only the ratios of peak to continuum from sample and standard must be measured.

The advantage of this technique is that it is relatively simple to use, can be employed with only energy-dispersive spectrometers, and has produced relatively accurate results. The disadvantages arise from the uncertainties about the Kramers relation for continuum radiation and thickness plus the need for standards that are either thin or that can be thin-sectioned. Finally, the sample must be mounted on a thin, uniform support. Preparing such supports and attaching thin sections to them in such a way as to produce flat samples with good thermal and electrical contact often proves troublesome. Nevertheless, several workers have reported good results using this method.[50]

3. BASIC AND BICEP

In our laboratory we have employed two correction techniques, BICEP and BASIC.[51,53] BICEP is essentially a modification of Colby's thin-film model[54] and was designed primarily to calculate ratios of elements present. The use of ratios is often useful in biological studies, since the volume analyzed in a thin film may be difficult to determine, and the use of ratios of atomic percent is comparable to mole ratios frequently employed in biological studies. The computer program is extensive and is usually run on an IBM 360. The data correction procedure is based on the atomic number correction according to Colby[54] [Chapter XII, equation (25)]. In this formulation, equation (15), the concentration C_A of the element of interest is proportional to the ratio of corrected characteristic x-ray intensity from a thick standard of known composition, represented by the numerator, and the corrected characteristic x-ray intensity from the sample, represented by the denominator

$$C_A = \frac{k_A R_A \int_{E_L}^{E_0} (Q/S_A)\, dE}{R^* \int_{E_L}^{E_0} (Q/S^*)\, dE + \eta_s \int_{E_L'}^{E_L} (Q/S^*)\, dE} \tag{15}$$

In this equation, A denotes the element of interest, the asterisk denotes values for the thin-film sample, R is the backscatter loss factor, Q is the ionization cross section, S is the stopping power, η_s is the electron backscattering coefficient for the substrate material. In addition, E is the energy of the electron, E_0 is the energy of the impinging electrons, E_c is the critical excitation potential, E_L is the mean electron energy at the substrate–specimen interface, and E_L' is the mean energy of electrons backscattered from the substrate as they leave the upper surface of the specimen. Figure 6 shows a diagrammatic representation of electron paths through a thin film placed on a thick substrate. Electrons may be backscattered from the film (trajectory 1 in Figure 6) or enter the thin film with energy E_0, and after traveling through the thin film enter the substrate with diminished energy E_L (trajectory 2 in Figure 6). To account for this the expression $\int_{E_L}^{E_0} (Q/S^*) \, dE$ is integrated from E_0 to E_L. Some electrons having passed through the thin film will be backscattered by the substrate (trajectory 3 in Figure 6). To account for this, an additional term $\eta_s \int_{E_L'}^{E_L} (Q/S^*) \, dE$ is added and integrated from E_L to E_L', the energy at which the electrons leave the thin film. The values for E_L and E_L' are calculated from the range equation of Andersen and Hasler[56] (see Chapter III):

$$\rho t = 0.032(A/Z)(E_0^{1.68} - E_L^{1.68}) \tag{16}$$

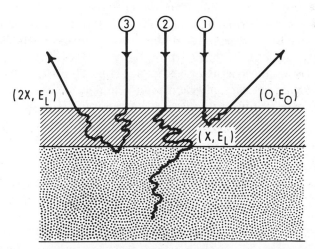

FIGURE 6. Electron paths through a thin sample mounted on an electron-opaque substrate. Film thickness X, accelerating voltage E_0. (Adapted from Colby.[55])

However, this equation requires the density and thickness or the product of the two. The correction procedure is of limited use for absolute values of concentration if ρ, t or ρt is not known, although it is still useful for the calculation of ratios.

In the BICEP method absorption in sample and standard is corrected according to the PDH equation (Chapter IX), with the exception of the σ coefficient. This coefficient accounts for the voltage effect on depth of x-ray production. It is formulated as $4.5 \times 10^5/E_0^{1.65}$ in the correction for standards. However, σ for the sample is taken as $4.5 \times 10^5/[E_0^{1.65} - (E_L')^{1.65}]$ to compensate for the fact that the x-rays will not be generated as deeply as in the standard and therefore will suffer less absorption before exiting from the surface of the sample. Characteristic fluorescence in sample and standard is corrected by the method of Reed.[56] Continuum fluorescence is not considered. To use the program it is necessary to measure peak and background intensities for all elements present in amounts as large as 1 or 2% from samples and standards. One element can be calculated as a ratio with respect to any other, and one element can be calculated by difference. In practice, the organic matrix is assumed to be all carbon and is calculated by difference. The values that must be read into the BICEP program are peak and background intensities for all elements and standards, the chemical symbols for the elements analyzed, the substrate material, counting times, accelerating potential, and takeoff angle Ψ.

The method has the following advantages. It uses conventional thick standards, it is based on the well-tested ZAF correction procedure (Chapter IX), and the accuracy is about $\pm 10\%$. The main disadvantage arises from the need to know ρt. Errors in ρt will influence the concentration calculated for each element. Another disadvantage is that, unlike the Hall technique, in which only one element need be measured, BICEP requires that all elements present in concentrations greater than 1 or 2% be measured. Finally, the program can only be run on a large computer facility.

In order to circumvent the necessity of knowing ρt, another procedure was formulated as follows. The characteristic signal from the substrate is a function of the energy of the impinging electrons. Thus, if one could measure the x-ray intensity of the characteristic radiation from the substrate beneath the sample, one would know the energy of the electrons that have traversed the sample to strike the substrate, i.e., E_L of trajectory 2 in Figure 6. E_L would be determined by the density and thickness of the sample through which the electron had traveled. Thus ρt could be calculated from the electron range [equation (5)].

In other words, when the electron beam is placed on the substrate, characteristic x-rays will be generated according to the relation

$$N_C = F(E_0 - E_c)^{\text{exp}} \tag{17}$$

where N_c is the intensity of characteristic radiation from the substrate, F is the efficiency of x-ray production,[57] Chapter IX. The value of the exponent is determined by the overvoltage ratio E_0/E_c. At 22 kV for a silicon substrate this exponent is estimated to be 1.63 from extrapolation of Green's data.[57] This value of the exponent was later confirmed by Monte Carlo calculations of Kyser and Murata.[58] When the electron beam is then moved to a portion of the substrate covered by the sample, characteristic radiation will be generated according to

$$N_{C'} = F(E_L - E_c)^{\exp} \tag{18}$$

where N_c' is the intensity of characteristic radiation from the substrate covered by the sample, and the value of the exponent is the same as in equation (17) for a thin section. The ratio of these two equations permits calculation of E_L:

$$\frac{N_C}{N_{C'}} = \frac{F(E_0 - E_c)^{\exp}}{F(E_L - E_c)^{\exp}} \tag{19}$$

and

$$E_L = E_c + \left(\frac{N_{C'}}{N_C}\right)^{1/\exp}(E_0 - E_c) \tag{20}$$

Thus this calculation can provide an internal compensation for variations in density or thickness.

The intensity of characteristic radiation from the substrate will be attenuated by absorption as it passes through the specimen lying on top of the substrate. In the BASIC method, absorption from this source is corrected by an empirical relationship,

$$\{\exp[-(\mu/\rho)\rho t \csc \Psi]\}^{0.8} \tag{21}$$

Equation (21) was tested with a series of aluminum films of known ρt and found to give optimum results.[53]

In order to use the method, one measures the characteristic and background radiation from the exposed substrate. The next step is to move the beam to the regions to be analyzed and measure the characteristic and background intensity at the site. Next, the peak and background intensities for each of the elements present at the site are analyzed. As with BICEP, one element can be analyzed by difference and one by ratio to another element. Peak and background intensities are measured from conventional, thick standards. The atomic number, fluorescence, and absorption corrections are calculated as in BICEP. The only data read into the computer program are peak and background intensities for exposed substrate, substrate covered by sample, chemical symbols of the elements analyzed, peak and background intensities for the elements analyzed, counting times, accelerating voltage, and takeoff angle, Ψ.

The advantages of this procedure are the intrinsic compensation for variations in ρ and t, the relative accuracy, $\pm 12\%$, the ability to handle up to 12 elements, and the use of thick standards in the calculation. Eshel[61] has shown that corrections for density and thickness are of primary importance in the analysis of thick sections. The empirical nature of the correction for absorption of substrate radiation by the overlying sample, and the necessity of analyzing virtually all the elements at a selected site, are disadvantages of the method. In addition, sizable computer facilities are needed. However, in practice, elements present in low concentrations can be assigned to the organic matrix and need not be measured.

III. SUMMARY

A variety of techniques are available for the preparation of biological materials for analysis by electron probe, microanalysis, and scanning electron microscopy. At present, however, it is not possible to predict the effects any particular preparative procedure will have on any given sample. This places the burden of testing the procedure for its validity in any analysis squarely on the shoulders of the analyst. At the minimum the technique chosen should be tested to discover: (1) to what extent it preserves natural morphology; (2) whether the elemental composition of the sample is altered during preparation; and (3) whether the natural distribution of elements in the sample has been distorted. A major effort is often required to establish these factors. However, the results of analysis are of dubious value until these factors are known.

Conventional methods of qualitative and semiquantitative analysis are applicable to properly prepared biological samples. Several well-tested quantitative methods are available for the analysis of thick biological samples. For thin biological samples, each of the available quantitative methods that has been tested extensively has been found to have sets of characteristic advantages and disadvantages. Nevertheless, the analysis of thin biological samples can be expected to be accurate to about 10%. Thus one is faced with the unhappy situation of having accurate quantitative methods whose utility may be limited by uncertainties associated with the preparation of materials for analysis.

REFERENCES

1. J. R. Coleman and A. R. Terepka, in *Principles and Techniques of Electron Microscopy; Biological Applications* (M. A. Hayat, ed.), Vol. IV, Van Nostrand-Rheinhold, New York (1974).
2. M. J. Ingram and C. A. M. Hogben, *Anal. Biochem.*, **18**, 54 (1967).
3. C. Lechene, F. Morel, M. Guinnebault, and C. deRouffignac, *Nephron*, **6**, 457 (1969).
4. F. Morel, C. deRouffignac, D. Marsh, M. Guinnebault, and C. Lechene, *Nephron*, **6**, 553, (1969).
5. C. deRouffignac, C. Lechene, M. Guinnebault, and F. Morel, *Nephron*, **6**, 643 (1969).
6. C. Lechene, in *Microprobe Analysis Applied to Cells and Tissues* (T. A. Hall and W. Kaufmann, eds.), Academic Press, London (1974), p. 351.
7. A. Boyde and V. R. Switsur, in *X-Ray Optics and X-Ray Microanalysis* (H. H. Pattee, Jr., V. E. Coslett, and A. Engstrom, eds.), Academic Press, New York (1963), p. 499.
8. A. M. Langer, I. B. Rubin, and I. J. Selikoff, *J. Histochem. Cytochem.*, **20**, 723 (1972).
9. A. M. Langer, I. B. Rubin, I. J. Selikoff, and F. D. Pooley, *J. Histochem. Cytochem.*, **20**, 735 (1972).
10. K. G. Carroll and J. L. Tullis, *Nature*, **217**, 1172 (1968).
11. S. L. Kimzey and L. C. Burns, *Ann. N.Y. Acad. Sci.*, **204**, 486 (1973).
12. E. J. Barrett and J. R. Coleman, in *Proceedings of the 8th National Conference on Electron Probe Analysis*, EPASA (1973), p. 60A.
13. J. R. Coleman, J. R. Nilsson, R. R. Warner, and P. Batt, *Exptl. Cell Res.*, **74**, 207 (1973).
14. J. R. Coleman, J. R. Nilsson, R. R. Warner, and P. Batt, *Exptl. Cell Res.*, **76**, 31 (1973).
15. J. R. Coleman and A. R. Terepka, *J. Histochem. Cytochem.*, **20**, 401 (1972).
16. T. F. Anderson, in *Physical Techniques in Biological Research* (A. W. Pollister, ed.), Academic Press, New York (1965), p. 319.
17. A. Boyde and C. Wood, *J. Microscopie*, **90**, 221 (1969).
18. E. R. Lewis and M. K. Nemanic, in *SEM/1973 Proceedings of the 6th Annual SEM Symposium* (O. Johari and I. Corvin, eds.) (1973), p. 767.
19. K. R. Porter and V. G. Fonte, in *SEM/1973 Proceedings of the 6th Annual SEM Symposium* (O. Johari and I. Corvin, eds.) (1973), p. 683.
20. R. R. Warner and J. R. Coleman, in *Microprobe Analysis Applied to Cells and Tissues* (T. A. Hall and R. Kaufmann, eds.), Academic Press, London (1974), p. 249.
21. R. R. Warner and J. R. Coleman, *J. Cell Biol.*, **54**, 64 (1975).
22. A. Boyde and P. Echlin, in *SEM/1973 Proceedings of the 6th Annual SEM Symposium* (O. Johari and I. Corvin, eds.), (1973), p. 759.
23. P. Echlin and R. Moreton, in *SEM/1973 Proceedings of the 6th Annual SEM Symposium* (O. Johari and I. Corvin, eds.) (1973), p. 325.
24. L. I. Rebhun, in *Principles and Techniques of Electron Microscopy Biological Applications* (M. A. Hayat, ed.), Vol. II, Van Nostrand-Rheinhold, New York (1972), p. 3.
25. F. D. Ingram, M. J. Ingram, and C. A. M. Hogben, *J. Histochem. Cytochem.*, **20**, 716 (1972).
26. H. P. Zingsheim, *Biochem. Biophys. Acta*, **265**, 339 (1972).
27. R. S. Morgan, J. Liebedzik, and E. W. Whyte, in *SEM/1973 Proceedings of the 6th Annual SEM Symposium* (O. Johari and I. Corvin, eds.) (1973), p. 205.

28. J. Philibert and R. Tixier, in "Proceedings of the 8th National Conference on Electron Probe Analysis," EPASA (1973), p. 27A.

29. J. E. Wergedal and D. J. Baylink, *Am. J. Physiol.*, **226**, 345 (1974).

30. A. W. Wachtel, M. E. Gettner, and L. Ornstein, in *Physical Techniques in Biological Research* (A. W. Pollister, ed.), Academic Press, New York (1966), p. 173.

31. A. R. Spurr, *J. Ultrastr. Res.*, **26**, 31 (1969).

32. A. K. Christensen, *J. Cell Biol.*, **51**, 772 (1971).

33. T. C. Appleton, *Micron*, **3**, 101 (1971).

34. R. S. Thomas, in *Advances in Optical and Electron Microscopy* (R. Barer and V. E. Coslett, eds.), Vol. III, Academic Press, London (1969), p. 99.

35. H. Schraer and W. Hohman, in *Biological Calcification. Cellular and Molecular Aspects* (H. Schraer, ed.), Appleton Century Crofts, New York (1970), p. 313.

36. S. J. B. Reed, in *Microprobe Analysis* (C. A. Anderson, ed.), Wiley, New York (1973).

37. C. A. Friskney and C. W. Haworth, *J. Appl. Phys*, **38**, 3796 (1967).

38. G. E. Pfefferkorn, in *SEM/1973 Proceedings of the 6th Annual SEM Symposium* (O. Johari and I. Corvin, eds.) (1973) p. 757.

39. P. Echlin and P. J. W. Hyde, in *SEM/1972 Proceedings of the 5th Annual SEM Symposium* (O. Johari, ed.) (1972), p. 137.

40. C. A. Andersen, in *Methods of Biochemical Analysis* (D. Glick, ed.), Vol. XV, Interscience, New York (1967), p. 147.

41. F. D. Ingram, J. J. Ingram, and C. A. M. Hogben, in *Proceedings of the 8th National Conference on Electron Probe Analysis*, EPASA (1973), p. 62A.

42. B. Lehrer and I. Rubin, *J. Histochem. Cytochem.*, **20**, 722 (1972).

43. C. Lechene, in *Electron Probe Analysis as Applied to Cells and Tissues*, (T. A. Hall and R. Kauffman, eds.), Academic Press, London (1974).

44. T. A. Hall, in *Quantitative Electron Probe Analysis* (K. F. J. Heinrich, ed.), NBS Special Publication 298 (1968), 298.

45. D. J. Marshall and T. A. Hall, in *X-Ray Optics and Microanalysis* (R. Castaing, P. Deschamps, and J. Philibert, eds.) Hermann, Paris (1966), p. 374.

46. T. A. Hall, in *Physical Techniques in Biological Research* (G. Oster, ed.) (1971), Vol. I, Part A, p. 157.

47. H. A. Kramers, *Phil. Mag.*, **46**, 836 (1923).

48. J. Philibert and R. Tixier, IRSID, Met Phy. 805-RT/CC (1972).

49. E. Lifshin, M. F. Ciccarelli, and R. B. Bolon, in *Proceedings of the 8th National Conference on Electron Probe Analysis*, EPASA (1973), p. 29A.

50. A. T. Marshall, in *Principles and Techniques of Electron Microscopy; Biological Applications* (M. A. Hayat, ed.), Vol. IV, Van Nostrand–Rheinhold, New York (in press.)

51. R. R. Warner and J. R. Coleman, *Micron*, **4**, 61 (1973).

52. R. R. Warner, Ph.D. Thesis, University of Rochester (1972).

53. R. R. Warner and J. R. Coleman, in *Electron Probe Analysis as Applied to Cells and Tissues* (T. A. Hall and R. Kaufman, eds.), Academic Press, London (1974).

54. J. Colby, in *Advances in X-Ray Analysis* (J. B. Newkirk, G. R. Mallett, and H. G. Pfeiffer, eds.), Vol. 11, Plenum Press, New York (1968), p. 287.

55. C. A. Andersen and M. F. Hasler, in *X-Ray Optics and Microanalysis* (R. Castaing, P. Deschamps, and J. Philibert, eds.), Hermann, Paris (1966), p. 310.

56. S. J. B. Reed, *Brit. J. Appl. Phys.*, **16**, 913 (1965).

57. M. Green, in *X-Ray Optics and X-Ray Microanalysis* (H. H. Pattee, V. Coslett, and A. Engstrom, eds.), Academic Press, New York (1963), p. 185.

58. D. F. Kyser and K. Murata, in *Proceedings of the 8th National Conference on Electron Probe Analysis* (1973), p. 28A.

59. C. A. Andersen, in *Methods of Biochemical Analysis* (D. Glick, ed.), Interscience Publishers, New York (1967), p. 147.

60. J. K. Koehler, in *Advances in Biological and Medical Physics.* (J. H. Lawrence and J. W. Gofman, eds.), Academic Press, New York (1968), p. 1.

61. A. Eshel, *Micron*, 5, 41 (1974).

XIV

ION MICROPROBE MASS ANALYSIS

J. W. Colby

Whereas the remainder of this book is concerned with electron optical instruments, this chapter will deal with the ion microprobe, one of the newest and perhaps one of the most powerful analytical instruments. It is capable of providing an *in situ* mass analysis of a microvolume of a solid sample and provides an extremely powerful analysis technique complementary to electron microprobe analysis. Such a local analysis is quite useful to the study and fabrication of solid state devices and to the studies of basic diffusion technology, growth kinetics, redistribution phenomena, surface catalysis, and various mineralogical and metallurgical processes. The ion microprobe has also been uniquely utilized for the dating of individual grains from lunar samples.

Factors that distinguish the ion microprobe from the electron microprobe are: (1) better detection sensitivities, which typically may be less than one part per million; (2) greater sensitivity to the light elements, such as boron, carbon, oxygen, and even hydrogen; (3) ability to be utilized for isotopic analyses; and (4) in-depth element profiling capability. Whereas the electron microprobe is nondestructive, the ion microprobe is destructive, thus affecting sample choice and handling. The basic design and use of these instruments will be discussed below.

J. W. COLBY—Bell Telephone Laboratories, Allentown, Pennsylvania.

I. BASIC CONCEPTS AND INSTRUMENTATION

In secondary ion mass spectroscopy (SIMS), of which ion microprobe mass analysis (IMMA) is a subcategory, a beam of ions (the primary beam) is accelerated, focused, and impinged on a sample to be analyzed. As the energetic primary ions strike the surface of the sample, their kinetic energy is transferred to the sample (target) atoms, causing them to be eroded away. A portion of the sputtered atoms become ionized positively or negatively and these ions then are collected and analyzed by suitable mass spectrometers. The majority of the sputtered ensemble comes off as neutral atoms, and the probability that ions will remain ions and hence be detected is related in a complicated way to the work function of the target[1,2] and to the kinetic energies of the ions. The work function, which influences the secondary ion yield, may be considerably altered by the choice of primary bombarding species[3,4] or by otherwise altering the surface chemistry.

From the above brief summary, it should be apparent that ion microprobe technology uniquely combines several other analytical disciplines and as a consequence the instrumentation is more sophisticated and expensive. First, the vacuum must be considerably better than that required for electron microprobes because surface chemistry[1,2] (hence yield) may be considerably altered by adsorbed residual contaminants. A means must be provided for introducing various gases into the ion source, depending on the analysis being performed. If lateral distribution information is desirable, a system must be provided for image formation. A sufficiently high mass-resolving power must be provided to assure adequate separation of the isotpes being analyzed. This latter consideration usually results in a compromise in sensitivity and will also be discussed in more detail later. Typical count rates obtained from the ion microprobe may be on the order of 10^7 cps with a microfocused beam for a matrix element, while backgrounds are on the order of 10 cps. These high count rates require special counting techniques with very low detector dead times. There are basically two distinctly different approaches to the design of such ion microprobe instrumentation, with a third approach under development.

II. ION MICROSCOPE

The first instrument capable of providing true microanalysis by ion bombardment was described initially by Castaing and Slodzian.[5,6] Several deficiencies in this early instrument were corrected in a later instrument by Castaing and Slodzian[7,8] and the resulting instrument became commercially available. This instrument, manufactured by Cameca, is illustrated schematically in Figure 1. The primary ion source is a hollow cathode duoplasmatron capable of bombarding the sample with either positive or negative ions. Primary ion energies are typically 5.5–14.5 keV and the ions are usually $^{16}O_2^+$, $^{16}O^-$, and $^{40}Ar^+$, though other primary ions can also be employed. No provision is made for mass analysis of the primary beam. The vacuum system employs oil diffusion pumps and liquid nitrogen-cooled baffling.

The primary beam is focused with a double condenser lens system that provides beam diameters from ~30 to ~400 μm. A set of electrostatic deflection plates following the second condenser lens serves to position the primary ion beam on the observed area, corrects any astigmatism in the primary beam, and can also be used to rapidly scan the beam in a raster over the sample surface. The sample is held at 4.5 kV, the polarity depending on whether positive or negative secondary ions are being analyzed.

An immersion lens at the sample forms a real secondary ion image of the area of the sample surface being eroded by the primary ion beam. This image is formed at the entrance to the magnetic prism used for mass analysis. The ions entering the prism are deflected through 90° and are directed toward an electrostatic mirror. This mirror reflects the lower energy ions back through the second half of the magnetic prism, whose chromatic aberration just cancels that of the first sector of the prism assembly, so that the final image is achromatic. The potential of the mirror is externally adjustable and provides high-energy discrimination. Placed in front of the mirror are externally adjustable slits which can be used to

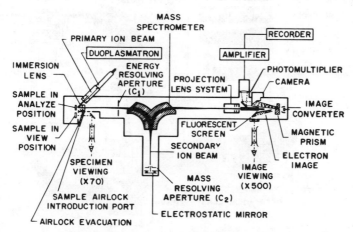

FIGURE 1.   Schematic of ion microscope manufactured by Cameca (from Morabito and Lewis[50]).

increase the mass resolving power to ~1000 (to be discussed later). Such an increase in mass resolving power decreases the transmission of the system and thus reduces sensitivity.

The mass-analyzed image, upon leaving the prism, is post-accelerated and projected on an image converter for viewing, or into a standard Daly detector[9] for digital output. A mechanical aperture can be positioned in front of the detector so that only a portion of the image is counted. This selected area analysis is essential for in-depth profiling. A digital data acquisition system was later added by Rouberol et al.,[10] which significantly increased the instrument's utility. A computer system is currently being integrated into the ion microscope.

III. ION MICROPROBE

The second type of instrument is illustrated in Figure 2. This type of instrument was originally designed by Liebl[11] and was first offered commerically by Applied Research Laboratories.[12] The primary ion source again is a hollow cathode duoplasmatron from which can be extracted either positive or negative ions. The primary ions can be accelerated up to

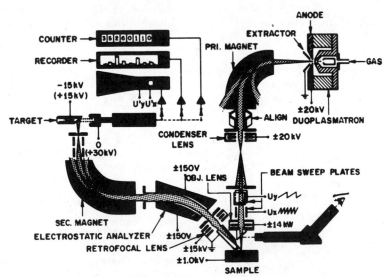

FIGURE 2.   Schematic of ion microprobe mass analyzer manufactured by Applied Research Laboratories (from Robinson[15]).

22.5 keV and are mass-analyzed prior to striking the sample. A two-stage electrostatic lens system following the 90° primary magnet focuses the beam onto the sample surface; the beam can be as small as 1.5 μm in diameter. A set of deflection plates between the lenses is used to raster the beam over the sample surface.

The secondary ions from the sample are accelerated to 1.5 keV and enter a tandem double-focusing mass spectrometer for analysis. The energy pass band of such a system is relatively large ($\sim$100 eV), leading to a high instrument transmission. No entrance slit is required in this system; hence instrument sensitivity is not degraded by a loss of signal intensity due to a slit.

The mass-analyzed secondary ions are detected by a modified Daly detector[13,14] system similar to that described by Bernhard *et al.*[13] Typical pulse pair resolution of the detector system is less than 30 nsec, thus minimizing dead-time losses. The detected secondary ions can be used to form images of the eroded surface or can be processed digitally in a multichannel analyzer or computer system. The images are formed by rastering the small ($\sim$1.5 μm) primary ion beam over the sample surface. The secondary ion current obtained from the sample is used to modulate the intensity of a cathode ray tube whose electron beam is being rastered in synchronism with the primary ion beam. This method of image formation is completely analogous to that in the SEM or EPMA. An "electronic aperture" is employed in this type of instrument to selectively view or analyze any portion of the rastered area. The electronic aperture consists of gating the counting electronics such that they are "on" only when the beam is within a predetermined area. This area can be made as small or large as desired and can be moved to any part of the rastered area.

A set of deflection plates placed between the exit of the mass spectrometer and the detector system can be used for isotope ratioing. Such a technique is quite useful for in-depth profiling or isotope dating. The mass-analyzed secondary ion beam is caused to rapidly scan repetitively over a slit. The output of the detector is gated to two separate scalers which accumulate counts for each of two isotopes.

The ion microprobe is currently manufactured by three companies, as shown in Table I. There are at this time, to the author's knowledge, two U.S. companies developing a third type of instrument based on a quadrupole mass spectrometer but these are not commercially available at this time. The latter instruments have the practical advantages of being lower in cost and simpler to operate and maintain. The availability and cost of these instruments is given in Table I. Since it is not the purpose of this chapter to compare instruments in detail, only a brief synopsis is given in Table II of the instrumental features and differences of the various com-

TABLE I

Type	Manufacturer	Cost (1974)	Availability
Ion microscope	Cameca Instruments, Inc.	~$240,000	Yes
Ion microprobe	Applied Research Labs. Inc.	~$240,000	Yes
Ion microprobe	AEI Scientific Apparatus, Inc.	~$300,000	Yes
Ion microprobe	Hitachi	~$175,000	Yes
Quadrupole	Applied Research Labs, Inc.	<$100,000	1975
Quadrupole	Etec Corp.	<$100,000	1975

TABLE II. Ion Microprobe Mass Analysis Instrumentation Features

	ARL	AEI	Hitachi	Cameca
Type	Microprobe	Microprobe	Microprobe	Microscope
Beam size, μm	1.5–300	2–300	2–300	30–300
Mass analysis of primary ions	Yes	Yes	Optional	No
Angle of incidence, deg	90	45	90	~45
Vacuum system	Ion pump	Ion pump	Oil diffusion	Oil diffusion
Sample chamber pressure, Torr	1×10^{-8}	1×10^{-9}	5×10^{-7}	1×10^{-8}
Cold surface surrounding sample	Yes	No	No	No
Direct viewing of sample during analysis	Yes	No	Yes	No
Mass spectrometer	Double focusing	Double focusing	Double focusing	Prism–mirror prism
Mass resolution (maximum)	1500	3,000 electrical 10,000 photographic	600	1000
Detector system	Electrical Daly type	Electrical/ photographic	Electrical Daly type	Electrical/ photographic imaging
Imaging	Point	Point	Point	Direct
x–y resolution, μm	~1	2	2	~1

mercially available instruments. More complete discussions of these instruments can be found in the literature.[15,16]

IV. PRODUCTION OF IONS

There appear to be two distinct processes for the production of ions during sputtering. These are the "kinetic"[17–21] process and the "chemical"[1–4] process. Both processes can be operating virtually all the time sputtering is taking place; however, they can differ considerably in magnitude. The kinetic process is predominant when elemental surfaces are bombarded by chemically inert ionic species. These impinging ions penetrate into the solid surface and are scattered much like electrons before coming to rest at some mean depth. During the scattering process lattice bonds are broken, and electrons can be ejected into the conduction band, leaving the host atoms in an excited metastable state. As these ions leave the solid, the probability of their escaping as ions depends on their kinetic energy and on the surface barrier potential that they must overcome. Typically under inert ion bombardment the greater majority of ions are neutralized by recombination with conduction band electrons and are thus lost as far as analysis is concerned.

In the chemical process, the surface chemistry is altered usually by bombardment with a chemically reactive species such as oxygen or nitrogen.[1–4] However, the reactive species may also be present in the sample itself, in the residual vacuum or may be introduced through a controlled leak.[22,23] The net effect is to form chemical compounds that are strongly bonded through chemisorption of these electronegative gases. These compounds raise the surface work function, and hence the barrier height, which an electron must overcome to be available for neutralization of a positively charged ion. In other words chemical compound formation has the net effect of reducing the number of conduction electrons available for neutralization of positive ions.

If the sample is bombarded with electropositive elements,[3,4] the work function is reduced and those elements with high electron affinities have an optimum environment for a high negative ion yield. This effect has been demonstrated by Andersen,[3,4] although very little work has actually been done with electropositive primary sources. In actual practice, the majority of ion microprobe work is done using oxygen as a primary source and the predominant mechanism for ion production is the chemical process. This has the net effect of greatly enhanced and stable ion yields, making possible trace analysis and quantitative analysis.

Because of the complexity of the sputtering and ion production processes, there are some problems associated with interpretation of results which will be dealt with in more detail later after discussing the sputtering mechanism and quantitative techniques. One effect, however, which can be understood from the above description of the chemical process arises when one attempts to interpret ion images. As noted above, if an inert gas is used as a primary source, a variation in the ion intensities in the images may not reflect a variation in chemical or elemental content but only variations in oxidation potential or in the work function. Variations with depth can be caused by variation in vacuum conditions, residual gas pressure, or even variations in the primary source gas. If polycrystalline targets are being analyzed, differences in work function due to differences in grain orientation,[24] differences in oxidation rates due to orientation,[25] and to a lesser extent differences in channeling[24] may affect the yield. Hence ion images may require a great deal of care in interpretation, perhaps more so than bulk chemical analyses or in-depth profiles.

The majority of analyses probably are done with positive secondary ions, although there are certain cases where negative secondary ions are preferred. The predominant species in the positive ion mode are the singly charged ions of the elements present. There are in addition doubly and triply charged ions from these same species, although their signals are considerably less intense. There are also dimer and other higher molecular ions present in the spectrum, which may be quite abundant. Particularly when using a reactive gas such as oxygen, many oxides of the major matrix elements (which may not be stable in nature) are found in the spectrum. If the vacuum system is not clean, hydrocarbons can be seen in the spectrum, thus considerably complicating matters. The use of ion pumping and a cold dome or cold surface around the sample considerably reduces the amount of high-molecular-weight hydrocarbons in the vicinity of the sample and so is quite helpful in reducing the complexity of the spectra.

V. SPUTTERING

It is pertinent at this time to discuss in some detail the phenomenon of sputtering, since it affects the rate at which the sample is destroyed and is the source of secondary ions. Sputtering yields, the number of sputtered atoms per incident ion, are a function of sample composition, primary ion species, incident ion energy, angle of incidence of the primary ion beam, and crystal orientation.

Other conditions being equal, the sputtering yield appears to be related to the concentration of electrons in the d orbital.[26] To illustrate this behavior, the sputtering yield S is plotted as a function of atomic number[27] in Figure 3a and the number of electrons[28] in the outermost shell is plotted versus atomic number in Figure 3b. There is not an exact correspondence, but the trend is definite and more or less explains the sputtering behavior across the periodic table. This relationship can be connected to the variation in range of primary ions with the degree to which the electronic structure

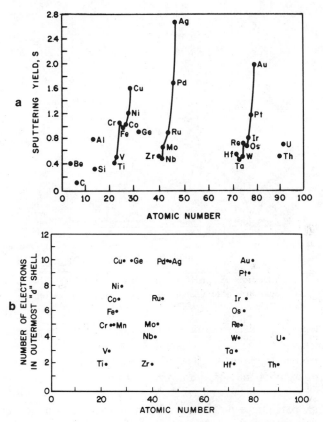

FIGURE 3.   (a) Variation of sputtered yield with atomic number of target material for argon ions incident at 400 eV (from Carter and Colligon[26]). (b) The number of electrons in the atomic d shell as a function of atomic number (from Carter and Colligon[26]).

of the matrix is open. The more open the structure (less filling of d orbitals), the greater the range and the less probability (in the hard sphere sense) that the damage sequence will be propagated back to the surface, causing atoms to be dislodged.

In single-crystal materials, variation can occur due to crystal orientation.[24] This is related to the degree to which incident primary ions can be channeled in the single crystal. That is, along certain crystallographic directions, the lattice is relatively open, and impinging ions can travel down these channels virtually unimpeded (suffering only small-angle scattering) for very large penetration depths. Under such conditions, it would be

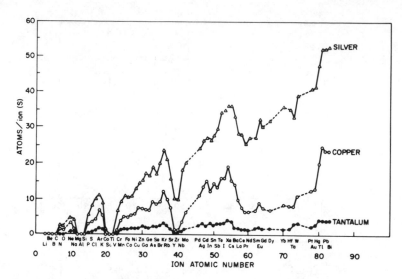

FIGURE 4.   Variation of sputtering yield with atomic number of the primary bombarding ions incident at 45 keV (from Carter and Colligon[26]).

expected that very little damage sufficient to dislodge atoms could be propagated back to the surface layers.

Variation in the sputter yield with primary ion species also shows an interesting periodic behavior. The work of Almen and Bruce[29] is shown in Figure 4, where it can be seen that yields increase fairly steadily within a given group, reaching a maximum for the inert species of the group, and then fall rather abruptly almost to zero. This pattern then repeats itself but increases in absolute magnitude as the mass of the primary ion increases.

For a given ion species, angle of incidence, and target, the sputter yield increases parabolically with primary energy over the low-energy range[27,30–32] and then increases nearly linearly up to approximately 10 keV.[26–33] At higher energies, the yield goes through a maximum and then decreases as the ions penetrate further into the lattice, dissipating their energy too deeply to affect surface layers. This behavior is illustrated in Figure 5a and 5b.

Almen and Bruce[33] and Wehner[34] have also measured the sputtering yield as a function of angle of incidence of the primary beam. Their results are reproduced in Figure 6, where it can be noted that the yield relative to that for normal incidence increases with the angle of incidence up to approximately 45°. Beyond this angle the yield falls rapidly. The yield for nonnormal incidence can be approximated as[35]

$$S_\theta \cong S_0/\cos\theta \qquad\qquad (1)$$

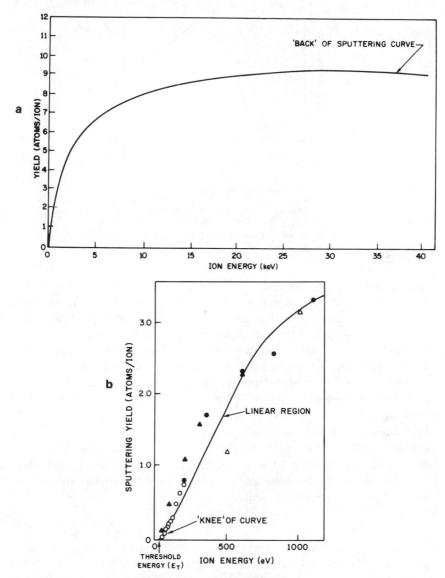

FIGURE 5. (a) Variation of sputtering yield with energy for argon ions incident on copper (from Carter and Colligon[26]). (b) Low-energy variation of sputtering yield for argon primary ions incident on copper. ●, Bader *et al.*[30]; ○, Henschke and Derby[31]; ▲, Laegrid and Wehner[27]; △, Keywell[32] (from Carter and Colligon[26]).

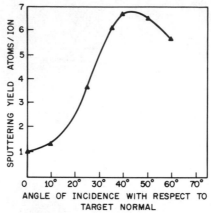

ANGLE OF INCIDENCE WITH RESPECT TO
TARGET NORMAL

FIGURE 6.   Variation of sputtering yield with angle of incidence (from Carter and Colligon[26]).

where S_θ is the sputtering yield at an angle of incidence θ (relative to the target normal) and S_0 is the sputter yield at normal incidence. The probable reason for this behavior is that at normal incidence, ions penetrate more deeply into the solid and the direction of momentum transfer is unfavorable to dislodge atoms from the surface. As the angle of incidence is increased, however, momentum transfer becomes more favorable and the penetration depth normal to the surface decreases. As still higher angles are approached the beam is just grazing the surface and a larger number of the incident ions are elastically scattered from the surface.

Knowledge of the sputtering yield is fairly important in microprobe analysis, particularly in obtaining depth profiles, because it allows one to calculate the sputtering rate, which in turn provides the depth calibration for the profile. It has been shown by Long[36] that the eroded volume can be related to the primary ion current by

$$i_p = n_a ev/tS \qquad (2)$$

where i_p is the primary ion current, n_a is the number of atoms per unit volume, e is the electronic charge, v is the volume of specimen eroded, t is the sputtering time, and S is the sputter yield. The number of atoms per unit volume is $N_0\rho/A$, where N_0 is Avogadro's number, ρ is the density, and A is the mean atomic weight. Letting $v = za$, where z is the layer thickness eroded and a is the *uniformly* eroded area, (2) becomes

$$z/t = \dot{z} = (1/N_0 e)(jAS/\rho) \qquad (3)$$

where $\dot{z}$ is the sputtering rate and j is the current density i_p/a. For current density in mA/cm² and sputtering rate in Å/sec, equation (3) becomes

$$\dot{z} = 1.04jAS/\rho \qquad (4)$$

Thus if S is known, j can be measured and the sputtering rate accurately calculated. It can also be noted from equation (4) that the sputtering rate for a given target and accelerating potential is a function of only the current density and hence sputtering rate can be controlled through varying the current density. This can be accomplished by varying either the primary beam current and/or the effective erosion area (e.g., by rastering the beam over larger or smaller areas). In this regard, it should be noted that when rastering the primary ion beam, the current density is the ratio of the primary beam current (measured in a Faraday cup) to the effective rastered area (measured by examining the eroded crater in a calibrated optical or scanning electron microscope or by Tallysurf measurement). By measuring the eroded depth either through Tallysurf measurement, by SEM, or by interference microscopy, the sputtering rate and hence the sputtering yield can also be determined. It can also be measured by sputtering through known layer thicknesses.

Although there are several models available to calculate the sputtering yield directly,[4,29,37,38] they are in general rather complicated or are only applicable for primary ions of the inert gases. For these other bombarding ionic species, saturation values (the amount of trapped primary ions in the target)[39] may no longer be negligible and the trapped ions may in fact alter the properties of the target material.[40–42] It is therefore best to measure the sputter yield directly by one of the above techniques and then use equation(4) to calculate the sputtering rate for the effective current density being employed. Sputtering will be discussed later in regard to obtaining useful in-depth profiles.

VI. QUALITATIVE ANALYSIS

In the ion microprobe the sputtered secondary ions are collected and analyzed by a mass spectrometer for their mass-to-charge ratio m/e. A typical mass spectrum from silicon is shown in Figure 7a. The m/e ratio is usually sufficient to permit identification of the isotopes or elements present in the sample. However, secondary ion mass analysis of specimens that contain many elements results in mass scans with peaks at nearly every integer mass unit. Signals are obtained for the elements and all of their isotopes, as well as oxides, hydrides, etc. In Figure 7b, a spectrum obtained from commercial grade (98%+) manganese is shown. The spectrum is complex, particularly in the range of masses 50–70. Note that the manganese signal is not even the most prominent, despite the fact that it is the major constituent. The very large iron and chromium peaks are due to surface contamination of the manganese by the stainless steel holder used during polishing. This thin surface contamination layer was not detectable by optical microscopy or electron probe microanalysis. The presence of this contamination layer further illustrates the care that must be taken in ion microprobe analysis.

A number of interferences occur. These interferences are particularly difficult in the case of isotopically pure elements, such as V and Mn, where a signal is obtained at only one mass window. The peaks at 51 (V) and 55 (Mn) suffer interferences from chromium hydride. The peaks from 56 to 60 can contain signal contributions from Si_2 dimer. Chromium, titanium, and vanadium oxides interfere with the peaks for zinc (64, 66, 67, 68, 70), making a zinc analysis difficult. The production of oxide series [^{52}Cr, $^{68}(CrO)$, $^{84}(CrO_2)$, $^{100}(CrO_3)$] makes analyses at high mass numbers (>75) quite difficult. The problem of interferences is severe, and the difficulties that are encountered with complicated specimens must be recognized.

In many ion microprobe mass analyzers, the mass resolving power[43] is quite low, and is typically between 200 and 1500, depending on the setting of slits or energy filters. The mass resolving power as used here is a measure of the instrument's ability to resolve adjacent masses and is defined as the ratio of the mass at which the measurement is being made to the "mass distance" ΔM between adjacent mass units at 5% of the peak height (10% valley definition). These definitions are illustrated in Figure 8. If two or more components have the same isobar (components having the same integer isotopic mass but differing slightly in their true isotopic masses), they cannot be distinguished from one another unless the mass resolving power is sufficiently high. Although the isotopes are usually referred to in integer numbers, the true mass differs slightly from this integer mass, and this small difference is referred to as the mass defect. If the mass resolving power is greater than the ratio of mass to mass defect, then the two isotopes can be distinguished from one another. As an example, doubly charged silicon has isotopes occurring at 14, 14.5, and 15, while nitrogen has isotopes at 14 and 15. The true isotopic masses for doubly charged silicon are 13.992911, 14.492851, and 14.991644, and those for nitrogen are 14.0075263, and 15.0048793. The two isotopes at mass 14 differ by 0.0146153; hence, to distinguish between the two requires a mass resolving power of at least 14/0.0146153 or 960. While this is possible with

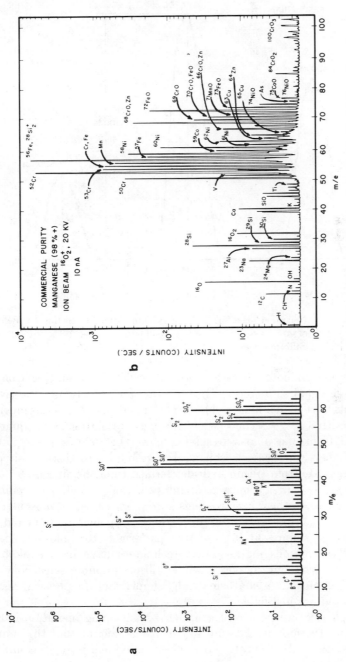

FIGURE 7. (a) Typical mass spectrum for silicon semiconductor bombarded with oxygen at 18.5 keV. (b) Secondary ion mass scan (positive ions) of commercial purity manganese (98%+). Primary ion beam conditions, $^{16}O_2^+$, 20 kV, 10 nA.

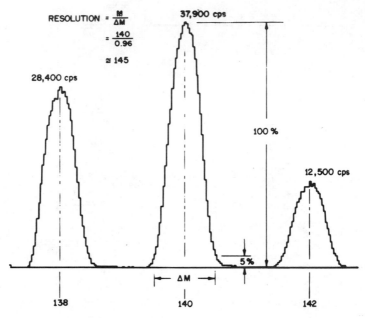

FIGURE 8.   Recorded mass spectrum of gallium dimer illustrating the definition of mass resolving power and illustrating the MIA pattern of the gallium dimer.

the earlier instruments, there is a loss in sensitivity resulting from the reduction in signal as a consequence of using finer resolving slits. (Newer instruments have higher mass resolving power and do not suffer as much of a loss in sensitivity.[44]) Second, while the mass resolution is adequate for the lighter elements, a mass resolving power of 1000 is inadequate to distinguish between heavier isobars. For example, to distinguish phosphorus (mass 31) from silicon hydride (formed from the silicon 30 isotope and hydrogen) requires a mass resolving power of ~3500. To circumvent these problems, other spectroscopists[45,46] measured "molecular ion abundance (MIA) patterns" for the various components considered, and then did a least squares fit of these MIA patterns to the unknown spectra.

For these reasons mathematical techniques have been exploited to reduce the isobaric interferences, while still maintaining sensitivity. The techniques employed are mathematically simple, and the physical assumptions involved are justifiable.

The physical assumptions on which the stripping operation is based are: (1) that the ion currents are sufficiently stable, or that the counting rates are statistically accurate, (2) that the individual isotopes combine in a linearly additive fashion, and (3) that a redundant system can be obtained. Also, it is required that the "MIA pattern" for all components be known, or that a procedure for calculating them be available. The "MIA pattern" expresses the relative relationship between isotopes for a particular component, e.g., commercially available silicon always has isotopes occurring at masses 28, 29, and 30, and with relative abundances of 0.9227, 0.0468, and 0.0305. This natural isotopic abundance pattern can be obtained directly from mass abundance tables. MIA patterns for more complex species can be calculated by binomially expanding the relative isotopic abundances[47] of the individual components. For example, gallium has two isotopes, at masses 69 and 71, whose relative abundances are 0.602 and 0.398, respectively. To calculate the MIA pattern for the dimer of gallium Ga_2, one observes that the mass-69 isotope can combine with another mass-69 isotope to give a dimer of mass 138, also the mass-69 isotope can combine with a mass-71 isotope to yield a mass of 140. Similarly, the mass-71 isotope can combine with the mass-69 isotope or another mass-71 isotope, giving a second mass of 140 or a mass of 142, respectively. The relative abundances occur as the products of the various combinations, giving rise to a simple binomial expansion. For example, the MIA pattern for gallium dimer is calculated as

$$(a + b)^2 \tag{5}$$

where a and b are the relative abundances of the gallium isotopes at masses 69 and 71, respectively. Thus the isotopes at masses 138, 140, and 142 for the gallium dimer would have relative abundances of 0.362, 0.497, and 0.158, respectively. That this is so can be seen from Figure 8, which is the recorded mass spectrum for gallium dimer.

In Table III measured abundances (as measured on an ion microprobe mass spectrometer) are compared to the calculated abundances for silicon dimer. The slight differences are due to inaccuracies in the measurements. This technique can be extended to other combinations, such as oxides of silicon, by a similar procedure. Hence the relative abundances of a combination can be easily calculated independent of any assumptions other than that the relative abundances of the individual components are accurately known.

Since the MIA pattern can be calculated, it is now possible to show how the components can be separated. Let us suppose that we have a simple ternary mixture whose components have some common isotopes. For example, we might choose a mixture of Si_2, Fe, and CaO, and let us call the total intensity (sum of intensities for all isotopes) for each component $I(Si)$, $I(Fe)$, and $I(CaO)$. Now if we are concerned only with those isotopes from mass 56 through mass 60, Fe has isotopes at masses 56–58, Si_2 has isotopes at masses 56–60, and CaO also has isotopes at masses

TABLE III. MIA Pattern for Silicon Dimer

Isotope	Calculated abundance	Measured abundance
56	0.85138	0.851
57	0.08636	0.0864
58	0.05847	0.0578
59	0.00285	0.00327
60	0.00093	0.00109

56–60. Let us call the measured intensities at each mass Y_{56}, Y_{57}, Y_{58}, Y_{59}, and Y_{60}, and the relative abundance of each component as r_{56}^{Fe}, r_{57}^{Fe}, r_{58}^{Fe}, r_{56}^{Si}, etc. Then the measured intensity at mass 56 is the sum of the products of the relative abundance and total intensity of each of the three components, i.e.,

$$Y_{56} = r_{56}^{Fe}I(Fe) + r_{56}^{Si_2}I(Si_2) + r_{56}^{CaO}I(CaO) \tag{6}$$

Similarly, we have

$$Y_{57} = r_{57}^{Fe}I(Fe) + r_{57}^{Si_2}I(Si_2) + r_{57}^{CaO}I(CaO) \tag{7}$$

$$Y_{58} = r_{58}^{Fe}I(Fe) + r_{58}^{Si_2}I(Si_2) + r_{58}^{CaO}I(CaO) \tag{8}$$

$$Y_{59} = 0 \qquad\qquad + r_{59}^{Si_2}I(Si_2) + r_{59}^{CaO}I(CaO) \tag{9}$$

$$Y_{60} = 0 \qquad\qquad + r_{60}^{Si_2}I(Si_2) + r_{60}^{CaO}I(CaO) \tag{10}$$

In the last two expressions, the first terms are zero because there are no isotopes of iron at masses 59 and 60. These equations can be conveniently rewritten in matrix notation as

$$[Y] = [R] \cdot [I] \tag{11}$$

where $[Y]$ is the vector of measured intensities, $[I]$ is the vector of total intensities, and $[R]$ is the matrix of isotopic abundances. Conventional matrix algebra then yields the solution as

$$[I] = [R]^{-1} \cdot [Y] \tag{12}$$

If it is desirable at this point, the contribution at each mass can now be obtained for each component, by multiplying the total intensity for that component by the relative isotopic abundance for the component.

Since the matrix $[R]$ is nonsquare (by definition) its inverse cannot be obtained in the usual manner; however, the solution (in the least-squares sense) can be obtained from

$$[I] = ([R]^T \cdot [R])^{-1} \cdot [R]^T \cdot [Y] \tag{13}$$

That is, the vector $[I]$ is equal to the inverse of the product of $[R]$ transpose and $[R]$, times the product of $[R]$ transpose and the vector $[Y]$. Having obtained the vector of intensities $[I]$, one can now premultiply this vector by the matrix of relative abundances, to obtain a vector of calculated intensities at each mass $[Y']$. The agreement between $[Y]$ and $[Y']$ is a measure of the goodness of fit, and can be calculated from the residuals. That is, the standard error σ_0 is given by

$$\sigma_0 = \left[\frac{\sum (Y_i - Y_i')^2}{N_1 - N_2} \right]^{1/2} \tag{14}$$

where N_1 is the number of equations and N_2 is the number of unknowns. The standard error for each of the components I_i is given by

$$\sigma_i = \sigma_0 \sqrt{D_{ii}} \tag{15}$$

where D_{ii} is the ith term on the principal diagonal of the matrix $([R]^T \cdot [R])^{-1}$.

These calculations are completely independent of method of excitation, measurement technique, or the matrix material. They assume *only* that the MIA patterns are available, and that the components are linearly additive. Numerous examples[45,46] of the applicability of this latter condition are contained in the mass spectrometry literature. These principles have been combined in a computer program written in FORTRAN IV for use on an IBM 360 computer.

The utility of this program is illustrated in Tables IV and V. In Table IV, the calculated intensities do not agree well with the measured intensities, and the sigma is fairly large as a consequence of assuming that the entire $m/e = 28$–30 spectra is due only to silicon. In Table V, much better agreement between measured and calculated intensities is obtained by attributing part of the spectrum to SiH. There is also a significant reduction in the sigma.

The use of such a stripping technique is particularly advisable prior to attempting to make ion probe data quantitative and is absolutely

TABLE IV

Mass	I(meas.)	I(calc.)	Si
28	225.00	225.11	225.11
29	13.50	11.42	11.42
30	7.50	7.44	7.44
		Total	243.97
		Sigma	1.60

TABLE V

Mass	I(meas.)	I(calc.)	Si	SiH
28	225.00	225.00	225.00	0.0
29	13.50	13.50	11.41	2.09
30	7.50	7.54	7.44	0.11
			Total 243.85	2.20
			Sigma 0.05	

required when attempting to interpret spectra such as that shown in Figure 7. As mentioned earlier, when using a reactive gas as a primary source of ions and because of the omnipresent molecular ions and hydrides, mass spectra of even simple samples become quite complicated.

VII. QUANTITATIVE ANALYSIS

There are currently several models for quantitative ion microprobe analysis, the most successful and widespread being that of Andersen.[1-4, 48-51] The predominant use of oxygen as a primary ion source has enabled high, *stable* ion yields to be obtained and as a consequence literally made quantitative analysis possible. This value of the use of oxygen cannot be underestimated and is perhaps one of the most significant contributions to the field of ion microanalysis.

In 1965, Long[36] developed relationships to estimate minimum detectability limits, precision, volume consumed, primary ion current necessary to erode at a specific rate, and, in terms of basic measurable parameters, a means of obtaining quantitative analyses. However, some of the parameters were difficult to estimate and were ambiguous, such as the instrument transmission, which can range from $\sim 1\%$ to $\sim 100\%$ depending on how it is estimated. The absolute ionization yields (the ratio of the number of positive or negative ions ejected to the number of sputtered atoms) are difficult to measure and furthermore are matrix dependent, so that such yields measured from one element or compound cannot be used for analysis in another system. Werner[52] extended these calculations, putting measurements on a relative basis (i.e., measuring the isotopic intensity of an unknown element relative to an isotopic intensity from the matrix). As a result, the instrument transmission factor dropped out of the calculation and absolute ionization efficiencies were replaced with relative ionization efficiencies. Werner was still using argon as a primary ion source, however, and uncertainties due to varying vacuum conditions and unstable yield detracted somewhat from his approach. This same approach was taken up by Morabito and Lewis,[53] whose equations are basically the same as those given by Werner.[52] Their relative yields are specific to a particular instrument and would have to be adjusted empirically for any other ion probe and would depend on operating conditions.

A simpler approach has been taken[54-56] which can be employed whenever the species being analyzed are in dilute solution in a host matrix, such as boron, phosphorus, arsenic, etc., doping in silicon, or trace elements in a mineral. The method is based on known standards and it can be shown that only one standard is necessary for each element being analyzed. If the ratio of the secondary ion intensity from the analyte to the secondary ion intensity from the matrix is plotted as a function of atomic concentration, the resulting plot is linear, as shown in Figure 9 for boron in silicon. In this case, the $^{11}B^+$ isotope is ratioed to the $^{28}Si^{2+}$ isotope and plotted as a function of concentration. The boron concentrations were determined by electrical resistivity measurements, Hall measurements, and electron microprobe measurements (at the higher concentration). For most dopants or analytes, electron probe analysis of the standards is preferred over electrical measurements for the simple reason that electrical techniques measure the net uncompensated electrical activity of the sample, which may or may not represent the true composition. It might also be noted that the technique of Werner[52] (and of Morabito and Lewis[53]) essentially reduces to this technique after one empirically adjusts their factors for the host matrix and combines all of their coefficients into a single multiplier.

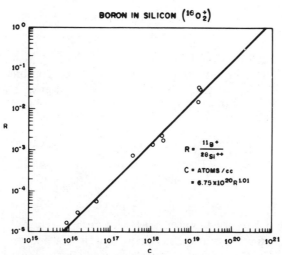

FIGURE 9.   Calibration curve for boron in silicon showing linearity of boron intensity with concentration (from Colby[54]).

More sophisticated and generally applicable techniques for quantitative analysis have been suggested by Andersen,[1-4,51] Andersen and Hinthorne,[48-50] Schroeer *et al.*,[57] Jurella,[58] Joyes and Toulouse,[59] and Blaise and Slodzian.[60,61] However, only the model of Andersen has been applied to a large variety of compounds, alloys, and minerals and found to be generally applicable.

Andersen's model relies on the use of reactive gases as primary ion sources, to control the surface chemistry and hence enhance and stabilize the secondary ion yields.[1-4] The model assumes that the region being sputtered resembles a dense plasma in local thermal equilibrium. This plasma is composed of the excited atoms and ions that are being sputtered. Under equilibrium conditions the energy states of all particles within the excited volume can be described thermodynamically and the ion-to-atom ratio of each element can be calculated by the use of the Saha–Eggert ionization equation[62-64]:

$$K_{n+} = \left(\frac{2\pi}{h^2}\frac{m_{M+}m_{e-}}{m_{M0}}kT\right)^{3/2}\frac{B_{M+}B_{e-}}{B_{M0}}e^{-E/kT} \tag{16}$$

where h is Planck's constant, k is Boltzmann's constant, m is the mass, T is the absolute temperature, B is the internal partition function, and E is the dissociation energy of the species in question, and in this case is the first ionization potential of the atom. K_{n+} is the dissociation constant of the dissociation reaction of the ionization process assuming thermal equilibrium,

$$K_{n+} = n_{M+}n_{e-}/n_{M0} \tag{17}$$

where n_{M+}, n_{M0}, and n_{e-} represent the concentrations per unit volume of the singly charged positive ions and atoms of element M and of the electrons, respectively. Equations (16) and (17) can be combined to give the ratio of the number of positive ions to neutral atoms. These combined equations can be transformed to

$$\log\frac{n_{M+}}{n_{M0}} = 15.38 + \log\frac{2B_{M+}}{B_{M0}} + 1.5\log T$$
$$- \frac{5040(I_p - \Delta E)}{T} - \log n_{e-} \tag{18}$$

where n_{M+} and n_{M0} are the atom densities in two adjacent charge states, B_{M+} and B_{M0} are the internal partition functions for these charge states, 2 is the partition function of a free electron, I_p is the ionization potential of the lower charge state, ΔE is the ionization-potential depression due to

Coulomb interation of the charged particles, and n_{e-} the electron density. If the ionization potentials, internal partition coefficients, and ionization-potential depression are known, the ratio of the number of singly charged ions to neutral atoms is a function of the electron temperature and density. The ionization-potential depression ΔE can be calculated from the Debye–Hückel model[65] and the partition function and ionization potentials for many elements are available in the literature. If the atomic concentrations of two elements in a sample are known, then an equation such as (18) can be written for each of the elements and the pair solved simultaneously for the electron density and temperature. These can then be used to calculate the concentrations of the other elements in the sample. A more complete discussion of the technique is available in the literature[48–51] and is combined in a computer program referred to as CARISMA.

The model has been criticized primarily because it has not been accurate for all systems, and because it predicts temperatures that are considered unrealistic (e.g. 10^4 °K). These temperatures, while high, are not inconsistent with thermal spike temperatures. Inaccuracies in the final results can be due to inadequate internal partition functions and ionization potentials as required by equation (18), and because in many cases the energy spread of some of the secondary ions sputtered from the sample may be too large to all be detected, even with an instrument bandpass of 100 eV. Instrument improvements for data collection may also be necessary for improved quantitation. Recent work by other investigators[66–69] with completely different instruments has confirmed the CARISMA type of analysis for several examples.

If one is attempting to analyze multiphase alloy structures, or diffusion zones with large concentration gradients, in which second phases are formed, the analysis is especially difficult. The use of standards or other empirical techniques would be highly complicated. It should be obvious that the "absolute" ionization yield or "relative" ionization yield for a given ionic species varies markedly depending on the matrix. When the matrix itself contains several phases, any analytical procedures that rely on relative yields cannot work, including the standards approach. But when working with dilute solutions of trace analytes in a known matrix, all of the techniques work. If standards are not available, Andersen's model can be employed to calculate a correction factor.

Statistical precision can be calculated for ion probe analysis in the same manner as for x-ray intensity data. That is, the precision[36] can be related to the standard counting error σ, which is the square root of the total number of counts N, as

$$CV = 100\sigma/N = 100\sqrt{N}/N \tag{19}$$

where CV is the coefficient of variation. For the precision to be better than 1%, N must exceed $10^4/(CV)^2$ counts. The minimum detectability limit is inversely proportional to the square root of the product of the total intensity N and the peak-to-background ratio N/B. The proportionality constant can be determined from known standards as described by Ziebold.[70] (Also see Chapter XII.)

VIII. DEAD-TIME LOSSES

Dead-time losses can occur in the detector system or in the pulse counting electronics. In either event, the net effect is a loss in count rate, and as a consequence may affect attempts at quantitative analysis. Counting rates are usually corrected for dead-time losses by the following equation:

$$N = N'/(1 - N'\tau) \tag{20}$$

where N is the "true" count rate, N' is the "observed" count rate, and τ is the dead time. This equation is valid for $N'\tau \leq 0.1$.

To determine the dead time when it is unknown, two different techniques can be employed. In the first technique, it can be noted that in the absence of dead time, and for all other parameters remaining constant, the counting rate is proportional to the incident primary ion current, i.e.,[71]

$$N = \alpha j \tag{21}$$

where j is the incident current and α is a proportionality constant.

Combining equations (20) and (21), we obtain

$$\alpha j = N'/(1 - N'\tau) \tag{22}$$

or

$$N'/j = \alpha - \alpha\tau N' \tag{23}$$

Hence by varying the incident primary ion current and recording the intensity of a given isotope as a function of ion current and then plotting observed intensity per unit ion current versus observed intensity, a straight line is obtained whose Y intercept is α, the proportionality constant, and whose slope is $\alpha\tau$. The dead time calculated in this way for the ARL IMMA was 32 nsec and for the Cameca was 140 nsec.

These dead times are relatively small. However, for an observed counting rate of 12.5 million counts per second and a dead time of 32 nsec, the true counting rate is at least 21 million counts per second, or almost a

2:1 difference.* These are typical of the counting rates obtained on the ^{30}Si isotope normally used for the quantitative analyses for a beam current of 30 nA.

A second method that can be employed to obtain the dead time is to measure the intensity of two different isotopes (if available) or the isotope ratio. The ratio of the "true" counting rates to each other is given simply by the ratio of the two isotope abundances. Thus

$$N_1 = N_1'/(1 - N_1'\tau) \quad \text{and} \quad N_2 = N_2'/(1 - N_2'\tau) \quad (24)$$

or

$$\frac{N_1}{N_2} = \frac{N_1'/(1 - N_1'\tau)}{N_2'/(1 - N_2'\tau)} = R \quad (25)$$

N_1/N_2 is the ratio of isotope abundances, and

$$\frac{R(1 - N_1'\tau)}{(1 - N_2'\tau)} = \frac{N_1'}{N_2'} \quad (26)$$

or

$$\tau = \frac{N_1' - RN_2'}{N_1'N_2'(1 - R)} \quad (27)$$

If the measured isotope ratio is R', where $R' = N_1'/N_2'$, then

$$\tau = (R' - R)/N_1'(1 - R) \quad (28)$$

The dead time from equation (28) can then be used in equation (20) to correct counting rates for dead-time losses. In practice, the first method yields more consistent results and is preferred. The second method can be employed where precision is not necessary, or to determine if the dead time is significant. A major problem with the second method may arise if interfering mass peaks are present, which may alter significantly the measured isotope ratio. Thus the first method described above is preferred.

IX. IN-DEPTH PROFILING

In addition to trace element microanalysis, one of the most useful features of the ion microprobe mass analyzer is its ability to do in-depth

---

*Actually this correction would not apply for these dead times and count rates. When the product of count rate and dead time is large, other terms in the series must be included,[72] i.e., the observed count rate N' is equal to $N/[1 + N\tau + (N\tau)^2/2! + (N\tau)^3/3! + \ldots]$, which is a series approximation to $Ne^{-N\tau}$, the actual correction to be applied. This is completely discussed in Ref. 72.

profiles.[54,55,73–79] However, the nature of the ion microprobe imposes several limitations on the technique of in-depth profiling which must be taken into account.

Such profiling is accomplished by continually eroding away the region of interest on the sample while monitoring one or more isotopes of interest. If the erosion rate is known accurately and remains constant with time, then a profile of intensity versus depth can be obtained. Such an in-depth profile is extremely useful in the semiconductor industry, to determine doping profiles and reaction or alloying zones.

The primary ion beam has an intensity distribution that is ideally Gaussian. That is, the current density in the beam is Gaussian, so that a static beam impinging on a sample at some angle θ will erode a crater having a cross section that is a skewed Gaussian, as shown in Figure 10. This is a purely geometric treatment assuming that the sputtering rate is a function only of the current density in the primary beam. As the angle of incidence becomes normal to the specimen, the crater becomes more Gaussian. If the entire secondary ion output from this crater is detected as a function of depth, it can be readily noted that the resulting profile will not accurately reflect the true profile, due to the continuous contribution from the side walls of the crater. If the secondary ion signal is mechanically apertured so that only the signal coming from the central portion of the crater is detected, the resulting profile will better approximate the true profile. However, as seen in Figure 11, which is a computer-generated plot

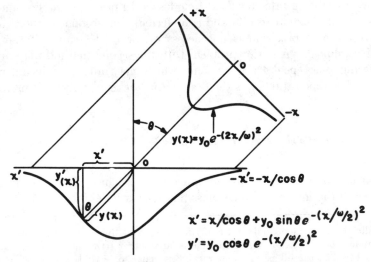

FIGURE 10.   Schematic of crater eroded by ion beam having a Gaussian energy profile, incident at nonnormal incidence (from Colby[54]).

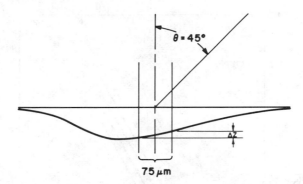

FIGURE 11. Computer-generated plot of crater eroded by a primary ion beam (250 μm FWHM) incident at 45°, and illustrating the volume sampled with a 75-μm aperture (from Colby[54]).

of such a crater for 45° incidence, the aperture represented by the pair of vertical lines about the center still intercepts ions from a region having a finite thickness ΔZ. This condition also represents a loss in resolution, which is not constant but is a function of the depth at which it is measured. This loss in resolution is a function of the depth at which it is measured D, the angle of incidence of primary beam, the full-width at half-maximum of the beam (beam diameter) W, and the diameter of the aperture A. Figure 12 is a plot of the resolution-to-depth ratio versus the aperture-to-beam size ratio for several angles of incidence. For example, from Figure 12, for a beam diameter W of 250 μm, an aperture A of 50 μm, and an angle of incidence of 45°, $A/W = 0.2$, and $R/D = 0.017$. Thus at a depth of 1 μm, the in-depth resolution is ~170 Å, while at 2 μm, the resolution is ~340 Å, etc. That is, at a depth of 1 μm, secondary ions are being collected from a region whose thickness is 170 Å, and at 2 μm is 340 Å.

It has been suggested[22,23] that these geometric limitations can be avoided if the primary beam is also apertured. However, such aperturing only makes the side walls steeper, by eliminating the low-current-density tails, but does not alter the central portion of the eroded crater from which the signal is extracted. A second technique has been employed by Guthrie and Blewer,[80,81] who defocus the objective lens such that the primary beam has a current density cross section that is nearly flat. Such a technique, while appearing to work quite well, requires careful calibration and meticulous adjustment of the objective lens.

The technique most commonly employed is to raster the primary ion beam across the surface as is done in the EPMA–SEM. This provides a region that is uniformly eroded in the center and is flat to within the tolerance with which it can be measured (~50 Å). If an aperture is now

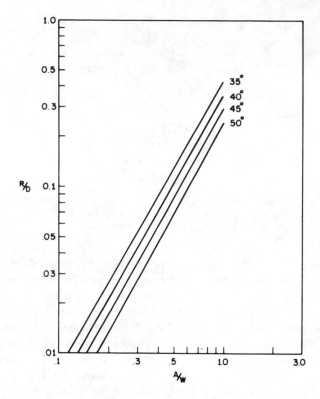

FIGURE 12.   Curves showing loss in resolution R as a function of depth D for variations in beam size W and aperture diameter A and for various angles of incidence.

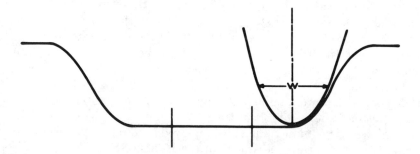

FIGURE 13.   Schematic showing flat-bottomed crater eroded by rastered primary ion beam and area sampled by aperture (from Colby[54]).

employed as shown in Figure 13, a profile is obtained that much more closely approximates the true profile. Either a mechanical aperture or electronic aperture can be employed with equal success. Electronic aperturing[82] is done by gating the detector electronics on and off in synchronism with the rastered beam such that counts are being obtained only when the beam is in the central portion of the raster. Sputtered ions from the sides of the crater are thereby rejected. The difference between profiles obtained with a static beam, a rastered beam, and a rastered beam with electronic aperturing can be seen in Figure 14. This is an in-depth-profile of an ~300-Å layer of titanium beneath other layers of copper and nickel and on top of silicon dioxide. Another example of the difference between profiles obtained with a static beam and a rastered, electronically apertured beam is shown in Figure 15.[83] This represents the profile obtained with and without aperturing from boron implanted into silicon at 50 keV.

A second effect that limits resolution close to the surface (and also surface analysis) is related to the dynamics of sputtering and implantation. A typical unnormalized profile is shown in Figure 16. Near the surface the secondary ion intensity is fairly high. The intensity passes through a minimum, then rises rapidly to a steady state value. This behavior can be explained quite easily in the following manner. On the surface is a native oxide,[1,2] which may be ~30 Å thick. The secondary ion yield from an oxide (strongly bonded, low neutralization probability) is known to be quite high. As this oxide is sputtered away, the yield drops to that of the virgin metal.[1,2] However, during this time, oxygen (from the primary ion beam) has been continuously implanted into the sample, eventually reaching a steady state or saturation value.[1,2,84] As the saturation value is being approached, the sputtering front is continuously moving inward until the sputtering front has caught up with the implantation front. This is illustrated in Figures 17 and 18, which are from Carter et al.[84]

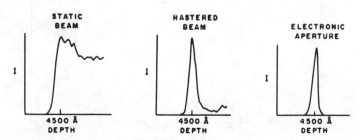

FIGURE 14.   In-depth profiles of titatium-48 for: static beam, no aperture; rastered beam, no aperture; and rastered beam, electronic aperture (from Colby[54]).

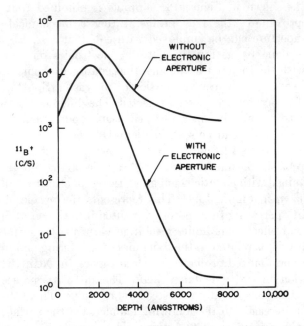

FIGURE 15.   In-depth profiles of boron implanted in silicon with and without electronic aperture (from Andersen *et al.*[83]).

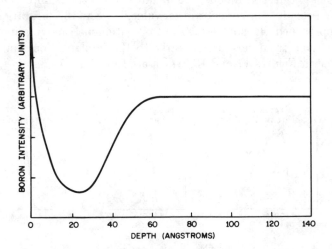

FIGURE 16.   Typical response of detector to boron uniformly distributed in silicon showing the variation in intensity until saturation is reached.

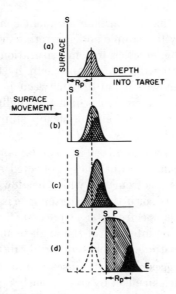

FIGURE 17. Steps in the ion collection process showing at (a) the implant profile due to the primary beam, at (b) and (c) the implant concentration builds up as the target surface moves inward, and at (d) the final implant concentration depth profile reached at equilibrium (from Carter *et al.*[84]).

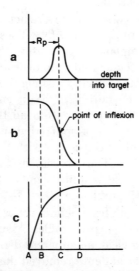

FIGURE 18. Point-to-point correspondence between: (a) instantaneous implantation profile, (b) implant concentration profile, and (c) collection curve (from Carter *et al.*[84]).

This effect implies that as ion bombardment continues, the two processes of ion collection by the sample and sputtering proceed simultaneously and that as the collection increases to some saturation value, the secondary ion yield is also increasing due to enhancement by the collected oxygen. There is thus a continuous competition between implant collection (enhanced yield) and sputtering erosion, which eventually leads to a collection saturation (by the sample) and hence a steady state yield.[84] This effect has recently been verified by Lewis *et al.*[85]

Such an effect can be minimized if secondary ions from the analyte and from the matrix can be sequentially collected extremely rapidly, as with a quadrupole mass spectrometer, which has low hysteresis and hence can be rapidly multiplexed, or if the sputtering rate is made extremely slow for a conventional mass spectrometer. Under such conditions, the ratio between the secondary ion intensity of the analyte and the matrix is much more nearly invariant with depth for a uniformly distributed analyte.

It has also been suggested that the use of an oxygen leak close to the sample will minimize such an effect by continuously replenishing the surface oxide.[22,23] Moreover, it has been suggested that the oxygen leak would minimize anomalous effects that sometimes occur at interfaces. While this may be true, it would seem to be no more effective than the ratioing techniques and does not improve the depth resolution. However, the use of the oxygen leak does improve the secondary ion yield, which improves the sensitivity and so is still desirable.

The saturation effect discussed above also acts to limit depth resolution in another way, according to Schulz *et al.*[86] These authors measured the projected range of 20-keV implanted boron atoms in silicon as a function of primary beam energy and found that the measured range increased with primary beam energy while the range straggling (i.e., implantation distribution) remained relatively constant with beam energy. This is shown in Figure 19. They interpret these results to mean that there is a change in sputtering yield at the beginning of sputtering, until a saturation of implanted ions is obtained. This is consistent with the treatment and experimental results of Smith.[40–42] The increase in the final sputtering yield over the initial sputtering yield is quite significant and can indeed cause significant errors in in-depth profiles.

The following relations are from Schulz *et al.*[86] and form the basis for correction of ion microprobe in-depth profiles. Assuming a constant erosion rate from a mean sputtering yield, an apparent depth z^* is obtained instead of the true depth z. If the total depth removed during sputtering is t, then for $z \geq R_x$, where R_x is the range of the primary ion,

$$z = z^*(1 + \delta R/t) - \delta R \qquad (29)$$

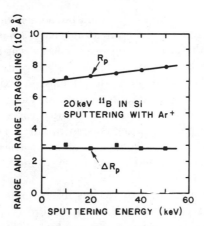

FIGURE 19.   Measured range R_p and range straggle ΔR_p versus energy of primary argon ion beam for 20-keV implanted boron in silicon. (From Schulz et al.[86])

and

$$\Delta z = \Delta z^*(1 + \delta R/t) \qquad (30)$$

To obtain δR, an actual and apparent depth of some characteristic feature (e.g., interface, projected range for implantation,) must be measured. This is done by measuring depth or range R^* as a function of primary ion beam energy, as shown in Figure 19, and extrapolating to zero primary ion beam energy to obtain the actual or true depth R. Then for any given primary ion beam potential

$$\delta R = (R^* - R)/(1 - R^*/t) \qquad (31)$$

As an example from Figure 19 (from the work of Schulz et al.,[86]) R (at zero energy) is 690 Å and at 50 keV, R^* is 790 Å. The total depth sputtered was 5000 Å, so $\delta R = 120$ Å from equation (31). This value of δR can then be employed in equation (29) to correct the depth scale at each point. In this case, the total correction amounted to approximately 2.4% but would be larger for smaller sputtering yields. Therefore, for accurate work δR should be taken into account. The sputtering yield, independent of saturation effects, is a function of primary ion beam energy, so the correction described above is essentially a combined correction taking into account both effects. However, since the sputtering yield decreases nonlinearly at the lower primary ion beam energies, linear extrapolation to zero energy would tend to overestimate the "true" depth slightly and probably accounts for the small error between the results of Schulz et al.[86] and the theoretical value.

The remaining problem to be dealt with in profiling and perhaps the least understood is the knock-on or recoil phenomenon. As shown in Figure 20, an increasing number of boron atoms have been "knocked" deeper into the sample as the primary ion beam energy was increased. Although such knock-on effects can in principle be calculated by Monte Carlo techniques,[87] no classical or analytical technique exists as yet for these calculations. It is also believed that direct knock-on by primary ions is unlikely due to their limited range, the limited range of the boron atoms themselves, and the relatively low concentration of boron atoms. In this case it is more likely due to some repeated low-energy recoil processes, since several such events would be necessary to move the boron atoms as deeply as they have been moved.[88]

It is also possible that there has been some radiation-enhanced diffusion of boron even though the temperature of the sample was not appreciably increased above room temperature during the sputtering. It has been shown previously that radiation-enhanced diffusion can take place at low temperatures. The results are also consistent with the increasing amount and extent of the damaged layer with increasing primary ion beam energy.

It has also been shown by Andersen *et al.*[83] that knock-on or recoil phenomena can be partially accounted for by assuming that the measured profile is the convolution of the true profile and the implanted primary ion distribution. A simple Gaussian distribution was convoluted with the known true profile and the resulting profile was shown to agree reasonably well

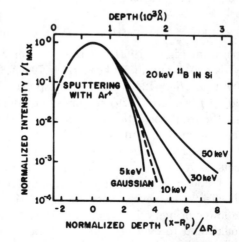

FIGURE 20.   Effect of sputtering energy on the apparent boron distribution in the low-concentration tail of the distribution, illustrating the effects of knock-on (from Schulz *et al.*[86]).

with the measured profile, using values of range straggling from Domeij *et al.*[89] Deconvolution[90] techniques are currently being developed that will empirically remove the effects of knock-on or recoil.

From the work of Schulz *et al.*[86] and Andersen *et al.*[83] optimum conditions for obtaining meaningful in-depth profiles are: (1) as low an accelerating potential as is practical from the standpoint of counting statistics and sputtering rate; (2) near grazing incidence of the primary ion beam; and (3) use of a primary ion species having a fairly high mass-to-charge ratio.

X. APPLICATIONS

The ion microprobe mass analyzer has been applied to a variety of problems in the fields of mineralogy,[91,92] semiconductor technology,[54,55,73–77,90,93] metallurgy,[94] nuclear fuels,[95,96] and particle analysis.[97] Its chief published uses, however, have been in the field of geology and semiconductors. A few applications will be described that illustrate the potential uses of the instrument in the various fields.

In the semiconductor industry, dopants used in typical transistors may be uniformly distributed, diffused, or implanted and are generally present in low concentration (less than 10^{19} atoms/cm^3 or 0.02 wt. %). Such low concentrations do not lend themselves to analysis by electron microprobe techniques, particularly when the dopant is boron. Electrical techniques may be utilized, but they only provide information about the net electrically active impurity level. To fully understand and improve transistor performance, it is often necessary to measure the individual dopant levels and profiles.

For instance, when boron is implanted into silicon, the resulting boron profile is Gaussian, as shown in Figure 21. If phosphorus is diffused into the same structure after the boron has been implanted, the boron is "pushed" deeper into the silicon as a result of vacancy flow as shown in Figure 22. This pushout can considerably alter a transistor's performance by changing the position of the junction (position where the phosphorus and boron profiles intersect) relative to where it would be without this cooperative diffusion effect. The opposite effect, or pullback, can occur if arsenic is employed instead of phosphorus. A composite of the boron profile before and after pushout and the phosphorus profile is shown in Figure 23.

Another effect that can occur due to a different mechanism is shown in Figure 24. In this case, the phosphorus was diffused into the silicon first, followed by a boron implant. The sample was then annealed for 15 min

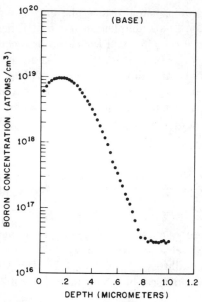

FIGURE 21.   Profile of boron implanted in silicon at 50 keV. Typical base profile.

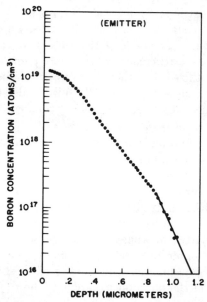

FIGURE 22.   Profile of boron implanted in silicon (same sample as in Figure 21) followed by diffusion of phosphorus (emitter formation) showing change in shape of profile as a consequence of cooperative diffusion.

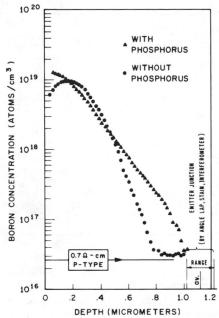

FIGURE 23.   Composite of Figures 21 and 22 showing how the tail of the boron distribution is pushed out and the peak of the distribution is sucked toward the surface by cooperative diffusion effects.

at 900°C to remove the damage caused by ion implantation. Normally for this time and temperature, boron would not diffuse appreciably. However, due to the electric field set up by the phosphorus, the boron is attracted toward the surface, resulting in a considerable profile shift. Such anomalous diffusion is not completely understood and so it is extremely useful to be able to characterize a transistor in this way.

Extensive tables[98] exist for calculating the range and range straggling of ion implantations, which are based on LSS theory.[99] While LSS theory[99] has been reasonably successful in calculating these range parameters at least qualitatively, the actual values depend to a great extent on the electron stopping power used, and it has been shown that the tables are inaccurate and consistently overestimate the range and range straggling. The ion microprobe offers an excellent technique therefore for calibrating the ion implantation curve, as shown in Figure 25, where the range of implanted boron atoms is plotted as a function of implantation energy. The boron was implanted into (111) silicon tilted 7° off orientation.

When doped silicon is oxidized, impurity segregation occurs at the moving oxide interface. If the dopant is boron, it is rejected from the silicon

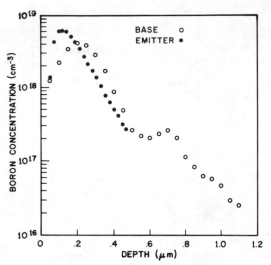

FIGURE 24.   Effect of emitter suck showing how electrical field from phosphorus sucks the peak of the boron distribution toward the surface. In this case the phosphorus was diffused prior to boron implantation. Boron redistribution occurred as a result of 15 min anneal at 900°C.

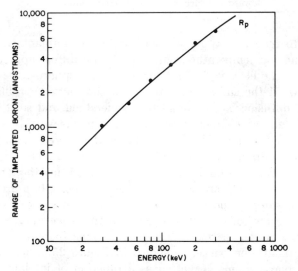

FIGURE 25.   Range R_p as a function of ion implant energy for boron implanted into silicon. Range (points) was determined by ion microprobe and solid line is LSS theory with modified electron stopping power.

into the growing oxide. Such a segregation results in a lowering of the boron surface concentration, as shown in Figure 26. The ratio of the boron concentration in the oxide at the interface, to the boron concentration in the silicon at the interface, is called the segregation coefficient. The segregation coefficients and diffusion coefficients are extremely important for predicting surface dopant concentrations, and dopant profiles in semiconductors. Segregation effects have been studied in detail by Colby and Katz,[93] who found both a crystal orientation and temperature dependence. Their results are summarized in Figure 27. They also obtained diffusion coefficients by applying a diffusion model to their data; the resulting diffusivities are shown in Figure 28.

The ion microprobe has been extensively applied to the field of geology to analyze earth minerals, meteorites, and lunar rocks.[91] In these cases, the ion microprobe was employed primarily for the local analysis (individual grains) and excellent detection sensitivity for trace elements. One particularly unique analysis, however, was its use in dating lunar mineral phases from the Apollo flight.[92] The $^{207}Pb/^{206}Pb$ ratio was measured on a number of different phases and this ratio was then used to date these phases. It was found that some phases differed appreciably from others but that the composite gave an apparent age of 4.1 billion years, in good agreement with other results.

The ion microprobe has also been employed to analyze particulate material[97] and for metallurgical problems. One particular such use was the study of the distribution of potassium in extruded tungsten wire. Both the lateral and in-depth distributions are shown in Figure 29, which also illustrates the extremely fine resolution in depth and the lateral resolution

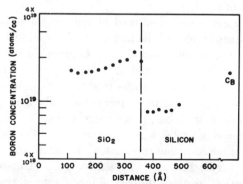

FIGURE 26. Redistribution of boron in uniformly doped silicon upon thermal oxidation: (100) crystal orientation, 1100°C. The segregation coefficient is the ratio of the boron concentration in the oxide at the interface to the concentration in silicon at the interface.

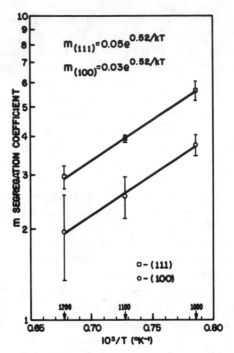

FIGURE 27. Segregation coefficient m for boron in silicon versus crystal orientation and oxidizing temperature for dry oxidation.

of $\sim$1.0 μm that can be obtained with the scanning ion probe without an appreciable loss of sensitivity.

While the applications described above are by no means exhaustive, they should provide an idea of the instrument's capabilities and utility. The IMMA, electron microprobe, and the SEM combined in a single laboratory provide an extremely potent analytical capability for microanalysis and failure analysis. The ion microprobe is still in its infancy. The IMMA is expected to become an even more powerful analytical tool when coupled to a computer and more laboratories begin to use it. The ion probe today is where the electron probe was ten years ago. Considerable debate is expected to continue for some time yet over quantitative interpretation of results. Another major difficulty with the ion probe, caused by its relatively low mass resolution, is the number of interferences that occur in fairly complex steels and minerals, making both identification of species present and quantitative interpretation difficult at best. This should not detract from the instrument's acceptance, however, but should be treated as a

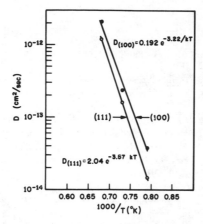

FIGURE 28.   Diffusivity for boron in silicon versus crystal orientation and temperature.

challenge, as was the EPMA. The stripping procedure described here and the CARISMA or standards approach are beginnings and permit one to obtain useful results that are otherwise unobtainable.

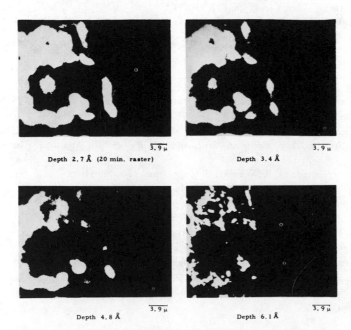

FIGURE 29.   Variation in areal distribution with depth for potassium in tungsten. Tungsten wire fracture surface, $^{39}K^+$ ion images.

REFERENCES

1. C. A. Andersen, in *Third National Conference on Electron Microprobe Analysis*, EPASA, (1968), p. 27.
2. C. A. Andersen, *Int. J. Mass Spectrom. Ion Phys.* **2**, 61 (1969).
3. C. A. Andersen, in *Fourth National Conference on Electron Microprobe Analysis* (1969), p. 19.
4. C. A. Andersen, *Int. J. Mass Spectrom. Ion Phys.*, **3**, 413 (1970).
5. R. Castaing, B. Jouffrey, and G. Slodzian, *C. R. Acad. Sci. (Paris)*, **251**, 1010 (1960).
6. R. Castaing and G. Slodzian, *C. R. Acad. Sci (Paris)*, **255**, 1893 (1962).
7. R. Castaing and L. Henry, *C. R. Acad. Sci. (Paris)*, **255**, 76 (1962).
8. G. Slodzian, *Ann. Phys.*, **9**, 591 (1964).
9. N. R. Daly, *Rev. Sci. Instr.*, **31**, 264 (1960).
10. J. M. Rouberol, P. H. Basseville, and J. -P. Lenoir, *J. Radioanal. Chem.*, **12**, 59 (1972).
11. H. J. Liebl, *J. Appl. Phys.*, **38**, 5277 (1967).
12. C. F. Robinson, H. J. Liebl, and C. A. Andersen, in *Third National Conference on Electron Microprobe Analysis*, EPASA (1968), p. 26.
13. F. Bernhard, K. H. Krebs, and I. Rotter, *Z. Phys.*, **16**, 103 (1961).
14. A. C. Tyrrell, R. G. Ridley, and N. R. Daly, *Int. J. Mass. Spectrom. Ion Phys.* **1**, 69 (1968).
15. C. F. Robinson, in *Microprobe Analysis* (C. A. Andersen, ed.), Wiley—Interscience, New York (1973), p. 507.
16. H. J. Liebl, *Anal. Chem.*, **46**, 22A (1974).
17. P. Joyes and R. Castaing, *C. R. Acad. Sci. (Paris)*, B **263**, 384 (1966).
18. P. Joyes and J. F. Hennequin, *J. Phys.*, **29**, 483 (1968).
19. P. Joyes, *J. Phys.*, **29**, 774 (1968).
20. P. Joyes, *J. Phys.*, **30**, 224 (1969).
21. P. Joyes, *J. Phys.*, **30**, 365 (1969).
22. B. Blanchard, N. Hilleret, and J. Monnier, *Mat. Res. Bull.* **6**, 1283 (1971).
23. B. Blanchard, N. Hilleret, and J. B. Quoirin, *J. Radioanal. Chem.*, **12**, 85 (1972).
24. M. Bernheim, *Rad. Effects*, **18**, 231 (1973).
25. M. Bernheim and G. Slodzian, *Int. J. Mass Spectrom. Ion Phys.* **12**, 93 (1973).
26. G. Carter and J. S. Colligon, *Ion Bombardment of Solids*, American Elsevier, New York (1968), p. 313.
27. N. Laegried and G. K. Wehner, *J. Appl. Phys.*, **32**, 365 (1961).
28. D. Rosenburg and G. K. Wehner, *J. Appl. Phys.*, **33**, 1842 (1962).
29. O. Almen and G. Bruce, *Nucl. Inst. Methods*, **11**, 279 (1961).
30. M. Bader, F. C. Witteborn, and T. W. Snouse, NASA Report TR-R-105 (1961).
31. E. B. Henschke and S. E. Derby, *J. Appl. Phys.*, **34**, 2458 (1963).
32. F. Keywell, *Phys. Rev.*, **97**, 1611 (1955).
33. O. Almen and G. Bruce, *Nucl. Inst. Methods*, **11**, 257 (1961).
34. G. K. Wehner, *J. Appl. Phys.*, **30**, 1762 (1959).
35. V. A. Molchanov and V. G. Tel'Kovskii, *Bull. Acad. Sci., USSR, Phys. Ser.*, **26**, 1381 (1963).
36. J. V. P. Long, *Brit. J. Appl. Phys.*, **16**, 1277 (1965).
37. P. Sigmund, *Phys. Rev.*, **184**, 383 (1969).
38. K. Kanya, K. Hojou, K. Kuya, and K. Tuki, *Jap. J. Appl. Phys.*, **12**, 1297 (1973).

39. J. L. Whitton, G. Carter, J. M. Baruah, and W. A. Grant, *Rad. Effects*, **16**, 101 (1972).
40. H. J. Smith, *Rad. Effects*, **18**, 55 (1973).
41. H. J. Smith, *Rad. Effects*, **18**, 65 (1973).
42. H. J. Smith, *Rad. Effects*, **18**, 73 (1973).
43. J. Roboz, *Introduction to Mass Spectrometry*, Interscience, New York (1968), p. 11.
44. B. M. Colby and C. A. Evans, Jr., *Appl. Spectrosc.* **27**, 274 (1973).
45. L. Fraser Monteiro and R. I. Reed, *Int. J. Mass Spectrom. Ion Phys.*, **2**, 265 (1969).
46. S. L. Grotch, *Anal. Chem.*, **43**, 1362 (1971).
47. J. Roboz, *Introduction to Mass Spectrometry*, Interscience, New York (1968), p. 290.
48. C. A. Andersen and J. R. Hinthorne, *Science*, **175**, 853 (1972).
49. C. A. Andersen and J. R. Hinthorne, in *Seventh National Conference on Electron Probe Analysis*, EPASA (1972), p. 39.
50. C. A. Andersen and J. R. Hinthorne, *Anal. Chem.*, **45**, 1421 (1973).
51. C. A. Andersen, in *Microprobe Analysis* (C. A. Andersen, ed.), Wiley—Interscience, New York (1973), p. 531.
52. H. W. Werner, in *Developments in Applied Spectroscopy*, Vol. 7A, Plenum Press, New York (1969), p. 239.
53. J. M. Morabito and R. K. Lewis, *Anal. Chem.*, **45**, 869 (1973).
54. J. W. Colby, in *Proceedings of 11th Annual Reliability Physics Symposium*, IEEE Cat. No. 73 CHO 755-9-PHY (1973), p. 194.
55. J. W. Colby, in *Eighth National Conference on Electron Probe Analysis* EPASA (1973), p. 6.
56. M. Croset, *J. Radioanal. Chem.*, **12**, 69 (1972).
57. J. M. Schroeer, T. N. Rhodin, and R. C. Bradley, *Surf. Sci.*, **34**, 571 (1973).
58. Z. Jurella, *Int. J. Mass. Spectrom. Ion Phys.*, **12**, 33 (1973).
59. P. Joyes and G. Toulouse, *Phys. Lett.*, **39A**, 267 (1972).
60. G. Blaise and G. Slodzian, *C. R. Acad. Sci.*, *(Paris)*, B **266**, 1525 (1968).
61. G. Blaise and G. Slodzian, *J. Physique*, **31**, 93 (1970).
62. M. N. Saha, *Phil. Mag.*, **40**, 472 (1920).
63. M. N. Saha, *Z. Phys.*, **6**, 40 (1921).
64. J. Eggert, *Z. Phys.*, **20**, 570 (1919).
65. H. W. Drawin, in *Reactions Under Plasma Conditions* (M. Venugopalan, ed.), Wiley, New York (1971), p. 94.
66. F. G. Rudenauer and W. Steiger, to be published.
67. R. Shimizu, T. Ishitani, and Y. Ueshimu, *Jap. J. Appl. Phys.*, **13**, 250 (1974).
68. A. Lodding, S. J. Larssun, J. -M. Goungout, L. G. Petersson, and G. Frostell, *Zeit. Naturforsch.* **29A**, 897 (1974).
69. L. G. Petersson, G. Frostell, and A. Lodding, *Zeit. Naturforsch.* **29C**, 417 (1974).
70. T. O. Ziebold, *Anal. Chem.*, **39**, 858 (1967).
71. K. F. J. Heinrich, D. Vieth, and H. Yakowitz, *Advances in X-Ray Analysis*, Vol. 9 (G. R. Mallett, M. J. Fay, and W. H. Mueller, eds.), Plenum Press, New York (1966), p. 208.
72. L. I. Schiff, *Phys. Rev.*, **50**, 88 (1936).
73. J. W. Colby, P. Miller, and A. D. Bridges, paper presented at ECS Meeting in San Francisco, California (1974).
74. W. K. Hofker, H. W. Werner, D. P. Oosthoek, and H. A. M. de Grefte, *Rad. Effects*, **17**, 83 (1973).
75. G. Schwartz, M. Trapp, R. Schimko, G. Batzke, and K. Rogge, *phys. status solidi*, **17**, 653 (1973).

76. C. A. Evans, Jr., and J. P. Pemsler, *Anal. Chem.*, **42**, 1060 (1970).

77. R. P. Gittins, D. V. Morgan, and G. Dearnaley, *J. Phys. D*, **5**, 1654 (1972).

78. H. Liebl, *J. Vac. Sci. Technol.*, to be published Jan. (1975).

79. C. A. Evans, Jr., *J. Vac. Sci. Technol.* to be published Jan. (1975).

80. J. W. Guthrie and R. S. Blewer, *Rev. Sci. Instr.*, **43**, (1972).

81. R. W. Blewer and J. W. Guthrie, *Surf. Sci.*, **32**, 743 (1972).

82. T. A. Whatley, C. B. Slack, and E. Davidson, in *Proceedings of the 6th International Conference on X-Ray Optics and Microanalysis* (G. Shinoda, K. Kohra, and T. Ichinokawa, eds.), Univ. of Tokyo Press, Tokyo (1972), p. 417.

83. C. A. Andersen, J. W. Colby, R. Dobrott, T. A. Whatley, and D. J. Comaford, paper present at 25th Pittsburgh Conference on Analytical Chemistry (1974).

84. G. Carter, J. N. Baruah, and W. A. Grant, *Rad. Effects*, **16**, 107 (1972).

85. R. K. Lewis, J. M. Morabito, and J. C. C. Tsai, *Appl. Phys. Lett.*, **23**, 260 (1973).

86. F. Schulz, K. Wittmaack, and J. Maul, *Rad. Effects*, **18**, 211 (1973).

87. T. Ishitani, Thesis, Osaka University (1972).

88. J. A. McHugh, *Rad. Effects*, **21**, 209 (1974).

89. B. Domeij, F. Brown, J. H. Davies, and M. McCargo, *Can. J. Phys.*, **42**, 1724 (1964).

90. J. W. Colby, paper presented at 24th Pittsburgh Conference on Analytical Chemistry (1973).

91. C. A. Andersen, J. R. Hinthorne, and K. Fredriksson, in *Proceedings of the Apollo 11 Lunar Science Conference* (1970), Vol. 1, p. 159.

92. K. Fredriksson, J. Nelen, A. Noonan, C. A. Andersen, and J. Rittinthorne, in *Proceedings of the Second Lunar Science Conference*, M.I.T. Press, Cambridge, Massachusetts (1971), Vol. 1, p. 727.

93. J. W. Colby and L. E. Katz, paper presented at ECS Meeting, San Francisco, California (1974).

94. J. M. Walsh, in *Ninth National Conference on Microbeam Analysis*, (1974), p. 51.

95. C. E. Johnson, in *Eighth National Conference on Microbeam Analysis*, (1973), p. 68.

96. I. Johnson, C. E. Johnson, C. E. Crouthamel, and C. A. Seils, *J. Nucl. Mater.*, **48**, 21 (1973).

97. J. A. McHugh and J. F. Stevens, *Anal. Chem.*, **44**, 2187 (1972).

98. W. S. Johnson and J. F. Gibbons, *Projected Range Statistics in Semiconductors*, Stanford Univ. Press, Palo Alto, California (1969).

99. J. Lindhard, M. Scharff, and H. Schiott, *Mat. Fys. Medd. Dan. Vid. Selsk.*, **33**, 1 (1963).

INDEX